Parasitology

Volume 119 Supplement 1999

Parasite adaptation to environmental constraints

EDITED BY
R. C. TINSLEY

CO-ORDINATING EDITOR
L. H. CHAPPELL

CAMBRIDGE
UNIVERSITY PRESS

University Printing House, Cambridge CB2 8BS, United Kingdom

One Liberty Plaza, 20th Floor, New York, NY 10006, USA

477 Williamstown Road, Port Melbourne, VIC 3207, Australia

314–321, 3rd Floor, Plot 3, Splendor Forum, Jasola District Centre, New Delhi - 110025, India

79 Anson Road, #06-04/06, Singapore 079906

Cambridge University Press is part of the University of Cambridge.

It furthers the University's mission by disseminating knowledge in the pursuit of education, learning and research at the highest international levels of excellence.

www.cambridge.org
Information on this title: www.cambridge.org/9780521005005

© Cambridge University Press 1999

This publication is in copyright. Subject to statutory exception and to the provisions of relevant collective licensing agreements, no reproduction of any part may take place without the written permission of Cambridge University Press.

A catalogue record for this publication is available from the British Library

ISBN 978-0-521-00500-5 Paperback

Cambridge University Press has no responsibility for the persistence or accuracy of URLs for external or third-party internet websites referred to in this publication, and does not guarantee that any content on such websites is, or will remain, accurate or appropriate.

Front Cover Illustrations: Upper: Sassendalen, Spitsbergen. Location for studies by Halvorssen *et al.* (1999) showing that transmission of parasitic nematodes in reindeer continues throughout the Arctic winter (see p. S3). Photograph by Matthew Tinsley. Lower: Sonoran Desert, southwestern U.S.A. Location for studies by Tinsley (this volume) based on a host-parasite system where transmission is restricted to <24 h each (see pp. S31–S56). Photograph by Richard Tinsley. Centre: Pentastomids, *Armillifer armillatus*, taken from the lungs of a gaboon viper, *Bitis gabonica*. The large, long-lived, blood-feeding worms cause little observable pathology; they evade immune surveillance and reduce inflammation by continuous secretion of a disguise of host-like surfactant that coats the entire cuticle (see pp. S89–S105). Photograph by John Riley

Background Illustration: Transmission electron micrograph of a section of a trypanosome-infected tsetse fly salivary gland showing the attachment junctions between parasite flagellum and the microvillar border of the salivary gland epithelium. Original micrograph, Dr L. Tetley, University of Glasgow.

Contents

List of contributions	vii

Overview: extreme environments — S1
Introduction — S1
   Extreme environments — S1
   Physical environmental conditions — S1
   Macro-environmental conditions — S3
   Host environmental factors — S4
   Evolutionary considerations — S5
Acknowledgements — S6
References — S6

Parasites and low temperatures — S7
Summary — S7
Effects of low temperatures on animals — S7
Effects of low temperatures on parasites — S8
Low temperatures and parasite life cycles — S8
   Diapause in nematodes — S9
   Arrested development in trichostrongyle nematodes — S9
   Infective larvae — S9
   Ectoparasitic arthropods — S9
Free-living stages and off-host survival — S10
   Globodera rostochiensis — S10
   Trichostrongylus colubriformis — S10
   Entomopathogenic nematodes — S11
   Ticks — S11
   Insect parasitoids — S12
   Other examples of cold tolerance in parasite free-living stages — S12
Low temperatures and parasitic stages — S13
Conclusions — S14
References — S15

Desiccation survival of parasitic nematodes — S19
Summary — S19
Introduction — S19
Terminology — S19
Overview of anhydrobiosis in parasitic nematodes — S20
Behavioural responses that enhance survival — S20
Physical attributes that enhance survival — S21
Morphological changes induced by desiccation — S23
Biochemical and molecular correlates of anhydrobiosis — S24
Rehydration — S26
Conclusions — S28
Acknowledgements — S28
References — S28

Parasite adaptation to extreme conditions in a desert environment — S31
Summary — S31
Introduction — S31
The life cycle adaptations of *Pseudodiplorchis americanus* — S32
   Adaptations to the constraints operating during host invasion: reproductive preparation — S32
   Adaptations to the constraints operating during host invasion: oviposition — S35
   Survival in the external environment: characteristics of transmission — S36
   Survival in the external environment: transmission efficiency — S36
   The transfer from external to internal environments: invasion — S38
   Internal environmental conditions: survival in the respiratory tract — S38
   Internal environmental conditions: migration through the alimentary tract — S38
   Constraints of external environmental conditions: direct and indirect effects on parasite development and reproduction — S39
   Constraints of internal (host) environmental conditions: effects on parasite survival — S40
Host-parasite population ecology — S42
Effects of extreme environmental perturbation on the host-parasite system — S42
   The environment — S42
   Assessment of environmental variation: rationale and interpretation — S43
   Host population ecology — S45
   Parasite population ecology — S46
   The outcome of environmental perturbation: recovery — S48
   The outcome of environmental perturbation: extinction — S49
Discussion — S50
   Extreme environmental perturbation — S50
   Effects on the host populations — S50
   Effects on the parasite populations — S50
   Factors contributing to extinction — S51
   Factors involved in recovery — S52
   Life cycle characteristics — S52
Conclusions — S53
Acknowledgements — S55
References — S55

The survival of monogenean (platyhelminth) parasites on fish skin — S57

Summary — S57
Introduction — S57
Fish skin — S58
  Structure — S58
    The epidermis — S58
    The dermis — S60
  Function — S60
Permanent attachment — S61
  The haptor — S61
    Margin hooklets — S61
    Hamuli — S65
    The adult haptor and the role of sclerites — S66
    Haptors with reduced sclerites — S68
    Sensory structures — S71
  The pseudohaptor — S72
Temporary attachment — S73
  Morphology of the anterior adhesive areas — S73
  Sensory structures associated with anterior adhesive areas — S75
  Function of adhesive pads — S75
  Feeding — S76
Susceptibility and defensiblity of fish skin — S77
Site specificity and migration — S81
Predation — S81
Summary and conclusions — S82
References — S84

Pentastomids and the tetrapod lung — S89

Summary — S89
Introduction — S89
  Macroparasites and lungs — S90
  The lung and pulmonary surfactant in endothermic tetrapods — S92
  Lung surfactant in ectothermic tetrapods — S93
  The lung and regulation of immunity — S94
The adaptations of pentastomids for life in the tetrapod lung — S94
  The pentastomid cuticle and its relation to glands which excrete E/S products — S94
  Pentastomid SPC-derived surfactant and the host immune response — S96
  Non-lipid E/S products of HG and FG — S99
  In vitro lipid production by *P. crotali* and a comparison with rattlesnake lung surfactant — S99
Functional considerations — S101
References — S102

Do parasites live in extreme environments? Constructing hostile niches and living in them — S107

Summary — S107
Introduction — S107
Parasites construct hostile niches — S107
Hostile niche and host specificity — S108
Parasites and their ever-evolving hostile environment — S108
Conclusion — S109
References — S109

Analysis of parasite host-switching: limitations on the use of phylogenies — S111

Summary — S111
Introduction — S111
Terminology — S112
Congruence and history — S112
  Colonization, speciation and extinction — S113
    Host-switching — S113
    Speciation and extinction — S113
    Other significant historical processes in host-parasitic associations — S114
    Colonization and phylogeny — S114
    Asynchronous cospeciation — S115
    Widespread parasites — S116
    Cospeciation with host assemblages — S117
    Reticulate evolution — S118
Historical reconstruction — S118
  Methodologies — S118
  Practical application of numerical methods — S118
  Host switching versus extinction — S119
  Assumed models of historical host-parasite associations — S120
  Timing — S120
Conclusions — S121
Acknowledgements — S121
References — S121

Digenean parasites of deep-sea teleosts: a review and case-studies of intrageneric phylogenies — S125

Summary — S125
Introduction — S125
  The deep-sea — S126
  Digeneans in the deep-sea — S129
Case-studies on the phylogeny of two digenean genera — S132
  Materials and methods — S132
    Gene amplification and sequencing — S132
    Sequence alignment — S132
    Phylogenetic reconstruction — S132
  Results and conclusion from molecular phylogenies — S133
    Lepidapedon-phylogeny — S133
    Hosts — S134
    Depth — S135
    Steringophorus-phylogeny — S135
    Hosts — S135
    Depth — S135
    General conclusions on molecular trees — S135
Discussion — S136
  Deep-sea digeneans — S136
  Digenea life-cycles — S136
  Putative intermediate host frequency at depth — S138
  Zoning at depth — S139
  Constraints of physical conditions — S140
    Temperature — S140

Pressure	S140	Are evolutionary transitions from shallow to deep (or vice versa) relatively rare?	S141
Light	S140		
Seabed composition	S140	Have most deep-sea digeneans radiated in the deep-sea?	S141
Currents	S140		
Concluding questions	S140	How much do we really know about the deep-sea digeneans?	S141
Is the deep-water fauna smaller than the shallow?	S140		
		Final conclusions	S141
Is the deep-water fauna less diverse than shallow?	S141	Acknowledgements	S141
		References	S141

List of contributions

1. *Overview: extreme environments* by R. C. TINSLEY — S1
2. Parasites and low temperatures by D. A. WHARTON — S7
3. Desiccation survival of parasitic nematodes by R. N. PERRY — S19
4. Parasite adaptation to extreme conditions in a desert environment by R. C. TINSLEY — S31
5. The survival of monogenean (platyhelminth) parasites on fish skin by G. C. KEARN — S57
6. Pentastomids and the tetrapod lung by J. RILEY *and* R. J. HENDERSON — S89
7. Do parasites live in extreme environments? Constructing hostile niches and living in them by C. COMBES *and* S. MORAND — S107
8. Analysis of parasite host-switching: limitations on the use of phylogenies by J. A. JACKSON — S111
9. Digenean parasites of deep-sea teleosts: a review and case-studies of intrageneric phylogenies by R. A. BRAY, D. T. J. LITTLEWOOD, E. A. HERNIOU, H. B. WILLIAMS *and* R. E. HENDERSON — S125

Overview: extreme environments

R. C. TINSLEY

Bristol University

INTRODUCTION

An important element in most approaches to the subject of parasitism is the consideration of environment. Parasites are set apart within animal ecology because they experience two environments, one the 'external' conditions and the other created by the living body of the host. As in any ecological system, external environmental conditions have a major influence on life history parameters: these conditions may be experienced directly by 'off-host' stages of a parasite or, to a greater or lesser extent, indirectly through the body of the host. However, uniquely in parasitic associations, the internal (host) environment has a dual influence on the physiological conditions encountered by parasites: first, the host buffers the external conditions (by homeostatic mechanisms moderating environmental fluctuations, by behavioural responses selecting appropriate habitat conditions, etc.), but second, the host creates a suite of hostile factors associated with immune defence. This living, reactive environment has no parallels elsewhere in free-living animal ecology: it has the characteristic of reacting specifically to kill the organisms within its boundaries.

The approach of this supplement is to examine environmental constraints affecting parasites at four levels. First, there is a focus on selected physical (abiotic) conditions that are hostile to life in general and a review of the specific adaptations that allow survival of highly specialized parasites. Second, particular macro-environments (ecosystem conditions) are considered in which a series of environmental factors may combine to jeopardize survival, for instance in deserts and the deep-sea. Third, there is a detailed examination of certain micro-environments (host conditions) which may not at first sight appear limiting but which actually present a major challenge to parasite exploitation (the skin of fishes, the lungs of tetrapods). Finally, this volume considers the outcome of constraints over an evolutionary timescale, including the concepts of the niche, the arms race and host immunity, adaptation, specificity and host-switching, and inferences from phylogeny.

Extreme environments

Major constraints are easily recognized where environmental conditions approach the limits at which life can be sustained. These conditions may be regarded intuitively as 'extreme'. However, the concept of an 'extreme environment' may be criticized as anthropocentric. Clearly, whilst specific conditions may exclude survival by a majority of animal groups, animals living in supposedly extreme environments appear to behave as if they regard their environment as 'normal'. Animals inhabiting the deep-sea (see Bray *et al.*, this supplement) illustrate the difficulty of defining conditions as 'extreme' and 'normal'. When removed from the high pressures and other constraints of their 'extreme environment' and brought to the surface, these animals are rapidly killed. Emperor penguins not only survive the Antarctic midwinter exposed to blizzards and to temperatures of $-50\,°C$, they also incubate their eggs during this period and endure over 3 months without feeding. Of course, this life style is 'normal', the product of a long process of evolutionary adaptation. If this view of 'extreme environments' is anthropocentric then it reflects an obvious fascination for both the scientist and layman. Perhaps the greater feat of endurance was that undertaken by the humans – a species singularly poorly-adapted to these conditions – who first undertook a midwinter expedition, subsequently called 'The worst journey in the world', in order to collect Emperor penguin eggs for studies concerned with development and evolution (Cherry-Garrard, 1922). For this species, the conditions are indisputably 'extreme'.

In the field of human endeavour, the concept of 'extreme' is often measured in terms of the proportion of attempts that fail. The highest levels of severity are reserved for cases where only a small minority of individuals, those exceptionally capable, succeed in a particular venture. Indeed, this may lead to the eulogizing of 'heroic failure' (as in the early exploration of polar regions). The popular view of extreme conditions evokes a significant degree of 'suffering' (ranging from discomfort to life-threatening dysfunction). In nature, of course, failures would not be a routine occurrence and it is unlikely that animals adapted to severe environmental conditions experience 'suffering': the conditions are 'normal' to them. For many biologists, the concept 'extreme' relates to physical and other environmental conditions where most life is excluded but some groups of organisms may routinely tolerate these conditions as a result of specific adaptations. Thus, the anthropocentric view may perhaps be distinguished by its focus on the survival of few *individuals*, whereas the scientific view emphasises the few *species* that tolerate extreme conditions. This latter feature leads to the recognition that extreme environments have reduced species richness (see Combes & Morand, this supplement).

Almost inevitably, an extreme environment comes to be defined in terms of its effects on the animals that inhabit it (or by the difficulties experienced by animals unable to inhabit it!) rather than by its intrinsic characters. Constraints imposed by the environment could be depicted by plotting along intersecting axes the environmental variables that determine survival of living organisms (in the manner of the niche concept). This would demonstrate a major part of the abundance and diversity of life accommodated within an irregular central 'multidimensional hyperspace' framed by well-recognized limits of temperature tolerance, oxygen and water availability, and so on. 'Extreme environments' would then be represented on the outer margins of space, as satellites where smaller peaks of abundance are created by certain combinations of conditions. Thus, marine environments characterized by high pressure, low temperature, lack of light, etc., are actually occupied by prolific life (in the deep-sea). The organisms that flourish under these conditions do not represent the tails of a normal distribution of adaptation extending from the central hyperspace; instead, they occupy their own 'normal distributions' and may show little or no overlap with the centre. Thus, these satellites may be more-or-less self-contained respectively within their own adaptive hyperspace. The separation of these central and peripheral hyperspaces emphasizes that it is inappropriate to compare putative 'normal' and 'extreme' environments by the same criteria. Each is constructed of different subsets of intersecting environmental gradients and their characteristic organisms are adapted to these conditions and should not be considered as displaced representatives from a different environment.

The term 'extreme environment' is therefore anthropocentric or, more precisely, represents a view from within the set of environmental parameters that accommodates the 'familiar' diversity of life. However, Bray and co-authors (this supplement) point out that, on the basis of its total volume, the abyssal ocean and its inhabitants could be considered to represent the most typical environment of this planet.

Given this rather artificial, but nonetheless practical, viewpoint, there are two distinct components of life in extreme environments. One group of organisms survives extreme conditions as a temporary measure, often as a dormant stage designed to resist hostile factors until the return of conditions in which its activity can resume. To continue the analogy outlined above, these organisms may make a short-term excursion from their central hyperspace (perhaps as dispersal propagules), or they become isolated from this central zone by unfavourable (sometimes seasonal) conditions. Their adaptations enable them to tolerate extreme conditions but, in their dormant state, they may not actually exploit this environment. By contrast, other organisms can maintain active life in these conditions which represent their natural zone. Transfer of these to different (supposedly less extreme) conditions may exceed their survival range and they die. The distinction between these types of environmental adaptation is discussed further by Perry (this supplement) within the categories 'resistance adaptation' and 'capacity adaptation'.

This supplement considers 'extreme environments' as part of its remit, but to become distracted by the problems of defining 'extreme' would destroy the inherent fascination of a wide range of parasite adaptations. The scope, therefore, is to consider the wider range of constraints that regulate parasite biology and to gain insight into the mechanisms by which parasites may respond to these factors. In association with this, the supplement aims to highlight less familiar examples, taken from natural host-parasite systems, to provide a glimpse of the wider zoological and evolutionary view of the diverse and often ancient lineages of animal parasites.

Physical environmental conditions

The nature of many parasite life cycles, where a transmission stage is released into the external environment, produces a requirement to survive in the absence of the host and often, therefore, to tolerate adverse conditions. Perry documents how the ability to survive different environmental stresses, including low and high temperatures, lack of oxygen, osmotic stress and dehydration, may be linked by common mechanisms. It follows from this that the animal phyla that are specialists in these mechanisms – including the nematodes, tardigrades and rotifers – can survive a range of the most severe stresses. For this supplement, two examples of extreme physical conditions have been chosen to illustrate these features with respect to parasite adaptation: low temperatures and desiccation. Reviews by Wharton and Perry document adaptations that fascinate both the biologist and the layman, permitting life to be sustained in circumstances where death should be inevitable. Wharton records how freeze-avoiding, supercooled potato cyst nematodes can survive temperatures as low as $-38\ ^\circ\text{C}$; larvae of a nematode parasite of reindeer survive freezing at $-80\ ^\circ\text{C}$; larvae of *Trichinella nativa* remain infective in host tissue at $-18\ ^\circ\text{C}$ for four years. The mechanisms are varied; some nematodes enter a state of anhydrobiosis, where there is no water to freeze, and this strategy overlaps with that involved in desiccation survival. Perry notes that where water content falls below about 20% there is no free water in cells. The 20% of bound water is involved in the structural integrity of macromolecules and macromolecular structures such as membranes. It is remarkable, therefore, that the water content of desiccated, anhydrobiotic nematodes may be only 1–5%, indicating that the bound water has been lost. Some plant parasitic nematodes, such as *Ditylenchus dipsaci*, demonstrate the most spectacular abilities for long-term survival, remaining viable after 23 years storage in dry plant material. These are the mechanisms that would aid the transport of living organisms through space!

Macro-environmental conditions

Temperature has an all-pervading influence on parasite life cycles, regulating rate-dependent processes of stages in the external environment and within ectothermic hosts. Only those internal parasites that exploit two classes of vertebrate animals – birds and mammals – are free of this constraint, during the period within their endothermic hosts. Even for these, temperature fluctuations affecting only part of the life cycle are sufficient to exert a powerful control on parasite population dynamics. Some parasite life cycles are adapted to this seasonal inhibition with a strategy that synchronises transmission with favourable environmental conditions and increased host availability. Thus, eggs of the sheep nematode *Nematodirus battus* enter diapause in autumn and overwinter on pasture; hatching of infective larvae requires chilling followed by increased temperatures in spring, and this synchrony enables *en masse* infection of lambs (see Wharton, this supplement). It is not unexpected that low temperatures, especially those approaching freezing point, should represent a major constraint on transmission. However, it is surprising that temperature-induced inhibition of parasite transmission may begin at temperatures as high as 10 °C. This is illustrated by representatives of three diverse groups of helminths: monogeneans, digeneans and nematodes. In species of *Discocotyle*, *Fasciola* and *Ascaris*, development of stages exposed to the external environment is extremely slow or ceases below 10 °C (indeed, in *Ascaris suum* this threshold is 15 °C) (Gannicott & Tinsley, 1998a, b; Ollerenshaw, 1974; Larsen & Roepstorff, 1999, respectively). In each case this affects the eggs, and hence the generation of infective stages, but also the intra-molluscan stages of the liver fluke. For *Discocotyle*, development and maturation of worms established on the fish host is also interrupted and output of eggs by previously-matured parasites is minimal. Transmission of *Ascaris* and *Fasciola* to the final hosts may continue from the pool of infective eggs and still-viable cysts that completed development before the decline in temperatures, but the rate will be greatly reduced. The remarkable outcome is that, in equable temperate climates as in Northern Europe, host-to-host transmission is almost eliminated for half of each year, from early November until late May. Since each of these parasites is responsible for significant disease under appropriate circumstances, the corollary is that transmission during the other half of each year is very successful in generating high worm burdens. The case study outlined by Tinsley, based in the Sonoran Desert of North America, illustrates the same effect: here, the threshold temperature for parasite development and reproduction is 15 °C and, in exact parallel, all life cycle progress is precluded between October and April each year. The extent of this inhibition is counterintuitive. An evolutionary biologist might predict the strong selective advantage of life cycle adaptations that would enable exploitation of a greater proportion of each annual cycle.

Given this documentation of temperature-dependent regulation of transmission, it is fascinating to encounter data demonstrating continued transmission under exceptionally severe conditions. In the high Arctic, Svalbard reindeer (*Rangifer tarandus platyrhynchus*) experience a long cold winter: typically the upper metre of ground is frozen and snow-covered from October to the end of May or early June. Transmission of trichostrongyle nematodes would be expected to be low. However, Halvorsen *et al*. (1999) recorded an increase in total abundance of infection over the winter, indicating that reindeer continue to ingest infective larvae from snow-covered pasture. They considered it unlikely that there is significant larval development below freezing point, but the infective stages have a relatively high survival rate at these temperatures so transmission probably involves larvae that developed to the infective stage during the preceding summer and autumn. One implication of the overwinter transmission in Svalbard is that progressively increasing worm burdens may induce disease in late winter when reindeer are also stressed by limited food availability and when their immune response may be reduced. This contrasts with the outcome well-documented for related parasites in domestic livestock where temperature constraints inhibit overwinter transmission: in these cases, hosts have respite from accumulation of infection at a time when environmental conditions may prejudice the ability to cope with disease. In ectothermic vertebrates, the temperature dependency of the immune response adds another component to the host-parasite interactions that are regulated by seasonal temperature cycles. This is explored in the paper by Tinsley. In the monogenean *Pseudodiplorchis americanus*, pathogenic effects (caused by blood-feeding) are parasite size dependent; growth inhibition during the period of low temperatures

Overview

therefore prevents any increase in disease effects. Parasite blood consumption is also temperature-dependent, so that energetic demands on the host are minimal during winter. However, coincidentally, temperature-induced suppression of the immune response allows worm burdens to survive, whereas an equivalent period at higher temperatures would lead to elimination of the entire infection. A special concern of this supplement is to document information such as this, indicating the complexity of host–parasite interactions in the natural environment.

Studies based in a hot desert (in Arizona) would be expected to demonstrate major environmental constraints on parasite biology, especially in a case where transmission occurs in water. The paper by Tinsley begins by documenting remarkable specializations associated with achieving synchrony of life cycle events with briefly favourable environmental conditions. Fortuitously, during a long-term study of the host and parasite populations, it became possible to follow the effects of a period of exceptional drought. The impact was evident in a succession of years of failed recruitment, leading to local extinction of previously-stable parasite populations. It might be concluded that this provides the most dramatic illustration of the power of extreme environmental conditions, leading to extinction despite finely-tuned adaptations. However, this paper also reviews the quantitative data now available for this host parasite interaction: aside from the effects of rare environmental perturbation, the major part of parasite pre-reproductive mortality occurs within the host, attributable to immune attack. This suggests the conclusion that the most severe environmental constraint, even in a desert, may actually be a feature of parasite life cycles generally – the conditions created by the host.

Host environmental factors

Alongside inclusion of constraints that are, to any biologist, immediately recognisable as extreme, this supplement aims to highlight other less obvious subjects where specific environmental conditions create very considerable difficulties for exploitation by parasites. Although at first sight the external surface of a fish presents an apparently rather unspectacular environment, its characteristics actually illustrate a series of major constraints affecting ectoparasites. Fish skin is wet, notoriously slippery, and swept by powerful water currents. The task is accentuated for those parasites that must also retain the ability to move – to find a sexual partner or new areas for feeding – as well as maintain secure attachment on the slime-covered surface of a fast-moving fish. Kearn presents a comprehensive review of the environment and the survival mechanisms of monogeneans that inhabit fish skin. This platyhelminth group has become specialized in the deployment of hard, proteinaceous hooks borne on a posterior disk-like haptor to pin themselves to the host. Kearn correlates the form and function of these hooks to the microscopic structure of the host's skin. Tiny marginal hooklets have blades that pierce the apical membrane of host epithelial cells and become embedded in the terminal web of tonofilaments just below the cell surface. The terminal web is sufficiently strong to resist tearing of the cell. The load is spread by employing multiple points of attachment to the epidermis, typically 14 or 16 hooklets, and this arrangement is very successful – as judged by its highly uniform organisation amongst monogeneans – for the attachment of small skin parasites (generally not exceeding 300 μm in body length). With increasing size, many monogeneans rely on a second series of hooks, the much larger hamuli, that may penetrate through the host epidermis and its collagenous basement membrane into the fibrous dermis beneath. Monogeneans employ two further mechanisms for attachment, often complementary to sclerite penetration: muscular suction and sticky secretions. The latter represent a particularly fascinating aspect of the biology of monogeneans. Kearn describes how, apparently against all odds, monogeneans employ adhesive secretions to attach to the mucus-covered and water-current-swept skin surface. In *Entobdella soleae*, two distinct secretions are discharged at the attachment site and their mixing generates adhesion reminiscent of that effected by the interactive components of commercial epoxy resins. The cement layer is 4 or 5 μm thick and penetrates between the microridges of the host's epidermal cells and between the microvilli of the parasite's adhesive pads. The bond is sufficiently strong to resist powerful water currents and yet, even more remarkably, the parasite is able to sever the adhesion instantly during its leech-like locomotion. Dissolution of the cement is mediated via the apical membrane of the adhesive pad tegument and a third secretion could provide the means of release.

These are problems encountered by marine engineers who require adhesives that will operate instantly underwater, but the parasites have taken this technology one stage further in producing a mechanism that can instantly dissolve the bond. The adaptations provide an exciting example of the way in which fundamental research may have important commercial applications.

Kearn's review also illustrates the reciprocal nature of host-parasite interactions. Fish skin appears highly vulnerable to parasite infection; however, those features that appear to increase its vulnerability – exposed living cells and proximity of blood vessels – may also be associated with powerful immune responses. Recent immunological studies suggest that the skin of fishes has a secretory defence system independent of the systemic system. This is likely to be a major factor influencing host specificity, site specificity and the regulation of infection levels. This theme, the environmental constraint created by host immunity, leads to consideration of the lethal conditions generated by the host immune response. Combes and Morand develop the idea that parasites themselves have created these hostile conditions within hosts by having driven the evolution of the immune system. They argue that parasites have responded to 'genes to kill' by 'genes to survive', and this has triggered selection of measures and countermeasures that form the basis of a continuing coevolutionary arms race.

Understanding of the interactions between the killing mechanisms of the host and the survival and manipulative strategies of the parasite contributes to a relatively well-reviewed area of parasitology, including the recent *Parasitology* supplement (Doenhoff & Chappell, 1997). This supplement does not intend to duplicate any of this well-established information. Instead, a paper by Riley focuses on an 'unconventional' case study involving an ancient parasite group, the phylum Pentastomida, whose adults infect the respiratory tract of tetrapod vertebrates, principally reptiles. Although the delicate pulmonary epithelium and rich blood supply might appear to provide a favourable environment for macroparasites, Riley points out that relatively few helminths have colonized vertebrate lungs to reside in or on the respiratory epithelium (aside from those, principally larval nematodes, that penetrate the lungs in transit to the gut). A range of protective mechanisms are involved in lung homeostasis. A key component is pulmonary surfactant, a complex mixture of phospholipids, neutral lipids and proteins. This has an essential biophysical role and is also crucially involved in protection of the lungs from infection. Lung-dwelling pentastomids are typically large (some are 5–10 cm in length), long-lived (in some cases many years), and feed on blood from the pulmonary capillaries. Surprisingly, they cause little observable pathology. Riley documents evidence suggesting that these parasites evade immune surveillance and reduce inflammation by secreting their own surfactant. This has a lipid composition very similar to that of lung surfactant and coats the parasite's cuticle with material that is therefore immunologically compatible with host secretions.

Most of the recent developments in parasite immune evasion and exploitation are based, for reasons of major medical importance and research funding, on a relatively small number of intensively studied human parasitic diseases, and most of these are studied, for reasons of practical expediency, in laboratory rodents. Riley's account provides a glimpse of interactions in the rest of the parasitological world; it should contribute to the prediction that there must be an enormous diversity of parasite adaptations to counter what is, arguably, the dominant environmental influence on parasites: the host immune response.

Evolutionary considerations

Several of the reviews emphasise evolutionary aspects of host-parasite interactions. Combes and Morand link the concept of the hostile niche with the specialization of parasites to survive in a limited number of these environments and, hence, the characteristic of host specificity. Kearn follows this idea in relation to the influence of the defensive responses of fish skin on the host ranges of monogeneans. The case studies employed in some papers also highlight the very long evolutionary association of particular host and parasite groups. For monogeneans, the association with their fish hosts may extend back to Palaeozoic times, before the modern lineages of fish had emerged. Pentastomids have the distinction, very rare amongst parasite groups, of a fossil record: this provides a tantalizing glimpse of forms present in the Cambrian. This evidence points to an enormously long period of natural selection and specialization.

It is logical, therefore, that further consideration of parasite responses to environmental constraints should take a phylogenetic perspective. Jackson pursues the concept of host specificity as a reflection of the constraints that restrict parasite occurrence in nature; he notes the potential for host specificity to change over evolutionary time, with new host lineages acquired and others lost. Changes of host by parasites are events of intrinsic evolutionary interest and may have had an important influence on present parasite distributions. Jackson critically evaluates the inference of host-switching from phylogenetic data.

The deep-sea environment presents exceptional constraints on life (at least, life as it is known nearer to the surface), including physical conditions such as high pressure, low temperature and lack of light. For parasites there are the additional factors of a sparse distribution of potential hosts, highly specialized food chains, and the characteristic that there is no 'platform' on which transmission can occur. Because of this lack of a surface on which parasite eggs may settle and develop and from which new hosts may be invaded, all life processes must take place 'on the run'. Bray and co-authors document the records of digeneans in the teleost fishes that inhabit the deep-sea and note the distinctiveness of this parasite fauna. They employ molecular phylogenies as a means of making inferences about the origins, routes and frequency of invasion of the deep-sea. They also review some of the adaptations for survival, including the nature of life cycles. This concluding paper fulfils one of the aims of this selection of studies, in providing a new, thought-provoking, insight into consideration of parasite adaptation. In case one might lightly dismiss the deep-sea fauna of hosts and parasites, Bray and co-authors emphasize that over half of the earth's surface is covered by the deep-sea with a depth of over 3200 m. There is enormous faunal diversity, including potential intermediate hosts such as molluscs, annelids and arthropods. The impression is clearly conveyed, as in some other areas of parasitology, of how little is still actually known.

Where harsh conditions are 'unrelenting', it might be predicted that animals regularly inhabiting such environments have appropriate adaptations to counter the constraints: the proof of their specializations must be assumed from their continuing survival. It might be argued that, by contrast, conditions that are really 'hostile' are those that occur sporadically, where life-threatening circumstances are unexpected (for instance, in temperate areas, a sudden period of freezing in late spring). Sometimes, the final challenge may follow the stress of already-difficult conditions: effects are exacerbated by existing weakness. A series of phrases in the English language express this concept, including 'the last straw breaking the camel's back'. In the scenario recorded by Tinsley (this supplement), conditions of sufficient severity to cause local parasite extinction could be attributed to a succession of unfavourable circumstances, the worst in a 35-year record of weather conditions ('a run of bad luck'). These are the conditions that test adaptations 'to the limit', and these events provide the basis for natural selection of genotypes most fit to cope with a recurrence of these circumstances.

Apart from its intrinsic interest, understanding the environmental constraints that limit parasite populations has wider importance. Removal of some of these constraints, especially those that limit transmission efficiency, has been a major contributory factor in the imbalance involved in parasite-induced disease in the medical and veterinary arenas.

ACKNOWLEDGEMENTS

I would like to thank all the contributing authors for their enthusiastic involvement. I am also grateful to those who acted as referees, including Dan Brooks, Wendy Gibson, Joe Jackson, Rod Page, Roland Perry, Robert Poulin, John Riley, Keith Stobart, Mark Viney and David Wharton, and especially to those who, to help achieve greater uniformity, refereed several manuscripts each. I express particular thanks to Les Chappell for his patient encouragement of this exercise and his efficiency in dealing with the contributions.

REFERENCES

CHERRY-GARRARD, A. (1922). *The Worst Journey in the World*. Chatto & Windus, London.

DOENHOFF, M. J. & CHAPPELL, L. H. (1997). Survival of parasites, microbes and tumours: strategies for evasion, manipulation and exploitation of the immune response. *Parasitology* **115**(suppl.), S1–S175.

GANNICOTT, A. M. & TINSLEY, R. C. (1998a). Larval survival characteristics and behaviour of the gill monogenean *Discocotyle sagittata*. *Parasitology* **117**, 491–498.

GANNICOTT, A. M. & TINSLEY, R. C. (1998b). Environmental control of transmission of *Discocotyle sagittata* (Monogenea): egg production and development. *Parasitology* **117**, 499–504.

HALVORSEN, O., STIEN, A., IRVINE, J., LANGVATN, R. & ALBON, S. (1999). Evidence for continued transmission of parasitic nematodes in reindeer during the Arctic winter. *International Journal for Parasitology* **29**, 567–579.

LARSEN, M. N. & ROEPSTORFF, A. (1999). Seasonal variation in development and survival of *Ascaris suum* and *Trichuris suis* eggs on pastures. *Parasitology* **119**, 209–220.

OLLERENSHAW, C. B. (1974). Forecasting liver fluke disease. In *The Effects of Meteorological Factors upon Parasites* (ed. Taylor, A. E. R. & Muller, R.), pp. 33–52. Symposia of the British Society for Parasitology **12**. Oxford, Blackwell Scientific Publications.

Parasites and low temperatures

D. A. WHARTON

Department of Zoology, University of Otago, P.O. Box 56, Dunedin, New Zealand

SUMMARY

Low temperatures affect the rate of growth, development and metabolism of parasites and when temperatures fall below 0 °C may expose the parasite to the potentially lethal risk of freezing. Some parasites have mechanisms, such as diapause, which synchronise their life cycle with favourable seasons and the availability of hosts. Parasites of endothermic hosts are protected from low temperatures by the thermoregulatory abilities of their host. Free-living and off-host stages, however, may be exposed to subzero temperatures and both freezing-tolerant and freeze-avoiding strategies of cold hardiness are found. Parasites of ectothermic hosts may be exposed to subzero temperatures within their hosts. They can rely on the cold tolerance adaptations of their host or they may develop their own mechanisms. Exposure to low temperatures may occur within the carcass of the host and this may be of epidemiological significance if the parasite can be transmitted via the consumption of the carcass.

Key words: Temperature, freezing, cold tolerance, diapause.

EFFECTS OF LOW TEMPERATURES ON ANIMALS

Temperature is an important environmental influence on the biology of all animals. The responses to temperature are complex but at its simplest temperature affects the rate at which life processes (growth, metabolism, reproduction etc.) proceed. Every animal has an optimum temperature at which these are at their maximum. As the temperature declines the rate of metabolism slows, due to the decrease in kinetic energy imparted to reactions. This is reflected in changes in the level of activity displayed by the animal. At low temperatures the animal first displays cold stupor, in which its motion becomes disorientated and normal processes are disrupted, and then cold coma, in which activity (and development) ceases altogether (Fig. 1). Proteins are not denatured by low temperatures (as they are by high temperatures) and thus the effects are potentially reversible. If the temperature falls below the melting point of its body fluids (usually just below 0 °C), however, the animal may freeze. This review will focus upon the response of parasites to subzero temperatures where they are exposed to freezing or the risk of freezing.

When an animal freezes water within its body undergoes a change in state from a liquid to a solid. This is obviously a dramatic event to occur within a living organism. Although it is not fully understood why freezing is usually lethal to cells, there are a number of proposed mechanisms (Storey & Storey, 1988; Lee, 1991; Karlsson, Cravalho & Toner, 1993). Death may result from the mechanical disruption, during ice formation, of the plasma membrane of cells or the membranes of organelles. Ice forming in a solution results in the concentration of salts in the unfrozen portion as water molecules join the growing ice crystals. If this occurs extracellularly this will raise the osmotic gradient between the intracellular and extracellular compartments and result in cellular dehydration, which may prove lethal. Despite these potentially lethal effects there are several groups of animals which can survive ice formation in their bodies.

Freezing-tolerant animals include: many species of terrestrial insects; intertidal invertebrates such as barnacles, bivalves and gastropods; some terrestrial hibernating amphibians and reptiles; centipedes, slugs and nematodes (Storey & Storey, 1996). The cold-tolerance adaptations of insects are the best understood. Freezing-tolerant insects produce ice-nucleating proteins which ensure freezing at a high subzero temperature (Duman *et al.* 1991); although the gut, and other sites, may also act as sites of ice nucleation (Block, 1995). Freezing at a relatively high temperature ensures that it is a gentle process, allowing the cells of the animal to adjust to the stresses involved and reducing the chances of intracellular freezing. Insects also produce sugars and polyols, such as trehalose and glycerol, which act as cryoprotectants. These depress the melting point and decrease the amount of ice formed at a given temperature, reducing cellular dehydration (Lee, 1991). Thermal hysteresis proteins are produced by freezing-tolerant insects which act to inhibit the recrystallization which may damage cells by the growth of ice crystals (Knight & Duman, 1986).

Freezing avoidance is the alternative strategy adopted by cold-hardy animals. Freeze-avoiding insects maintain their bodies in a supercooled (still liquid) state at temperatures below the melting point of their fluids. If freezing occurs (at the supercooling point) it is lethal. This involves removing or

Fax: 064 3 479 7584,
email: david.wharton@stonebow.otago.ac.nz

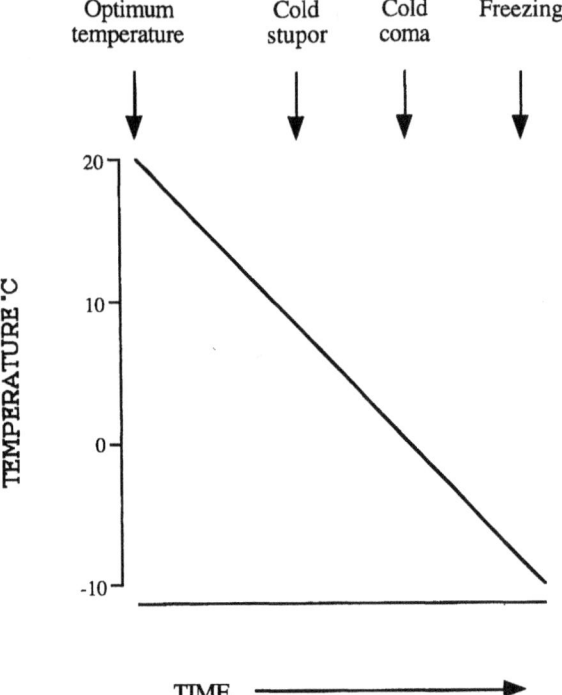

Fig. 1. The effect of decreasing temperature on a hypothetical animal (adapted from Vannier, 1994).

inactivating sources of ice nucleation, such as emptying the gut during winter and avoiding contact with external ice. Sugar and polyols are synthesized but in this case act as antifreezes, depressing the melting and supercooling points and masking or inactivating ice nucleators (Lee, 1991). Thermal hysteresis proteins are thought to stabilize the supercooled state by attaching to and inhibiting the growth of ice crystals (Duman et al. 1991). Antarctic fish are at constant risk of freezing since the temperature at which their blood melts is above that of the surrounding sea water and ice crystals in the water threaten to seed ice formation. They produce antifreeze proteins which attach to the surface of ice crystals thus preventing their growth and the freezing of the blood (Davenport, 1992).

Freeze-avoiding insects die when they freeze whereas freezing-tolerant insects do not. The third possibility is that the insect dies before it freezes. This is known as chilling injury or prefreeze mortality (Bale, 1993). Causes of chilling injury may include cold-induced phase changes in lipid membranes (Quinn, 1985), thermoelastic stress (McGrath, 1987), effects on protein structure and disruptions in metabolism which result in lethal imbalances (Storey & Storey, 1988).

EFFECTS OF LOW TEMPERATURES ON PARASITES

Endoparasites of birds and mammals gain protection from low temperatures by living within an endothermic host. The host generates its own heat and maintains a higher temperature within its body than that in the surrounding environment. This is achieved firstly by burning fuel (food) to generate heat through metabolism and secondly by mechanisms which reduce the loss of heat to the environment, such as insulation (fur, feathers and fat beneath the skin). These parasites will only be exposed to low temperatures if they leave the host during the free-living phase of their life cycle or if the host dies. Ectoparasites of endothermic hosts may still gain some thermal protection from their host if they are in close proximity to the skin and protected from the exterior by fur or feathers. This thermal protection will be less the further away the parasite is from the skin of its host and will be lost altogether if the parasite detaches from the host. Ectothermic animals (the majority of animals) are at the same temperature as their environment. A parasite of an ectothermic host thus experiences, and has to survive, low temperatures in the environment. If the host is exposed to subzero temperatures and is in danger of freezing, so are its parasites.

Low temperature may affect the host/parasite relationship in ways other than its potentially lethal effects on the parasite. The reproduction and activity of hosts will be low during cold seasons and they are thus likely to be unavailable for infection by parasites. The transmission of parasites will also be directly affected by low temperatures. The development of eggs, other free-living stages and of parasitic stages within an ectothermic intermediate host is related to temperature and development will not occur, or proceed slowly, at low temperature. The population dynamics and seasonal infection cycles of parasites are thus likely to be driven directly or indirectly by temperature both by its effects on parasite development and on the seasonal availability of hosts (Kennedy, 1975; Poulin, 1998).

The use of low temperatures is also important to parasitologists for the preservation of parasite cultures. The techniques used are often unrelated to the parasite's natural response to low temperatures (for example by freezing in liquid nitrogen) and are hence beyond the scope of this review.

LOW TEMPERATURES AND PARASITE LIFE CYCLES

In temperate and polar regions low temperatures are associated with seasonality and are more likely to be encountered during the winter. Parasites can improve their chances of locating a host and/or surviving the winter by a life cycle which is adjusted to this seasonal pattern. This allows the parasite to survive the cold season and to synchronize the production of infective stages both with conditions favourable to their survival and development and to the availability of hosts to infect. This is achieved by a variety of mechanisms.

Diapause in nematodes

The existence of a diapause in some species of parasitic nematode is indicated by the pattern of seasonal development and/or the response of a developmental event to temperature and other stimuli. Insects synchronize their life cycles with favourable seasons via a diapause which is hormonally regulated. Although hormone regulation in nematode diapause has yet to be demonstrated, there is experimental evidence for the existence of diapause mechanisms (Evans, 1987; Womersley, Wharton & Higa, 1998).

The eggs of some species of plant-parasitic nematodes show increased hatching after chilling, which may produce a pattern of invasion of host roots in the spring and maturation and egg production by adults in the summer (Evans & Perry, 1976). Different isolates vary in their responses, suggesting adaptation to local conditions, which may allow survival of a dry season rather than overwintering. There may be variation in the stimuli necessary to produce egg hatching even within a single population, allowing a proportion of eggs to hatch immediately and a proportion to overwinter and hatch in a second season. This improves the persistence of the population and reduces competition for feeding sites on the host (Jones, Tylka & Perry, 1998). Photoperiod is a more reliable seasonal indicator than temperature and there is evidence for a photoperiod-induced diapause in the eggs of *Globodera rostochiensis*, mediated by its effect on the host plant (Hominick, 1986).

A period of chilling is required for the hatching of the eggs of the animal-parasitic nematode *Nematodirus battus*, with infective larvae overwintering within the eggshell and hatching *en masse* in the spring (Evans & Perry, 1976). In Britain, eggs which have developed on pasture fail to hatch if kept at 20 °C but a mass hatch is produced if they are chilled at 5 °C before exposure to 20 °C. Conditions in autumn induce a diapause which is terminated by chilling over winter and a rise in temperature in spring. Infective larvae overwinter within the egg and hatch in spring ready to infect lambs. This response is so predictable that formulae have been derived which forecast the timing and severity of pasture contamination with hatched larvae (Thomas, 1974).

Arrested development in trichostrongyle nematodes

In some species and strains of trichostrongyle nematodes 3rd-stage infective larvae (L_3s) acquired by hosts during autumn do not immediately develop into adults but remain as 4th-stage larvae (L_4s) throughout the winter. This phenomenon is called hypobiosis or arrested development (Gibbs, 1986; Eysker, 1997). The L_4s resume development in the spring, in response, it is thought, to hormonal changes in the host at the start of the breeding season (Gibbs, 1986). Infective larvae are thus available on pasture at a time when new susceptible hosts (e.g. lambs and calves) start feeding. Low temperatures experienced by L_3s on pasture during autumn trigger the winter inhibition of L_4s. The process thus appears to be a diapause. The controlling mechanisms are unknown but the identification of a protein which is diagnostic for inhibited L_4s of *Ostertagia ostertagi* (Cross, Klesius & Williams, 1988) suggests an underlying molecular mechanism.

Infective larvae

To continue their life cycle parasites need to infect a host. The infective larva is thus in a state of developmental arrest and only resumes development once physiological stimuli are received which indicate entry into or the proximity of a suitable host. This ensures that development is prevented when conditions are likely to result in the death of the parasite (outside the host) and resumes when conditions are favourable (inside the host). This also has the effect of synchronizing the development of the parasite with the availability of hosts. A common physiological trigger which indicates entry into the intestine of a vertebrate is an increase in carbon dioxide concentration at an appropriate temperature and pH (Rogers, 1960; Lackie, 1975).

Ectoparasitic arthropods

Ticks (Ixodides) remain attached to their host for varying periods of time, depending on the species, but may detach after engorgement and be exposed to low temperatures off their host. *Ixodes uriae* is the predominant tick infecting seabirds on the Antarctic peninsula. It has a circumpolar distribution in both the Arctic and Antarctic and is also found in more temperate regions. It has a three-host life cycle which usually takes 4–5 years to complete in Antarctica due to the limited seasonal availability of hosts (Eveleigh & Threlfall, 1974). It has only occasionally been reported from the continental Antarctic, despite the presence of suitable hosts, and its distribution may be limited by its inability to tolerate very low temperatures off the host. *Ixodes uriae* can overwinter in all stages of its life cycle. It also shows behavioural diapause, which is a suspension of host-seeking behaviours, and morphological diapause: a delay in embryogenesis of the developing egg, metamorphosis of larvae and nymphs after engorgement or a delay in female oviposition (Lee & Baust, 1987). The ticks attach to hosts during nesting and engorgement is completed within 1 week, after which they detach (Eveleigh & Threlfall, 1974). They have access to their hosts for only a brief period of the year and it is likely that *I. uriae* must overwinter three times to complete its life cycle: as an egg or unengorged larva, an engorged larva and as an engorged nymph or pharate adult (Lee & Baust, 1987). Apart from the brief time spent feeding, the ticks are off the host and exposed to ambient temperatures.

Three-host ixodid ticks spend most of their time off the host and most of those of medical and veterinary importance in the USA overwinter off-host. The lone star tick, *Amblyomma americanum*, overwinters as a nymph in diapause, whilst the American dog tick, *Dermacentor variabilis* overwinters in any stage. The Eastern blacklegged tick, *Ixodes scapularis*, in the northern part of its range has a 2-year life cycle, overwintering as nymphs and adults. The ranges of these three species overlap and may partly be explained by their response to climatic factors such as low temperature (Burks *et al.* 1996).

Arthropod ectoparasites which remain attached to their host during winter will be protected against low temperatures. The Antarctic flea *Glaciopsyllus antarcticus* infects a number of migratory seabirds. No live fleas were found in nest material during the winter, suggesting that they overwinter on their hosts (Whitehead *et al.* 1991).

FREE-LIVING STAGES AND OFF-HOST SURVIVAL

The free-living stages of parasites and ectoparasites off their host are faced with low temperatures and the risk of freezing, whether their host is an ectotherm or an endotherm. In this section I will consider the cold-tolerance mechanisms of the free-living or off-host stages of a number of parasites.

Globodera rostochiensis

The potato cyst nematode *Globodera rostochiensis* evolved along with its potato host in the high Andes of South America. It is exposed to subzero temperatures in its natural alpine habitat and in many parts of its present range. After mating, the female becomes gravid, with 200–300 eggs being retained within her uterus. She then dies and her body is chemically hardened by a quinone-tanning process to form a cyst. *Globodera rostochiensis* overwinters as a 2nd-stage larva (L_2) within the eggshell within the cyst. This represents a complex situation for its cold tolerance since both the eggshell and the cyst wall could act as barriers to ice nucleation.

The eggshell prevents inoculative freezing in water (that is ice forming outside the egg from seeding the freezing of the contents) and allows the enclosed L_2 to supercool and survive temperatures as low as $-38\,°C$ (Wharton, Perry & Beane, 1993). Hatched L_2s, however, freeze by inoculative freezing and are killed. The larvae within the egg can thus use a freeze-avoiding strategy and supercool in the presence of external ice because the eggshell acts as a barrier to inoculative freezing. This interpretation is supported by studies using differential scanning calorimetry (DSC). During cooling DSC thermograms of cysts show three exotherms (production of heat), which indicate freezing events, and during warming three endotherms (absorption of heat), which indicate the melting of compartments of different composition (Wharton & Ramløv, 1995). These indicate the freezing and melting of three water compartments: the water surrounding the cyst (exotherm I, endotherm VI), the water between the cyst wall and the eggs (exotherm II, endotherm V) and the water contained within the eggs (exotherm III, endotherm IV) (Fig. 2). This explanation is supported by the disappearance of exotherm II after removal of eggs from the cyst and the merging of endotherms V and IV after heating cysts to 70 °C. Heating destroys the permeability barrier of the eggshell and allows mixing between these two compartments. The peak of the egg exotherm ($-38\,°C$) corresponds with the mean supercooling point of eggs ($-38.2\,°C$), determined by cryomicroscopy (Wharton *et al.* 1993; Wharton & Ramløv, 1995). The perivitelline space contains trehalose (Clarke & Hennessy, 1976), which is known to act as an antifreeze in some animals (Lee, 1991).

The eggshell is likely to act as a barrier to inoculative freezing of the eggs of many parasites. This may be of particular importance for parasites which spend all of the free-living phase of their life cycle within the egg and which are exposed to subzero temperatures. The walls of protozoan cysts, and of other encysted parasite stages, are also likely to act in this way. Retained cuticles or sheaths can also act as barriers to inoculative freezing (Wharton & Allan, 1989; Wharton & Surrey, 1994).

Trichostrongylus colubriformis

Trichostrongyle nematodes infect sheep and other ruminants. The infective L_3s can overwinter on pasture and are exposed to subzero temperatures in parts of their range. Wertejuk (1959) reported that a variety of species survived temperatures as low as $-28\,°C$ on Polish pastures. The L_2/L_3 moult is incomplete and the L_3 retains the cuticle of the previous stage as a sheath, which is shed upon ingestion by the host. This process is known as exsheathment but is actually an ecdysis (Wharton, 1986).

The apparent cold-hardiness strategy of the L_3s depends on the conditions under which they are exposed to low temperatures. If the surface water is removed and water loss prevented by covering with liquid paraffin the L_3s supercool, with a mean supercooling point of $-24.0\,°C$ (Wharton & Allan, 1989). They thus adopt a freeze-avoiding strategy under these conditions. The situation in which the nematodes are free of surface water and yet not

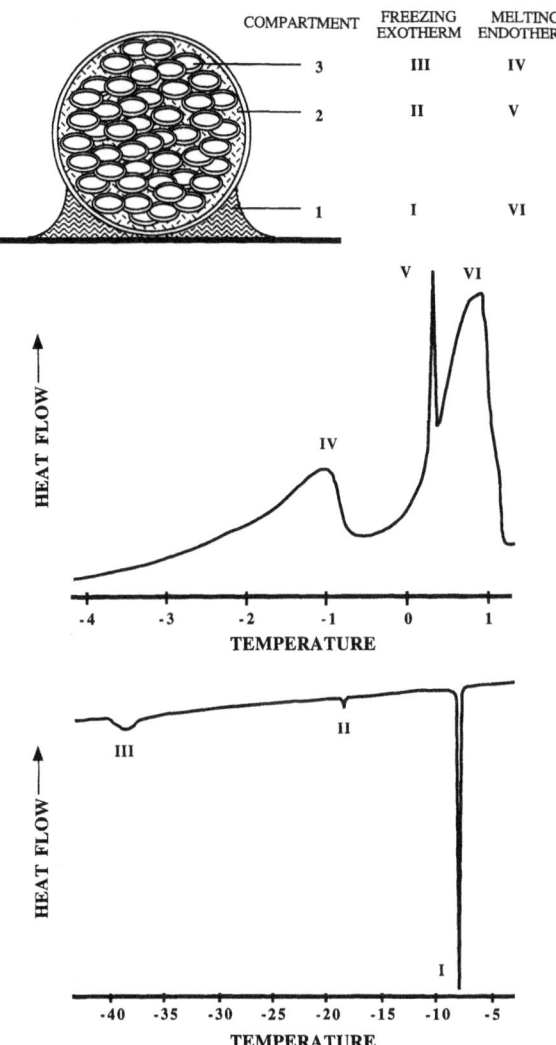

Fig. 2. The water compartments (top) and associated exotherms during freezing (middle) and endotherms during melting (bottom) in DSC thermograms of the cysts of *Globodera rostochiensis*. The compartments are: (1) the water surrounding the cyst, (2) the water between the cyst wall and the eggs and (3) the water within the eggs (redrawn from Wharton & Ramløv, 1995).

desiccated is, however, likely to be transient under natural conditions (Wharton, 1995). It is more likely that the nematode will continue to lose water and, if it is able to survive, enter a state of anhydrobiosis (see Perry, this volume). *Trichostrongylus colubriformis* L_3s can survive anhydrobiotically if they lose water slowly (Allan & Wharton, 1990). In a state of anhydrobiosis there is no water to freeze and hence the nematodes are also likely to be resistant to low temperatures.

Nematodes are essentially aquatic organisms, requiring at least a film of water for activity. Exposure to subzero temperatures is likely to occur whilst the nematodes are in contact with water and they are at greater risk of inoculative freezing than are terrestrial arthropods (Wharton, 1995). In water, the sheath can prevent inoculative freezing and allow the L_3 to supercool in the presence of external ice. A proportion of L_3s do freeze, however, but survive and thus show a degree of freezing tolerance (Wharton & Allan, 1989). In water the L_3s thus display both freeze avoidance and freezing tolerance. In arthropods these two strategies are often considered to be distinct (Zachariassen, 1985); although there have been reports of insects shifting their cold-tolerance strategy in successive seasons (Horwarth & Duman, 1984). If the sheaths of *T. colubriformis* L_3s are removed there is a shift from freeze avoidance to freezing tolerance but an overall decrease in cold tolerance (Wharton & Allan, 1989). This supports the suggested role of the sheath in reducing inoculative freezing.

Entomopathogenic nematodes

Studies on the cold tolerance of entomopathogenic nematodes of the genera *Steinernema* and *Heterorhabditis* have been initiated because of the need to develop a long-term storage method for these biological control agents of insect pests. Poor storage ability is a major impediment to their widespread commercial application (Friedman, 1990). The limited success of desiccation as a storage method (Womersley, 1990; Perry, this volume) has led to an interest in freezing as an alternative. Small quantities have been stored by freezing in liquid nitrogen (Popiel & Vasquez, 1991) but a technique for the storage of commercial quantities is more likely to be based on their natural cold-tolerance abilities.

The sheath of the infective larva of *Heterorhabditis zealandica* prevents inoculative freezing and allows the larva to supercool (Wharton & Surrey, 1994). Survival is, however, lower than would be predicted if each larva which did not freeze survived. They are thus chill tolerant (Bale, 1993) and suffer prefreeze mortality. The larvae are more cold tolerant if reared on an insect host (*Galleria mellonella*) than if grown in artificial culture, suggesting a nutritional component. Acclimation at low temperature, however, does not enhance their cold tolerance (Surrey, 1996).

Heterorhabditis bacteriophora, *Steinernema feltiae*, *S. anomoli*, *S. riobravis* and *S. glaseri* infective larvae are freezing tolerant and survive inoculative freezing from the surrounding water (Brown & Gaugler, 1996, 1998) but there is some evidence for prefreeze mortality in *H. bacteriophora* (Brown & Gaugler, 1996). Freezing tolerance in *H. bacteriophora* and *S. feltiae* was enhanced by acclimation at low temperatures and by the addition of glycerol (Brown & Gaugler, 1996).

Ticks

The Antarctic tick *Ixodes uriae* does not tolerate freezing but survives subzero temperatures by a freeze-avoiding strategy (Lee & Baust, 1987). Glycerol is present in *I. uriae* but at low levels and there is no seasonal variation in glycerol levels or changes in response to acclimation. This suggests that the synthesis of sugars or polyols as antifreezes does not play a role in the cold tolerance of this species (Lee & Baust, 1987). This tick appears to maintain its cold hardiness throughout the year, with the exception of a brief period during which feeding and the processing of a blood meal occurs. Engorged ticks have higher supercooling points than unengorged ticks, presumably because of ice nucleating agents in the blood meal. The processing and digestion of the meal must remove or mask these ice nucleators and thus restore the supercooling capacity (Lee & Baust, 1987).

The lower lethal temperatures of three North American hard ticks, *Amblyomma americanum*, *Dermacentor variabilis* and *Ixodes scapularis*, are much higher than their supercooling points. This indicates prefreeze mortality with chilling injury resulting in death before the supercooling point was reached (Burks *et al.* 1996). The lower lethal temperature was, however, much lower than temperatures recorded during winter in tick hibernaculae and they can thus survive temperatures encountered during winter. Inoculative freezing occurs at much higher temperatures, however, and results in lethal freezing, which may be the major cause of cold-related mortality in the field. This may indicate a preference for dry overwintering sites but ticks are susceptible to desiccation and require fairly moist sites for off-host survival (Needham & Teal, 1991). Microhabitat selection may thus have to balance the risk of desiccation with that of inoculative freezing (Burks *et al.* 1996).

The soft tick *Argus reflexus* is native to parts of Europe and the Near East but is an introduced species in Central Europe. It overwinters off its host, the rock pigeon. The hard tick *Ixodes ricinus* is native to Central Europe and parasitizes a range of mammals, birds and reptiles. Dautel & Knulle (1997) compared the cold hardiness of these two ticks, which were recorded as having been exposed to temperatures as low as $-13\ °C$ in an experimental attic. All postembryonic stages of *A. reflexus* had a marked supercooling capacity with supercooling points below $-20\ °C$ throughout the year. They did not survive freezing and thus employ a freeze-avoiding strategy. Engorgement did not appear to affect supercooling. The supercooling capacity of *I. ricinus* was very similar, although there was some evidence that engorgement elevated supercooling points in this species. Lower lethal temperatures were generally similar to the corresponding supercooling point for both species, indicating that there was little prefreeze mortality. Inoculative freezing in contact with ice resulted in increased mortality. The introduced species, *A. reflexus*, had a greater degree of cold tolerance than the native *I. ricinus*, despite the former being a native of areas where it is unlikely that it is exposed to very low overwinter temperatures. The supercooling capacity of *A. reflexus* may be a consequence of its ability to survive prolonged periods of starvation and low water availability (Dautel & Knulle, 1997).

Insect parasitoids

Insect parasitoids which overwinter outside their host may be exposed to subzero temperatures. The Braconid parasitoid, *Cotesa glomeratus* parasitizes a wide range of lepidopterous larvae but one of its most common hosts is the large white butterfly, *Pieris brassicae*. *Cotesa glomartus* overwinter as final instar larvae in diapause, with pupation delayed until spring. The larvae accumulate glycerol and have mean supercooling points between $-25\ °C$ and $-26\ °C$. This may indicate a freeze-avoiding strategy, but a decline in survival at subzero temperatures above the supercooling point indicates that they also suffer prefreeze mortality (Pullin, 1994). *Athrycia cinerea*, a parasitoid of the bertha armyworm, overwinters as pupae. The mean supercooling points of large puparia were $-21·1\ °C$ and of small puparia $-27·3\ °C$ (Turnock & Bilodeau, 1992). They also use a freeze-avoiding strategy, with some indication of prefreeze mortality. The hymenopteran egg parasite *Anastatus disparis* infects eggs of the gypsy moth. Larvae consume the entire egg contents and overwinter as mature larvae within the chorion of the host egg. The larvae are killed by freezing but have a mean supercooling point of $-28·8\ °C$ and thus have a freeze-avoiding strategy (Sullivan, Griffiths & Wallace, 1997).

Other examples of cold tolerance in parasite free-living stages

The freezing survival of larvae of the plant-parasitic nematode *Meloidogyne hapla* is enhanced after acclimation at low temperatures (Forge & MacGuidwin, 1990). The response is reversed if it is returned to higher temperatures (Forge & MacGuidwin, 1992*a*) and survival is enhanced at low water potentials (Forge & MacGuidwin, 1992*b*). It is not known whether this nematode survives by freezing tolerance or freeze avoidance. Forge & MacGuidwin (1992) suggest that in soil, the nematodes may desiccate rather than freeze due to freeze concentration effects. *Meloidogyne hapla* will freeze by inoculative freezing (Sayre, 1964); although this may not be the case at low water potentials in soil. The free-living Antarctic nematode *Panagrolaimus davidi* freezes by inoculative freezing at osmolalities which are much higher than those likely to occur by freeze concentration effects (Wharton & To, 1996).

The freezing survival of *Ditylenchus dipsaci* infective larvae is similar in contact or not in contact with water (Wharton, Young & Barrett, 1984), which may indicate freezing tolerance. Cryomicroscopy confirms that they freeze by inoculative freezing and are freezing tolerant (D. A. Wharton, unpublished observations). *Aphelenchoides ritzemabosi* also has some freezing tolerance ability (Asahina, 1959). There are other records of plant-parasitic nematodes surviving subzero temperatures (Miller, 1968; Antoniou, 1989) but the mechanisms by which they do so are unknown.

The eggs and hatched L_3s of the animal-parasitic nematode *Nematodirus battus* supercool in the absence of surface water. The degree of supercooling is increased after low-temperature acclimation (Ash & Atkinson, 1986), this is accompanied by an increase in trehalose concentration (Ash & Atkinson, 1983). Whether this nematode survives inoculative freezing in water is unknown but it might be expected that the eggshell would allow unhatched larvae to supercool in the presence of external ice, whereas hatched larvae would freeze.

Elaphostrongylus rangiferi L_1s, a nematode parasite of reindeer, survive freezing, at $-80\ ^\circ C$ or $-20\ ^\circ C$ in faeces and at $-20\ ^\circ C$ in water for 360 days (Lorentzen & Halvorsen, 1986). L_3s from a number of species of trichostrongyle nematodes have been reported to survive subzero temperatures under a variety of conditions (Wharton & Allan, 1989). The mechanisms by which these nematodes, and of others reported in the literature, survive are unknown but it is likely that if in contact with water they will freeze by inoculative freezing, unless this is prevented by the presence of a sheath or an eggshell. If free of surface water they may be able to supercool or to survive in a state of anhydrobiosis (Wharton, 1995). We clearly need data from a much wider range of species before we can determine the distribution of cold tolerance strategies in nematodes.

The cold tolerance mechanisms of the free-living stages of most other groups of parasites remain to be determined.

LOW TEMPERATURES AND PARASITIC STAGES

Parasites may be exposed to subzero temperatures within the host if the host is an ectotherm and is itself periodically exposed to these temperatures or within the carcass of an endothermic host after the host has died. The cold-tolerance strategy of the parasite may be dictated by that of its host. A parasite of a freeze-avoiding host can rely on the host's ability to supercool. The only requirement would be that the parasite did not initiate ice nucleation above the temperature at which nucleation is initiated within the body of the host itself. Since the parasite will always be smaller than its host the supercooling point of the parasite is likely to be below that of the host, as long as the parasite does not contain any ice-nucleating agents. A parasite of a freezing tolerant host, however, may also be subject to freezing unless it is able to inhabit parts of the host which do not freeze.

Wetanema sp. is a thelastomatid nematode which inhabits the hind gut of *Hemideina maori*, a large orthopteran insect which lives under rock slabs in the alpine zone of the South Island of New Zealand. The host is regularly exposed to temperatures below $-5\ ^\circ C$, and occasionally to as low as $-11\ ^\circ C$, and is freezing tolerant (Ramløv, Bedford & Leader, 1992; Sinclair, 1997). All stages of *Wetanema* are found within their host throughout the winter, suggesting that the nematode can survive freezing of its host in the field (Tyrrell et al. 1994). The gut of the host contains potent ice nucleators (Sinclair, Worland & Wharton, 1999) and hence *Wetanema* will experience freezing. In laboratory studies live nematodes were recovered from hosts which had been frozen to $-61\ ^\circ C$. The lower lethal temperature of the host is about $-11\ ^\circ C$ (Ramløv et al. 1992). *Wetanema* is thus able to survive much lower temperatures than its host. The freezing of *Wetanema* has also been observed directly on a microscope cold stage (Tyrrell et al. 1994). The nematode froze by inoculative freezing from the external medium, with freezing starting at the posterior end of the body and travelling forwards. Nematodes observed to freeze on the cold stage survived for six hours or more. The nematode is thus freezing tolerant and mirrors the cold tolerance strategy of its host; suggesting the coevolution of cold tolerance strategies of the parasite and its host.

There are few other reports of parasites surviving freezing within live hosts. The hymenopteran parasitoid *Syndipnus* sp. survives ice nucleation within its host the Arctic sawfly *Amauronematus* (Humble & Ring, 1985). The filarid nematode *Pelecitus fulicaeatrae* lives near the tendons of the ankles of its bird host and may be exposed to subzero temperatures during the winter within their host since they are located close to the surface of the skin and away from major blood vessels (Bartlett, 1992).

Many other freezing-tolerant hosts carry parasites. The parasite could avoid the risk of freezing within its host by being outside the host in the free-living phase of its life cycle during periods when the host is likely to freeze. However, the free-living stages of the parasite will be in a similar environment to that of the host and exposed to the same, or greater, risk of freezing. For parasites remaining in the host they may be exposed to freezing within the host's tissues. The wood frog, *Rana sylvatica* is widely distributed in N. America from the southern USA to above the Arctic Circle in Alaska. It overwinters under leaf litter where its surroundings often undergo extensive freezing. *Rana sylvatica* is freezing tolerant with about 65% of its body water being converted into ice and it survives temperatures as low as $-6\ ^\circ C$ (Layne & Lee, 1995). The freezing process is slow, often taking more than 24 h for completion. Survival is aided by the initiation of freezing at a high subzero temperature by inoculative freezing through the skin, or by nucleators such as bacteria in the gut, and by the production of glucose as a cryoprotectant (Storey & Storey, 1996; Lee & Costanzo, 1998).

Rana sylvatica has an extensive parasite fauna with a variety of protozoa, trematodes, cestodes and nematodes parasitizing sites including the intestine, lungs, liver, mesenteries, body cavity and musculature (McAllister *et al.* 1995; Yoder & Coggins, 1996). Survival of these parasites within the host during freezing may involve tolerating freezing themselves or survival in sites within the host which do not freeze. Ice formation in *R. sylvatica* appears to be restricted to the body cavity and extracellular spaces. Ice formation has been monitored in live animals using proton magnetic resonance imaging (Rubinsky *et al.* 1994). Freezing starts at the periphery and moves inwards. Water from the body organs appears to be sequestered into ice in the body cavity. This results in an osmotic dehydration of the organs (Storey & Storey, 1996; Lee & Costanzo, 1998). Parasites infecting the body organs could thus have less contact with ice than those in the body cavity but some ice does still form in the muscles, heart and other organs (R. Lee, personal communication). They may also be subjected to osmotically-induced dehydration in these sites.

The lungworm *Rhabdias ranae*, which parasitizes *R. sylvatica* and several other amphibian hosts, appears to be freezing tolerant (Woodhams *et al.* 2000). Live *R. ranae* were found in the lungs of frogs that survived freezing at -4 °C in the laboratory, and isolated lungworms tolerated *in vitro* freezing to the same temperature. Lungworms cooled on a cryomicroscope stage froze internally at -3 °C, shortly after contacting ice crystals in the suspension medium, whereas worms submerged in paraffin oil supercooled extensively (to -21 °C). Given its high susceptibility to ice inoculation, it is likely that *R. ranae*, like its host, survives exposure to subzero temperatures by tolerating internal freezing.

There are few reports on the cold tolerance of parasites within freeze-avoiding hosts. The hyperparasitoid *Tetrastichus galactopus* survives low temperatures within overwintering larvae of its host, the parasitoid *Cotesia glomeratus*, which can supercool but also shows some prefreeze mortality (Pullin, 1994). The egg parasitoid *Anaphes* sp. overwinters within eggs of its host, the carrot weevil *Listronotus oregonensis*. *Anaphes* sp. relies upon the supercooling of its host egg and freezing is fatal. Parasitized eggs synthesize glycerol and fructose which may enchance their supercooling capacity by acting as antifreezes (Hance & Boivin, 1993). Pupae of the cabbage root fly *Delia radicum* parasitized by the hymenopteran parasitoid *Trybliographa rapae* showed more variability in their supercooling points than unparasitized hosts, with more with high supercooling points and fewer with low supercooling points. This may indicate an adverse effect of the parasitoid on the cold tolerance of its host but parasitized hosts still have sufficient cold tolerance to survive temperatures they are likely to encounter in the field (Block, Turnock & Jones, 1987).

Parasites may also survive freezing within the frozen carcasses of their host. This is of biological relevance if new hosts are infected by consuming infective stages of the parasite when they eat the carcass. A number of parasites are transmitted in this way and hence survival of infective stages within the carcass of a frozen host may be an important epidemiological factor in some situations.

Trichinella spp. infect a variety of mammals and are transmitted by the ingestion of larvae encysted in the musculature. Arctic isolates of *Trichinella* (*T. nativa*) are more tolerant of freezing than are isolates from more temperate latitudes (Smith, 1984; Kapel *et al.* 1999). The ability to survive freezing at very low temperatures is enhanced by preconditioning at a higher (but still subzero) temperature (Smith, 1984, 1987). This may indicate an acclimation response, for example involving the production of cryoprotectants. Viable infective larvae have been recovered from arctic fox tissues after freezing at -18 °C for 4 years (Kapel *et al.* 1999). The mechanisms involved in the cold tolerance of this nematode are unknown.

Other reports of parasites surviving freezing within the host carcass are for *Pelecitus fulicaeatrae* a filarid nematode of the American coot (Bartlett, 1992), *Dirofilaria raemeri* a parasite of marsupials (Spratt, 1972) and *Anisakis simplex* within its fish intermediate host (Gustafson, 1953). *Anisakis* larvae infect a variety of fish and are unlikely to be exposed naturally to freezing. Although its ability to survive a degree of freezing stress (Gustafson, 1953; Wharton and Aalders, unpublished observations) is of no adaptive significance, it is important since fish and fish products which are to be consumed raw (e.g. as sushi) or lightly cooked (e.g. by cold smoking) are made safe for human consumption by freezing. It is therefore necessary to determine the temperature and time of freezing which will ensure the killing of larvae. This is also true for the use of freezing in this way for other parasites such as *T. spiralis* and *Echinococcus granulosus*.

CONCLUSIONS

In this review I have described how subzero temperatures affect parasites in a number of situations. Free-living and off-host stages are exposed to low temperatures in the environment outside the host. Parasites of ectothermic hosts may be exposed to freezing, or the risk of freezing, within the host. Finally, parasites may be exposed to freezing within the carcass of the host after it dies. These events could be important in determining the epidemiology of the parasite. Some parasites respond to low temperatures with strategies which synchronize their life cycle with favourable seasons and the availability of hosts. Parasites may take advantage of the cold tolerance adaptations of their host and they may have cold tolerance adaptations of their own. However, for most groups of parasites we have limited information on their cold tolerance mechanisms. Nematodes and ticks have been the most studied, but even for these groups a very limited range of species have been investigated. We have little or no information on the cold tolerance mechanisms of most other groups of parasites.

REFERENCES

ALLAN, G. S. & WHARTON, D. A. (1990). Anhydrobiosis in the infective juveniles of *Trichostrongylus colubriformis* (Nematoda: Trichostrongylidae). *International Journal for Parasitology* **20**, 183–192.

ANTONIOU, M. (1989). Arrested development in plant parasitic nematodes. *Helminthological Abstracts* **B58**, 1–19.

ASAHINA, E. (1959). Frost-resistance in a nematode, *Aphelenchoides ritzema-bosi*. *Low Temperature Science* **B17**, 51–62.

ASH, C. P. J. & ATKINSON, H. J. (1983). Evidence for a temperature-dependent conversion of lipid reserves to carbohydrate in quiescent eggs of the nematode, *Nematodirus battus*. *Comparative Biochemistry and Physiology* **76B**, 603–610.

ASH, C. P.J. & ATKINSON, H. J. (1986). *Nematodirus battus*: development of cold hardiness in dormant eggs. *Experimental Parasitology* **62**, 24–28.

BALE, J. S. (1993). Classes of insect cold hardiness. *Functional Ecology* **7**, 751–753.

BARTLETT, L. M. (1992). Cold-hardiness in *Pelecitus fulicaeatrae* (Nematoda: Filarioidea), a parasite in the ankle of *Fulica americana* (Aves). *Journal of Parasitology* **78**, 138–139.

BLOCK, W. (1995). Insects and freezing. *Science Progress* **78**, 349–372.

BLOCK, W., TURNOCK, W. J. & JONES, T. H. (1987). Cold resistance and overwintering survival of the cabbage root fly, *Delia radicum* (Anthomyiidae), and its parasitoid, *Trybliographa rapae* (Cynipidae), in England. *Oecologia* **71**, 332–338.

BROWN, I. M. & GAUGLER, R. (1996). Cold tolerance of steinernematid and heterorhabditid nematodes. *Journal of Thermal Biology* **21**, 115–121.

BROWN, I. M. & GAUGLER, R. (1998). Survival of steinernematid nematodes exposed to freezing. *Journal of Thermal Biology* **23**, 75–80.

BURKS, C. S., STEWART, R. L., NEEDHAM, G. R. & LEE, R. E. (1996). The role of direct chilling injury and inoculative freezing in cold tolerance of *Amblyomma americanum*, *Dermacentor variabilis* and *Ixodes scapularis*. *Physiological Entomology* **21**, 44–50.

CLARKE, A. J. & HENNESSY, J. (1976). The distribution of carbohydrates in cysts of *Heterodera rostochiensis*. *Nematologica* **22**, 190–195.

CROSS, D. A., KLESIUS, P. H. & WILLIAMS, J. C. (1988). Preliminary report: immunodiagnosis of pre-type II ostertagiasis. *Veterinary Parasitology* **27**, 151–158.

DAUTEL, H. & KNULLE, W. (1997). Cold hardiness, supercooling ability and causes of low-temperature mortality in the soft tick, *Argas reflexus*, and the hard tick, *Ixodes ricinus* (Acari, Ixodoidea) from Central Europe. *Journal of Insect Physiology* **43**, 843–854.

DAVENPORT, J. (1992). *Animal Life at Low Temperature*. New York and London: Chapman and Hall.

DUMAN, J. G., XU, L., NEVEN, L. G., THURSMAN, D. & WU, D. W. (1991). Hemolymph proteins involved in insect subzero-temperature tolerance: ice nucleators and antifreeze proteins. In *Insects at Low Temperatures* (ed. Lee, R. E. & Denlinger, D. L.), pp. 94–127. New York and London: Chapman and Hall.

EVANS, A. A. F. (1987). Diapause in nematodes as a survival strategy. In *Vistas on Nematology* (ed. Veech, J. A. & Dickson, D. W.), pp. 180–187. Hyattsville, Maryland: Society of Nematologists Inc.

EVANS, A. A. F. & PEERY, R. N. (1976). Survival strategies in nematodes. In *The Organisation of Nematodes* (ed. Croll, N. A.), pp. 383–424. London & New York: Academic Press.

EVELEIGH, E. S. & THRELFALL, W. (1974). The biology of *Ixodes* (*Ceratixodes*) *uriae* White, 1852 in Newfoundland. *Acarologia* **16**, 621–635.

EYSKER, M. (1997). Some aspects of inhibited development of trichostrongylids in ruminants. *Veterinary Parasitology* **72**, 265–283.

FORGE, T. A. & MacGUIDWIN, A. E. (1990). Cold hardening of *Meloidogyne hapla* second-stage juveniles. *Journal of Nematology* **22**, 101–105.

FORGE, T. A. & MacGUIDWIN, A. E. (1992*a*). Impact of thermal history on tolerance of *Meloidogyne hapla* second-stage juveniles to external freezing. *Journal of Nematology* **24**, 262–268.

FORGE, T. A. & MacGUIDWIN, A. E. (1992*b*). Effects of water potential and temperature on survival of the nematode *Meloidogyne hapla* in frozen soil. *Canadian Journal of Zoology* **70**, 1553–1560.

FRIEDMAN, M. J. (1990). Commercial production and development. In *Entomopathogenic Nematodes in Biological Control* (ed. Gaugler, R. & Kaya, H. K.), pp. 153–172. Boca Raton: CRC Press.

GIBBS, H. C. (1986). Hypobiosis in parasitic nematodes – an update. *Advances in Parasitology* **25**, 129–174.

GUSTAFSON, P. V. (1953). The effect of freezing on encysted *Anisakis* larvae. *Journal of Parasitology* **39**, 585–588.

HANCE, T. & BOIVIN, G. (1993). Effect of parasitism by *Anaphes* sp. (Hymenoptera, Mymaridae) on the cold hardiness of *Listronotus oregonensis* (Coleoptera, Curculionidae) eggs. *Canadian Journal of Zoology* **71**, 759–764.

HOMINICK, W. M. (1986). Photoperiod and diapause in the potato cyst-nematode, *Globodera rostochiensis*. *Nematologica* **32**, 408–418.

HORWARTH, K. C. & DUMAN, J. G. (1984). Yearly variations in the overwintering mechanisms of the cold-hardy beetle, *Dendroides canadensis*. *Physiological Zoology* **57**, 40–45.

HUMBLE, L. M. & RING, R. A. (1985). Inoculative freezing of a larval parasitoid within its host. *Cryo-Letters* **6**, 59–66.

JONES, P. W., TYLKA, G. L. & PERRY, R. N. (1998). Hatching. In *The Physiology and Biochemistry of Free-living and Plant-parasitic Nematodes* (ed. Perry, R. N. & Wright, D. J.), pp. 181–212. Wallingford & New York: CABI Publishing.

KAPEL, C. M. O., POZIO, E., SACCHI, L. & PRESTURD, P. (1999). Freeze tolerance, morphology, and RAPD-PCR identification of *Trichinella nativa* in naturally infected arctic foxes. *Journal of Parasitology* **85**, 144–147.

KARLSSON, J. O. M., CRAVALHO, E. G. & TONER, M. (1993). Intracellular ice formation: causes and consequences. *Cryo-Letters* **14**, 323–336.

KENNEDY, C. R. (1975). *Ecological Animal Parasitology*. Oxford: Blackwell Scientific Publications.

KNIGHT, C. A. & DUMAN, J. G. (1986). Inhibition of recrystallization of ice by insect thermal hysteresis proteins: a possible cryoprotective role. *Cryobiology* **23**, 256–262.

LACKIE, A. M. (1975). The activation of infective stages of endoparasites of vertebrates. *Biological Reviews* **50**, 285–323.

LAYNE, J. R. & LEE, R. E. (1995). Adaptations of frogs to survive freezing. *Climate Research* **5**, 53–59.

LEE, R. E. (1991). Principles of insect low temperature tolerance. In *Insects at Low Temperatures* (ed. Lee, R. E. & Denlinger, D. L.), pp. 17–46. New York and London: Chapman and Hall.

LEE, R. E. & BAUST, J. G. (1987). Cold-hardiness in the Antarctic tick, *Ixodes uriae*. *Physiological Zoology* **60**, 499–506.

LEE, R. E. & COSTANZO, J. P. (1998). Biological ice nucleation and ice distribution in cold-hardy ectothermic animals. *Annual Review of Physiology* **60**, 55–72.

LORENTZEN, G. & HALVORSEN, O. (1986). Survival of the first stage larva of the metastrongylid nematode *Elaphostrongylus rangiferi* under various conditions of temperature and humidity. *Holoarctic Ecology* **9**, 301–304.

McALLISTER, C. T., UPTON, S. J., TRAUTH, S. E. & BURSEY, C. R. (1995). Parasites of Wood Frogs, *Rana sylvatica* (Ranidae), from Arkansas, with a description of a new species of *Eimeria* (Apicomplexa, Eimeriidae). *Journal of the Helminthological Society of Washington* **62**, 143–149.

McGRATH, J. J. (1987). Cold shock: thermoelastic stress in chilled biological membranes. In *Network Thermodynamics, Heat and Mass Transfer in Biotechnology* (ed. Diller, K. R.), pp. 57–66. New York: United Engineering Center.

MILLER, P. M. (1968). The susceptibility of parasitic nematodes to sub-freezing temperatures. *Plant Disease Reporter* **52**, 768–772.

NEEDHAM, G. R. & TEAL, P. D. (1991). Off-host physiological ecology of ixodid ticks. *Annual Review of Entomology* **36**, 659–681.

POPIEL, I. & VASQUEZ, E. M. (1991). Cryopreservation of *Steinernema carpocapsae* and *Heterorhabditis bacteriophora*. *Journal of Nematology* **23**, 432–437.

POULIN, R. (1998). *Evolutionary Ecology of Parasites – From Individuals to Communities*. London: Chapman & Hall.

PULLIN, A. S. (1994). Cold tolerance of an insect parasitoid *Cotesia* (*apanteles*) *glomeratus* and a comparison with that of its host *Pieris brassicae* and a hyperparasitoid *Testrastichus galactopus*. *Cryo-Letters* **15**, 67–74.

QUINN, P. J. (1985). A lipid-phase separation model of low-temperature damage to biological membranes. *Cryobiology* **22**, 128–146.

RAMLØV, H., BEDFORD, J. & LEADER, J. (1992). Freezing tolerance of the New Zealand alpine weta, *Hemideina maori* Hutton (Orthoptera; Stenopelmatidae). *Journal of Thermal Biology* **17**, 51–54.

ROGERS, W. P. (1960). The physiology of the infective process of nematode parasites: the stimulus from the animal host. *Proceedings of the Royal Society of London* **B152**, 367–386.

RUBINSKY, B., WONG, S. T. S., HONG, J. S., GILLBERT, J., ROOS, M. & STOREY, K. B. (1994). 1H magnetic resonance imaging of freezing and thawing in freeze-tolerant frogs. *American Journal of Physiology* **266**, R1771–1777.

SAYRE, R. M. (1964). Cold-hardiness of nematodes. I. Effects of rapid freezing on the eggs and larvae of *Meloidogyne incognita* and *M. hapla*. *Nematologica* **10**, 168–179.

SINCLAIR, B. J. (1997). Seasonal variation in freezing tolerance of the New Zealand alpine cockroach *Celatoblatta quinquemaculata*. *Ecological Entomology* **22**, 462–467.

SINCLAIR, B. J., WORLAND, M. R. & WHARTON, D. A. (1999). Ice nucleation and freezing tolerance in New Zealand alpine and lowland weta, *Hemideina* spp. (Orthoptera; Stenopelmatidae). *Physiological Entomology* **24**, 56–63.

SMITH, H. J. (1984). Preconditioning of *Trichinella spiralis nativa* larvae in musculature to low temperatures. *Veterinary Parasitology* **17**, 85–90.

SMITH, H. J. (1987). Factors affecting preconditioning of *Trichinella spiralis nativa* larvae in musculature to low temperatures. *Canadian Journal of Veterinary Research* **51**, 169–173.

SPRATT, D. M. (1972). Aspects of the life cycle of *Dirofilaria raemeri* in naturally and experimentally infected kangaroos, wallaroos and wallabies. *International Journal for Parasitology* **2**, 139–156.

STOREY, K. B. & STOREY, J. M. (1988). Freeze tolerance in animals. *Physiological Reviews* **68**, 27–84.

STOREY, K. B. & STOREY, J. M. (1996). Natural freezing survival in animals. *Annual Review of Ecology and Systematics* **27**, 365–386.

SULLIVAN, C. R., GRIFFITHS, K. J. & WALLACE, D. R. (1977). Low winter temperatures and the potential for establishment of the egg parasite *Anastatus disparis* (Hymenoptera: Eupelmidae) in Ontario populations of the gypsy moth. *The Canadian Entomologist* **109**, 215–220.

SURREY, M. R. (1996). The effect of rearing method and cool temperature acclimation on the cold tolerance of *Heterorhabditis zealandica* infective juveniles (Nematoda: Heterorhabditidae). *Cryo-Letters* **17**, 313–320.

THOMAS, R. J. (1974). The role of climate in the epidemiology of nematode parasitism in ruminants. In *The Effects of Meteorological Factors Upon Parasites* (ed. Taylor, A. E. R. & Muller, R.), pp. 13–32. Oxford: Blackwell.

TURNOCK, W. J. & BILODEAU, R. J. (1992). Life history and coldhardiness of *Athrycia cinerea* (Dipt.: Tachinidae) in western Canada. *Entomophaga* 37, 353–362.

TYRRELL, C., WHARTON, D. A., RAMLØV, H. & MOLLER, H. (1994). Cold tolerance of an endoparasitic nematode within a freezing tolerant orthopteran host. *Parasitology* 109, 367–372.

VANNIER, G. (1994). The thermobiological limits of some freezing intolerant insects: The supercooling and thermostupor points. *Acta Oecologica* 15, 31–42.

WERTEJUK, M. (1959). Influence of environmental temperature on the invasive larvae of gastrointestinal nematodes of sheep. *Acta Parasitologica Polonica* 7, 315–342.

WHARTON, D. A. (1986). *A Functional Biology of Nematodes*. London & Sydney: Croom Helm.

WHARTON, D. A. (1995). Cold tolerance strategies in nematodes. *Biological Reviews* 70, 161–185.

WHARTON, D. A. & ALLAN, G. S. (1989). Cold tolerance mechanisms of the free-living stages of *Trichostrongylus colubriformis* (Nematoda: Trichostrongylidae). *Journal of Experimental Biology* 145, 353–370.

WHARTON, D. A., PERRY, R. N. & BEANE, J. (1993). The role of the eggshell in the cold tolerance mechanisms of the unhatched juveniles of *Globodera rostochiensis*. *Fundamental and Applied Nematology* 16, 425–431.

WHARTON, D. A. & RAMLØV, H. (1995). Differential scanning calorimetry studies on the cysts of the potato cyst nematode *Globodera rostochiensis* during freezing and melting. *Journal of Experimental Biology* 198, 2551–2555.

WHARTON, D. A. & SURREY, M. R. (1994). Cold tolerance mechanisms of the infective larvae of the insect parasitic nematode, *Heterorhabditis zealandica* Poinar. *Cryo-Letters* 25, 749–752.

WHARTON, D. A. & TO, N. B. (1996). Osmotic stress effects on the freezing tolerance of the Antarctic nematode *Panagrolaimus davidi*. *Journal of Comparative Physiology* B166, 344–349.

WHARTON, D. A., YOUNG, S. R. & BARRETT, J. (1984). Cold tolerance in nematodes. *Journal of Comparative Physiology* B154, 73–77.

WHITEHEAD, M. D., BURTON, H. R., BELL, P. J., ARNOULD, J. P.Y. & ROUNSEVELL, D. E. (1991). A further contribution on the biology of the Antarctic flea, *Glaciopsyllus antarcticus* (Siphonaptera, Ceratophyllidae). *Polar Biology* 11, 379–383.

WOODHAMS, D. C., COSTANZO, J. P., KELTY, J. D. & LEE, R. E. (2000). Cold hardiness in two helminth parasites of the freeze tolerant wood frog, *Rana sylvatica*. *Canadian Journal of Zoology* (in press).

WOMERSLEY, C. Z. (1990). Dehydration survival and anhydrobiotic survival. In *Entomopathogenic Nematodes in Biological Control* (ed. Gaugler, R. & Kaya, H. K.) pp. 117–137. Boca Raton: CRC Press.

WOMERSLEY, C. Z., WHARTON, D. A. & HIGA, L. M. (1998). Survival biology. In *The Physiology and Biochemistry of Free-living and Plant-parasitic Nematodes* (ed. Perry, R. N. & Wright, D. J.), pp. 271–302. Wallingford & New York: CABI Publishing.

YODER, H. R. & COGGINS, J. R. (1996). Helminth communities in the Northern Spring Peeper, *Pseudacris c. crucifer* Wied, and the Wood Frog, *Rana sylvatica* Le Conte, from Southeastern Wisconsin. *Journal of the Helminthological Society of Washington* 63, 211–214.

ZACHARIASSEN, K. A. (1985). Physiology of cold tolerance in insects. *Physiological Reviews* 65, 799–832.

Desiccation survival of parasitic nematodes

R. N. PERRY*

Entomology and Nematology Department, IACR-Rothamsted, Harpenden, Herts AL5 2JQ, UK

SUMMARY

The ability of certain species of parasitic nematodes to survive desiccation for considerable periods is a fascinating example of adaptation to the demands of fluctuating environments that occasionally can become extreme and life threatening. Behavioural and morphological adaptations associated with desiccation survival serve primarily to reduce the rate of drying, either to prolong the time taken for the nematode's water content to reach lethal low levels or, in true anhydrobiotes, to enable the structural and biochemical changes required for long-term survival to take place. Examples of these adaptations are reviewed, together with information on the factors involved in rehydration that ensure successful exit from the dormant state. Information on desiccation survival is central to effective management and control options for parasitic nematodes. It is also required to assess the feasibility of enhancing the longevity of commercial formulations of entomopathogenic nematodes, both before and after application; current research and future prospects for enhancing survival of these bio-insecticides are discussed.

Key words: Anhydrobiosis, desiccation, nematodes, survival.

INTRODUCTION

Two types of adaptation, defined originally for survival of temperature stress (Precht, 1958; Cossins & Bowler, 1987), are capacity adaptation and resistance adaptation. Capacity adaptation enables an organism to grow and reproduce under harsh environmental conditions that differ markedly from conditions required by the majority of species for continuation of their life-cycles. Resistance adaptation enables an organism to suspend development and survive environmental extremes until favourable conditions return, when growth and development can re-commence. Among species of nematodes, capacity adaptation is demonstrated by certain free-living species living in extreme environments, such as deserts or the terrestrial Antarctic, whereas resistance adaptation is an attribute of some parasitic and free-living forms that ensures survival in a fluctuating environment. For many parasitic nematodes, the requirement to persist in the absence of a host also necessitates survival of unfavourable conditions. The associated behavioural, morphological and biochemical mechanisms to withstand environmental extremes, often features of specific stages in the nematode life-cycle, contribute to the survival strategy of each species.

One aspect of nematode survival, the ability to withstand desiccation for periods considerably in excess of the duration of the normal life-cycle, has generated much research interest. In part, this has been engendered by the historical fascination of the ability, rare in the animal kingdom, to recover from extreme body-water loss. However, the effectiveness of management and control options for parasitic nematodes is often conditional on an understanding of the temporal factors involved in survival; this has also underpinned research which aims to understand the mechanisms of survival with the long-term aim of assessing the feasibility of disrupting dormancy and improving nematode control, especially of plant-parasitic species. More recently, the use of entomopathogenic nematodes as an environmentally acceptable method of controlling economically important insect pests has resulted in research aimed at not only enhancing the longevity of the nematodes in commercial formulations but also increasing survival after application (Glazer et al. 1999).

TERMINOLOGY

Dormancy can occur at most stages of the nematode's life. Keilin (1959) separated dormancy, involving lowered metabolism, from cryptobiosis, where no metabolism could be detected. In practice, it is frequently difficult to separate quiescence from cryptobiosis and many authors have used the term cryptobiosis in an arbitrary sense to indicate long-term survival, usually for years, of adverse conditions. Evans & Perry (1976) considered that the fundamental criterion for separating categories within dormancy should be the cause of arrest in development, rather than metabolic state, and considered that cryptobiosis should be viewed as the same kind of phenomenon as quiescence. Subsequently, Evans (1987) distinguished between dormancy affecting ontogenetic development and that affecting somatic development.

Dormancy is usually subdivided into two categories, diapause and quiescence. Diapause is a state of arrested development whereby development

* E-mail: roland.perry@bbsrc.ac.uk

does not occur until specific requirements have been satisfied, even if favourable conditions return. Quiescence is a spontaneous reversible response to unpredictable unfavourable environmental conditions and release from quiescence occurs when favourable conditions return. Adverse environmental conditions and the states they induce include cooling (cryobiosis), high temperatures (thermobiosis), lack of oxygen (anoxybiosis), osmotic stress (osmobiosis) and the subject of this review, dehydration (anhydrobiosis). These divisions are somewhat artificial. Nematodes are essentially aquatic organisms but many of the environmental stresses involve removal or immobilization of water. Desiccation concentrates body solutes and increases internal osmotic stress, which also may influence the rate of water loss. Exposure to hyperosmotic conditions causes partial dehydration of a nematode. Freezing may involve dehydration through sublimation of water from the solid phase (Wharton, this volume). Thus, it is unsurprising that the ability to survive one type of stress is frequently associated with increased resistance to others and nematodes exhibit similar behavioural, structural and physiological mechanisms to enhance their survival of different conditions.

OVERVIEW OF ANHYDROBIOSIS IN PARASITIC NEMATODES

Soil-dwelling nematodes are protected by the soil from extremes of moisture loss. Even when the soil water potential falls below -1.0 MPa, the relative humidity in soil pores is still above 99% (Stirling, 1991). Species of plant ecto- and endoparasitic nematodes that attack roots growing in the upper and lower soil profiles may avoid a decrease in the availability of water in the upper layers by moving downwards in the soil. Other plant-parasitic species locate and invade roots of their recently germinated hosts near the soil surface, and some species climb plants when water films cover them. In general, soil-dwelling stages of animal-parasitic nematodes remain on the soil surface or associated vegetation until ingested by their hosts. Entomopathogenic nematodes move through the upper layers of the soil in search of prey or wait on the soil surface for the host to pass. Although some plant-parasitic nematodes, such as *Globodera rostochiensis*, that inhabit deep soil have mechanisms to endure desiccation, species, such as *Ditylenchus dipsaci*, inhabiting aerial parts of plants demonstrate the most spectacular intrinsic abilities to withstand severe desiccation. The ability to survive desiccation is often commensurate with a dispersal phase of the life-cycle and such dormant stages are frequently formed in response to food shortage.

It is difficult to compare the results of different workers on the survival attributes of various species of parasitic nematodes because the experimental conditions are rarely comparable. Only general conclusions can be drawn. In most examples of anhydrobiotic survival in nematodes, the individuals involved depend on mechanisms to reduce their rate of evaporative water loss. The majority of free-living stages of animal- and plant-parasitic nematodes show little intrinsic ability to control water loss and survive desiccation, being dependent on the environmental conditions of high relative humidity within soil pores or plant material to slow down or prevent water loss. Womersley (1987) grouped nematode anhydrobiotes into slow-dehydration and fast-dehydration strategists based on the water loss dynamics commensurate with survival. Slow-dehydration strategists depend on surrounding environmental conditions to control water loss and variations in the rate at which water is removed from a given nematode species and stage depend largely on the moisture loss characteristics of the soil in which it lives (Womersley, Wharton & Higa, 1998). Physical and behavioural adaptations to control the rate of water loss are frequently associated with fast-dehydration strategists such as the infective fourth stage larvae (L_4s) of *D. dipsaci* (Perry, 1977 *a, b*).

Research into the mechanisms involved in anhydrobiotic survival has focused on three main areas: (1) behavioural and physical attributes that enhance survival, primarily associated with control of water loss from the nematode; (2) biochemical adaptations that maintain functional and structural integrity at the cellular level; and (3) the importance of rehydration in relation to metabolic and morphological changes required for successful revival. These aspects will be discussed using selected examples of animal- and plant-parasitic nematodes whose direct life-cycles involve survival outside the host.

BEHAVIOURAL RESPONSES THAT ENHANCE SURVIVAL

The two frequently quoted behavioural responses by some species of nematodes to removal of water are coiling and clumping. Coiling reduces the surface area of the nematode that is exposed to drying conditions and there is some experimental evidence to show that coiling reduces the rate of water loss of *D. myceliophagus* (Womersley, 1978). When exposed to desiccation, the infective larvae (L_3) of *Trichostrongylus colubriformis* form tight coils (Wharton, 1981). In the semi-endoparasitic nematode, *Rotylenchulus reniformis*, a direct relationship between coiling and anhydrobiotic survival has been shown (Womersley & Ching, 1989) but neither coiling nor the retention of moulted cuticles as sheaths (see below) enable this species to survive long-term dehydration. Womersley *et al.* (1998) considered that, although coiling may be an indicator of successful induction of anhydrobiosis, it cannot be used to distinguish between quiescent and cryptobiotic forms.

The classic image of anhydrobiotic nematodes is clumps of coiled, desiccated L_4s of *D. dipsaci*, yet only a very few species aggregate in this way. Although L_4s of *D. dipsaci* can survive extreme desiccation for long periods (see below), dry conditions rarely last long in the soil and the aggregations of dried L_4s, termed 'eelworm wool', are more usually associated with infected bulbs (Ellenby, 1969) or inside bean pods (Hooper, 1971). In infected narcissus bulbs, set out to dry at the end of the growing season, development is arrested at the L_4 stage and hundreds of L_4s issue from the basal plate and the lower end of the bulb scales and aggregate before drying. The death rate is greater on the outside of the aggregations and there is some evidence (Ellenby, 1969) that the outer fragments are drier. The death of the peripheral L_4s apparently provides a protective coat which aids survival of the L_4s in the centre of the 'wool' by slowing their rate of drying: the so-called 'eggshell' effect (Ellenby, 1968a). Thus, the outer layers perform the same function as physical structures discussed in the next section.

In other species of plant-parasitic nematodes, aggregations occur under natural conditions within modified plant tissue such as galls. *Anguina amsinckia* and *A. tritici* are examples of species that induce galls in the host inflorescence. Within the galls induced by *A. amsinckia* are hundreds of desiccated adults and larvae of all stages, many of which are coiled. However, not all nematodes need to coil to survive drying. The galls induced by *A. tritici* contain tightly packed aggregates of second stage larvae (L_2s) only, each of which remains uncoiled when dry. L_2s of *A. tritici* can survive severe desiccation as individuals (Ellenby, 1969), so the combination of this intrinsic ability and the behavioural adaptation of clumping, plus the protection of the gall tissue, enables L_2s to survive many years in the dry state.

Mass movement or swarming is found in mycophagous nematodes such as *D. myceliophagus* and *Aphelenchus avenae*, probably in response to lack of food or toxic products from decaying hosts. Swarming leads to aggregation and coiling during subsequent dehydration in the absence of a host but, in these examples, aggregation is not a behavioural response to desiccation *per se* (Womersley *et al.* 1998).

PHYSICAL ATTRIBUTES THAT ENHANCE SURVIVAL

Protection from environmental extremes is afforded by physical structures such as moulted cuticles, which are retained as protective sheaths, eggshells and resistant cuticles. One main function of such physical attributes is to control the rate of water loss. However, as will become clear, the ability to control the rate of water loss does not, by itself, ensure survival.

Whether in soil or plant tissue, the majority of animal- and plant-parasitic nematodes experience dehydration stress as individuals and *in vitro* experiments aiming to examine survival attributes of species do not reflect the natural situation if percentage survival is assessed using clumps of nematodes, particularly as percentage survival is influenced by the size of the aggregation (Ellenby, 1969). However, care should be taken in extrapolating from the results of laboratory-based experiments, which examine the mechanisms involved in anhydrobiotic survival, to survival under field conditions. Frequently such experiments are interpreted as indicating considerably enhanced survival attributes, but the interpretations are not borne out by field data. In part, this may be due to the interaction of factors prevalent in natural environmental conditions. The development of *Ascaris* species occurs within the egg and the infective stage is protected by the eggshell until ingestion by the host. When exposed to desiccation, the eggs of several animal-parasitic nematodes lose water very slowly and the eggshell has been implicated in enabling the unhatched larvae to survive desiccation, the lipid layer providing the main permeability barrier to water loss (Wharton, 1980). However, the rate of water loss of unhatched larvae increases as an exponential function of increasing temperature (Wharton, 1979) and, although *Ascaris* eggs lose water very slowly relative to their surface-volume ratio (Wharton, 1979), they are sensitive to desiccation in the long term and egg mortality due to dehydration has been claimed to be responsible for the complete lack of transmission of *A. suum* under intensive indoor production systems (Roepstorff, 1997). On grass plots, high temperature in combination with severe dehydration in faecal samples may have contributed to the large mortality of *A. suum* (Larsen & Roepstorff, 1999).

Females of the plant-parasitic cyst nematodes become spherical (e.g. *Globodera* spp.) or lemon-shaped (*Heterodera* spp.) and, after death of the fertilized female, the cuticle becomes tanned to form a tough, brown cyst containing 100–500 eggs, each one containing a tightly coiled infective L_2. When exposed to desiccation, the permeability characteristics of the surface layers of the cyst wall of *G. rostochiensis* change as they dry faster than the rate at which water can be replaced from within the cyst, resulting in an effective barrier to further water loss (Ellenby, 1946). The eggshell also becomes differentially permeable as it dries resulting in a reduced rate of water loss of unhatched L_2s compared with free L_2s (Ellenby, 1968a). Ultimately, the unhatched L_2

becomes as dry as the hatched L_2 yet the former survives but the latter perishes; the rate of water loss is a decisive survival factor. The susceptibility of hatched L_2s of *G. rostochiensis* to environmental extremes is offset by a sophisticated host-parasite interaction whereby the L_2 does not hatch unless stimulated by host root diffusates, thus ensuring that a large population of infective L_2s are released close to susceptible roots. An initial phase in the hatching process is a change in the eggshell permeability induced directly by hatching factors in host diffusate (Jones, Tylka & Perry, 1998); alteration in eggshell permeability characteristics through the action of hatching factors results in an increased susceptibility of unhatched L_2s to dehydration stress (Perry, 1983).

Like the cyst nematodes, the root-knot nematodes (*Meloidogyne* spp.) are plant endoparasites but the female lays eggs into a gelatinous matrix consisting of an irregular meshwork of glycoprotein material. The gelatinous matrix shrinks and hardens when dried (Bird & Soeffky, 1972), thus exerting mechanical pressure on the eggs to inhibit hatching during drought conditions and ensuring that the infective L_2s are retained within the protection of the eggs and matrix. A third protective layer, which appears as an extracuticular subcrystalline layer in *M. charis* (Demeure & Freckman, 1981), also may function to slow the rate of water loss.

Although the cyst wall or gelatinous matrix and the eggshell enhance the survival of unhatched L_2s of cyst and root-knot nematodes, they do not result in unhatched L_2s of the different species being able to survive dehydration equally well. The ability to survive severe desiccation varies considerably between species and long-term anhydrobiosis seems to be associated primarily with those species, such as *G. rostochiensis*, that have a very restricted host range. For example, compared with *G. rostochiensis*, species such as *H. schachtii* hatch well in water without depending on host stimulation (Perry, Clarke & Hennessy, 1980) but withstand desiccation poorly (Ellenby, 1968b); the very wide host range of *H. schachtii* (some 218 plant species, including many weeds) ensures survival of populations until the main host crop becomes available.

Eggs of adult trichostrongyle nematodes that parasitize sheep and cattle are passed to the outside in faeces and also have to withstand environmental extremes. Waller & Donald (1970) considered that eggs of *Haemonchus contortus* and *Trichostrongylus colubriformis* will survive dehydration provided that development can proceed to the prehatch stage during drying, and before the embryo loses a critical amount of water. Under desiccating conditions, the eggshell of *H. contortus* is more permeable to water loss than that of *T. colubriformis*. The inner layer of the eggshell of *H. contortus* contains non-polar lipids of the hydrocarbon type, whereas the equivalent layer of *T. colubriformis* eggs contains either more polar unsaturated lipids or proteins (Waller, 1971); such physico-chemical differences between the eggshells of these and other species of nematodes may be correlated, in part, with differences in the ability of the eggshells to control water loss. The survival attributes of the infective larvae of *H. contortus* also have been examined experimentally. Under suitable environmental conditions, the first stage larva hatches from the egg and development proceeds to the L_2 and then to the infective third stage larva (L_3). The L_3 of *H. contortus* retains the cuticle of the previous stage as a sheath and development is arrested until exsheathment occurs in the rumen of the host. The ensheathed L_3 survives desiccation better than the exsheathed form: at 47% relative humidity ensheathed L_3 can survive for at least 4 weeks whereas the exsheathed L_3 perishes after 8 h (Ellenby, 1968c). On exposure to desiccating conditions, the sheath dries first and becomes increasingly impermeable, thus slowing down the rate of water loss of the enclosed L_3 and enabling it to survive (Ellenby, 1968c). Exsheathed L_3s of *T. colubriformis* will survive transfer to 0% relative humidity if they are first dried slowly at high humidity (Allan & Wharton, 1990).

Some plant-parasitic and entomopathogenic nematodes also retain moulted cuticles to protect infective stages. L_2s of *Rotylenchulus reniformis* hatch in the soil and the post-hatch moulting from L_2 to adult is completed without feeding (Gaur & Perry, 1991a), resulting in a decrease in body volume from L_2s to adults (Bird, 1983). The young adults are enclosed in the three cuticular sheaths from the previous stages and remain inactive in dry soil until favourable conditions return allowing movement and exsheathment. Gaur & Perry (1991b) showed that the exsheathed adults survived poorly compared to ensheathed adults, and the sheaths aided desiccation survival by slowing the rate of drying of the enclosed adult. However, the reduced rate of water loss only assisted individuals to survive for periods over which water loss was controlled; they showed no ability for prolonged survival once their water content had been reduced to less than 10%. A similar situation appears to occur with the entomopathogenic nematode, *Heterorhabditis megidis*, where the sheath surrounding the infective larvae slows down the rate of drying of the enclosed larvae but does not result in them surviving for extended periods (Menti, Wright & Perry, 1997). Thus, whilst control of water loss enables some species to enter anhydrobiosis, *R. reniformis* and *H. megidis* are examples of nematodes that show little intrinsic ability for anhydrobiotic survival; control of water loss merely prolongs the time taken for the nematode's water content to reach lethal low levels.

O'Leary & Burnell (1997) isolated mutant lines of *H. megidis* with an increased tolerance to desiccation at low humidities. The surface of the sheaths of mutant lines is more negatively charged than that of the wild-type and removal of the outer layer, possibly the epicuticle, resulted in loss of the mutant phenotype (O'Leary, Burnell & Kusel, 1998). A strongly negative charge on the epicuticle has been found in larvae of *Strongyloides ratti* and related to desiccation tolerance (Murrell, Graham & McGreevy, 1983). O'Leary *et al.* (1998) suggested that the presence of a strongly ionized or polar coat on the surface of nematodes could facilitate the maintenance of a film of water over the cuticle.

The retention of moulted cuticles is found in other species of soil-dwelling nematodes but their presence does not necessarily indicate a rôle in desiccation survival; a sheath or sheaths also may afford protection against antagonistic organisms such as pathogenic fungi (Timper, Kaya & Jaffee, 1991). Species of *Steinernema*, another genus of entomopathogenic nematodes, have ensheathed infective soil-dwelling infective larvae but there is no evidence that the sheath aids desiccation survival (Campbell & Gaugler, 1991; Patel, Perry & Wright, 1997). The sheath of *Steinernema* spp. fits very loosely and is readily lost during movement through the soil whereas the sheath of *Heterorhabditis* spp. is closely associated with the nematode's body; the sheath of *Steinernema* may have no rôle in protection of the infective larva. Why such genera, occupying similar ecological niches, differ in this respect is not understood.

Nematodes parasitizing the aerial parts of plants provide some of the best examples of extended anhydrobiotic survival, and they also are able to tolerate very rapid dehydration regimes and repeated cycles of dehydration and rehydration. For example, L_4s of *D. dipsaci* have been revived after being stored in dry plant material for 23 years yet the total duration of the life-cycle ranges from only 19 to 23 days at 15 °C (Evans & Perry, 1976). In general, the survival of uncoiled, individual L_4, L_3 and L_2 in *in vitro* experiments can be expressed in weeks, days and minutes, respectively; in all cases survival increased with increase in humidity, especially in adults where survival was for hours at humidities under 50% but for days at higher humidities (Perry, 1977a). L_3s lost water less slowly than L_4s but both lost water more slowly than L_2s and adults (Perry, 1977b). Thus, for this species, the slower dryers are the best survivors. The superior survival ability of L_4s appears to be linked to an intrinsic property of the cuticle to resist water loss. The cuticle of the L_4 dries more rapidly than deeper layers of the nematode and slows down the rate of water loss of internal, and perhaps more vital, structures (Ellenby, 1969; Perry, 1977b). The cuticular permeability barrier is heat labile and is destroyed by brief extraction with diethyl ether, indicating that an outer lipid layer, possibly the epicuticle, is involved (Wharton *et al.* 1988). The permeability barrier of the cuticle of *Anguina agrostis* was also considered to lie in the epicuticle and to be lipoprotein in nature (Preston & Bird, 1987; Bird & Zuckerman, 1989). Repeated cycles of desiccation and rehydration of L_4s of *D. dipsaci* resulted in a decrease in the percentage surviving each cycle; however, after the initial cycle the rate of drying of previously desiccated and revived individuals remained constant, irrespective of the number of cycles (Perry, 1977c). Thus, control of the rate of drying does not, of itself, guarantee survival, and death caused by repeated cycles of desiccation and rehydration is not associated with an altered ability to control water loss.

Although nematodes protected by cysts, eggshells, sheaths or impermeable cuticles may lose all their body water, the rate of water loss is much slower than that of unprotected individuals. However, only a few species are able to survive beyond the period during which water loss is controlled. With these species, additional biochemical adaptations are required for long-term survival of anhydrobiosis.

MORPHOLOGICAL CHANGES INDUCED BY DESICCATION

A slow rate of water loss appears to allow orderly packing and stabilization of structures to maintain functional integrity during desiccation. Experimental analysis of the water dynamics of individual L_4 of *D. dipsaci* exposed to 0% and 50% relative humidities demonstrated that water loss occurred in three distinct phases (Perry, 1977b). An initial rapid loss of water was followed by a period of very slow water loss before the third phase of rapid water loss to leave individuals with no detectable water content (Fig. 1). The first two phases are separated by a permeability slump during which the permeability of the cuticle, and hence the subsequent rate of water loss, is reduced (Perry, 1977b; Wharton, 1996). During the first phase, Wharton & Lemmon (1998), using freeze substitution techniques, observed rapid shrinkage of the cuticle, the lateral hypodermal cords and the muscle cells, followed by a slower rate of shrinkage during the second phase. The contractile region of the muscle cells appears to resist shrinkage until desiccation becomes severe during the third phase (Fig. 1). The mitochondria swell and then shrink during desiccation, which may indicate disruption of the permeability of the mitochondrial membrane (Wharton & Lemmon, 1998). A decrease in thickness of the hyaline layer, caused by shrinkage of its constituent muscle cells and epidermis, results in a decrease in diameter of L_4s of *D. dipsaci* that is of a much greater magnitude than the accompanying change in length, and which has not been observed in other nematodes (Wharton, 1996). For example, reduction in the rate of water loss of *Rotylenchus robustus* is achieved by controlled contraction of

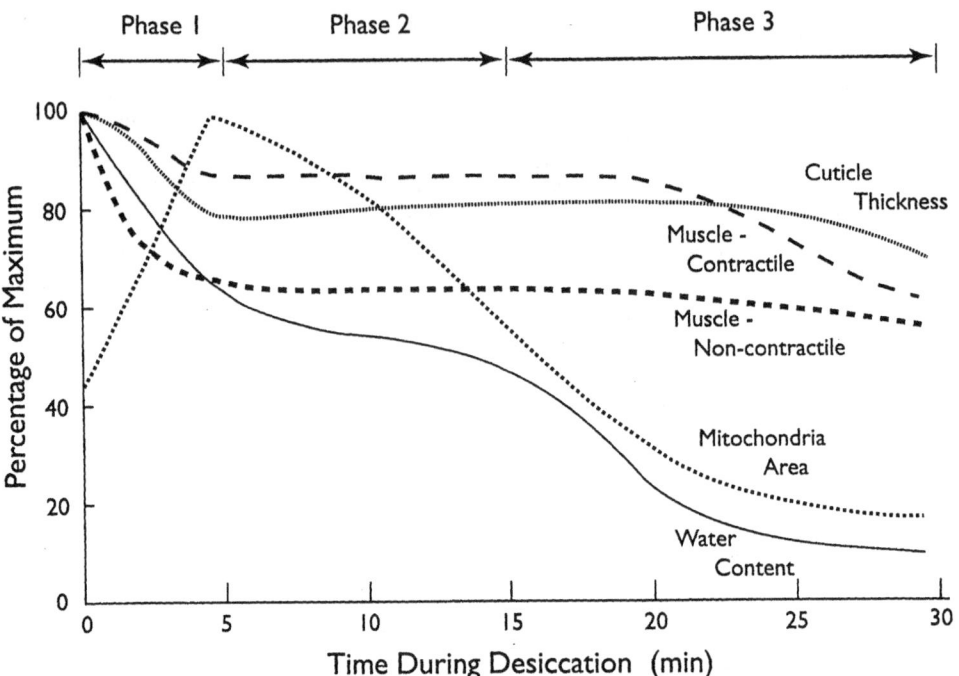

Fig. 1. Changes accompanying desiccation of L_4 of *Ditylenchus dipsaci* following placement of hydrated individuals in 50% relative humidity at time zero. Nematode water content data were calculated from Perry (1977b). Data for cuticle thickness, muscle region thickness and mitochondrial profile area were calculated from Wharton & Lemmon (1998). The three phases reflect differences in the rate of water loss (Perry, 1977b).

cuticular annuli resulting in decreased length, but not diameter, of the nematode (Rössner, 1973; Rössner & Perry, 1975). The large lipid reserves found in some nematodes, such as *D. dipsaci* and *A. tritici*, may prevent structural damage. Intestinal cells of *D. dipsaci* changed little during desiccation, possibly because the lipid droplets they contain resist shrinkage (Wharton & Lemmon, 1998).

Lee (1972) considered that the infective L_3 of *Nippostrongylus brasiliensis* became coated with a thin monolayer of lipid, obtained from the skin and hairs of the prospective host, that may reduce the rate of water loss from the nematode. Bird & Buttrose (1974) found differences between hydrated and desiccated L_2s of *A. tritici* in the structure of the external cortical layer of the cuticle. In desiccated L_2s, the outermost osmiophilic layer is doubled in thickness, indicating an increase in lipids. An outer layer is present in desiccated L_2s of *A. amsinkia* (Womersley et al. 1998) and in aggregations of *D. myceliophagus* (Perry, unpublished). However, the biochemical nature of these layers has not been investigated and it is not known whether they derive from the nematode or the host.

BIOCHEMICAL AND MOLECULAR CORRELATES OF ANHYDROBIOSIS

The biochemical adaptations associated with desiccation survival have been reviewed by Barrett (1991) and Womersley et al. (1998). At water contents below about 20%, there is no free water in the cells. This 20%, usually referred to as 'bound water', is involved in the structural integrity of macromolecules and macromolecular structures, such as membranes. The water content of desiccated, anhydrobiotic nematodes is estimated to be about 1–5%, so it is probable that the bound water has been lost although there is no experimental evidence that nematodes can survive the complete loss of structural water. Research on biochemical attributes of organisms that may be associated with anhydrobiosis has centred on molecules that might replace bound water and preserve structural integrity. Crowe & Crowe (1999) have recently presented a summary and supporting evidence for this 'water replacement' hypothesis in relation, primarily, to retention of membrane stability during desiccation. In nematodes, there is only limited information on biochemical mechanisms of desiccation survival. Most research has concentrated on the free-living mycophagous nematode, *Aphelenchus avenae*, but it is not certain that this is a useful 'model' for plant- and animal-parasitic nematodes capable of surviving anhydrobiotically.

The accumulation of the disaccharide, trehalose, the only naturally occurring non-reducing disaccharide of glucose, during water loss of anhydrobiotic nematodes has been reported frequently. In *A. avenae*, glycogen and lipid reserves are converted to trehalose and glycerol, respectively (Madin & Crowe, 1975). L_4 of *D. dipsaci* and L_2 of *Anguina tritici* also sequester trehalose, but not at the expense of lipid reserves; in these stages, other carbohydrates, such as myo-inositol and ribitol may be involved (Womersley, 1987). Lipid reserves are maintained at high levels in many anhydrobiotic nematodes and provide a food source for the nematode after revival and before they are able to feed on a host. The reported accumulation of glycerol in *Aphelenchus avenae* during desiccation (Madin & Crowe, 1975) was considered by Higa & Womersley (1993) to be an artefact due to the anaerobic conditions produced in large aggregates and not an adaptation to anhydrobiosis. There is no evidence that glycerol is preferentially synthesized during desiccation of *Anguina tritici* or *D. dipsaci* (Womersley & Smith, 1981; Womersley, Thompson & Smith, 1982). Womersley *et al.* (1998) considered that this is consistent with the fact that glycerol is highly fusogenic in dry membrane systems and, thus, would have an adverse effect on membrane stability during desiccation. Some nematodes, such as *Ditylenchus* (= *Orrina*) *phyllobius* accumulate no extra polyols during water loss yet can survive very rapid drying (Robinson, Orr & Heintz, 1984). Barrett (1991) considered it possible that such species may normally have large amounts of tissue polyols when active.

Several rôles for the involvement of trehalose in desiccation protection have been advanced and are detailed by Barrett (1991) and Crowe & Crowe (1999); only a brief summary is given here. Trehalose may replace bound water by attaching to polar side groups on proteins and phospholipids, thus maintaining the balance between hydrophilic and hydrophobic forces acting on the molecules and preventing their collapse. Preventing cross-linkage of molecules and fusion of membranes as bulk water is removed also preserves membrane stability. Stabilizing the membranes allows them to remain in a liquid crystalline phase and prevents a phase change to a gel state which would cause loss of the contents of cells and membrane vesicles during rehydration. Stabilization of molecules in the dry state also requires vitrification, which keeps membranes in a glass-like state to prevent a variety of deterioration processes (Levine & Slade, 1992; Crowe, Carpenter & Crowe, 1998). Trehalose also may prevent protein denaturation. Glucose reacts with the amino-acid side chains of proteins to form brown pigments called melanoidins. By contrast, trehalose does not react with proteins in this way and also appears to suppress this adverse reaction of other sugars with proteins (Loomis, O'Dell & Crowe, 1979). Trehalose can act as a free-radical scavenging agent to reduce random chemical damage (Barrett, 1991).

Synthesizing trehalose during dehydration may indicate preliminary preparation for a period in the dry state, but it does not necessarily mean that preservation of biological integrity and thus survival during subsequent severe desiccation is assured. Research on *Aphelenchus avenae* and *D. myceliophagus* illustrates this point. During desiccation preconditioning at 97% relative humidity, large aggregates (> 115 mg wet wt.) of *A. avenae* maxi-

mised their trehalose content after 72 h (Crowe & Madin, 1975), whereas small aggregates (< 10 mg wet wt.) achieved similar concentrations within the first 24 h (Higa & Womersley, 1993). However, only large aggregates survived direct transfer to extreme conditions such as 0% relative humidity; small aggregates needed further slow drying, by sequential transfer to successively lower humidities, to enable the nematodes to survive severe desiccation (Higa & Womersley, 1993). The rate of drying of large aggregates was probably reduced by the 'eggshell' effect mentioned previously in connection with survival of *D. dipsaci* in 'eelworm wool'. The research by Higa & Womersley (1993) contradicts the view that, once trehalose synthesis is complete, nematodes can survive further desiccation irrespective of the subsequent rate of water loss. It appears that, following trehalose synthesis, other, at present unknown, adaptations are required at the cellular and subcellular levels for nematode survival, and rate of drying still has to be controlled (Higa & Womersley, 1993). In contrast to *D. dipsaci*, individuals of *D. myceliophagus* survive desiccation poorly as individuals, even when dried at high humidities, and show no intrinsic ability to control water loss (Perry, 1977 *a*, *b*). When raised on different food sources and exposed to various desiccation regimes, aggregates of *D. myceliophagus* contained different amounts of trehalose (*ca.* 3–16% dry wt.), depending on treatment, yet the nematodes are unable to survive direct exposure to low relative humidity (Womersley & Higa, 1998). The survival of aggregates of *D. myceliophagus* was unrelated to their trehalose content, and elevated levels of trehalose did not enhance anhydrobiotic survival of this species.

In general, biochemical changes during drying of infective larvae of the entomopathogenic nematode, *S. carpocapsae*, parallel those observed in *A. avenae* and *D. myceliophagus*. When infective larvae were dried slowly at 97% relative humidity, glycogen and lipid reserves declined while the trehalose content increased from *ca.* 0·2% dry wt. in fully hydrated nematodes to a maximum of *ca.* 7%; however, the larvae were only able to survive at high relative humidities (Womersley, 1990). More research is needed to examine whether trehalose is implicated in the desiccation survival of other species or strains of entomopathogenic nematodes. Although steinernematids are not considered to be anhydrobiotes (Womersley, 1990), desiccation-tolerant strains have been isolated from a semi-arid region in Israel (Glazer *et al.* 1996; Solomon, Paperna & Glazer, 1999) but the adaptations enabling these strains to survive have not yet been investigated. Future work may focus on genetic transformation of entomopathogenic nematodes to improve their environmental tolerance (reviewed by Burnell & Dowds, 1996) and thus enhance survival of commercial formulations during storage and after foliar application. A transgenic approach would utilise the considerable information available from the *Caenorhabditis elegans* genome sequencing project (www.sanger.ac.uk/Projects/C_elegans/wormpep/) and has been used already by Gaugler, Wilson & Shearer (1997) to introduce a heat-shock protein gene, *hsp70*A, from *C. elegans* into *H. bacteriophora* to enhance thermotolerance. If trehalose is implicated in the survival of species and/or strains of entomopathogenic nematodes, then the use of genes for enzymes involved in the synthesis of trehalose, such as *tps* 1 coding for trehalose-6-phosphate synthase, may cause trehalose overproduction and enhanced survival (Vellai *et al.* 1999). Information on trehalose metabolism and its rôle in life-cycle physiology of animal- and plant-parasitic nematodes is also essential to evaluate possible novel control strategies. For example, if trehalose is important for the survival of animal-parasitic nematodes, enzymes of trehalose metabolism may offer molecular control targets as trehalose metabolism appears not to be important in mammals (Behm, 1997).

There is a shortage of information on other metabolic adaptations or changes involved in anhydrobiosis. The most detailed studies have been on L_4s of *D. dipsaci*. Barrett (1982) found that desiccation of these nematodes did not result in any appreciable denaturation of metabolic enzymes. There was no increase in the frequency of breaks in DNA obtained from desiccated L_4s compared with hydrated L_4s but as this also was observed in the desiccation-intolerant free-living nematode *Panagrellus redivivus*, it appears that DNA stability is a general feature of biological material and is not associated specifically with organisms able to enter anhydrobiosis (Barrett & Butterworth, 1985).

Completion of the sequencing of the 100 million base pair *C. elegans* genome provides a useful resource for the examination of the genetic induction of the survival forms in parasitic nematodes. The survival form of *C. elegans*, termed the 'dauer larva', represents a developmental arrest (Riddle & Albert, 1997) essentially similar to that found in some animal-parasitic nematodes, such as *Strongyloides ratti* which can switch between free-living and parasitic life-cycles in response to environmental cues (Viney, 1996). Dauer larvae are specialised L_3s enclosed by a dauer-specific cuticle and exhibit several characteristics including reduced metabolism, elevated levels of several heat shock proteins and an enhanced resistance to desiccation (Kenyon, 1997). They are formed, not in response to adverse environmental conditions acting on the L_3s but in response to specific factors acting on the L_1s and early L_2s. The factors initiating dauer formation are food availability, temperature and levels of a *C. elegans*-specific pheromone; details of the interaction of these factors and the *daf* genes (*daf* = dauer formation) involved have been reviewed by Riddle & Albert (1997). The information from this research may be relevant to other nematode groups as, in broad terms, diapause in plant- and animal-parasitic nematodes (Evans & Perry, 1976; Perry, 1989) and the formation of the infective larvae of entomopathogenic nematodes (Womersley, 1990) encompass developmental adaptations similar to dauer formation in *C. elegans*. Future research to investigate whether there are homologues of *daf* genes in parasitic nematodes will be an instructive first step.

REHYDRATION

Successful survival of desiccation requires not only completion of the induction into anhydrobiosis but also that changes during rehydration are ordered and controlled. Essentially, these changes reverse those that occur during drying but morphological and metabolic readjustments do not all occur at similar times (Fig. 2). The rate of rehydration by desiccated individuals when placed in water is very rapid and seems not to relate to the length of time they had been desiccated. L_2s of *G. rostochiensis* and L_4s of *D. dipsaci* took up water at the same rate (Ellenby, 1968a) and comparisons of all stages of *D. dipsaci* showed that, irrespective of the fact that they had been desiccated for different periods at 0% relative humidity, there were no differences in the rate of water uptake (Perry, 1977b). In all cases, the initial rate of rehydration was rapid with 50% water content being achieved in only a few minutes. The water content of L_4s of *D. dipsaci* increased logarithmically for up to 2·4 h of rehydration (Wharton, Barrett & Perry, 1985) whereas during rehydration of L_2s of *A. agrostis* cuticle permeability initially increased slightly followed by a sharp decrease in permeability between 1 and 8 h, after which there were two successive slower declines in permeability up to 24 h (Preston & Bird, 1987). These species are able to revive from the desiccated state on immediate transfer to water. With *Aphelenchus avenae*, successful revival depends on slow rehydration in saturated atmospheres (*ca.* 100% relative humidity) (Crowe, Hoekstra & Crowe, 1992). It appears that the fast-dehydration strategists are also fast-rehydration strategists and the slow-dehydration strategists are slow-rehydration strategists. Thus it is probable that the water dynamics of dehydration and rehydration are linked. The definitions of the two groups given by Womersley (1987) may be extended to characterize one group that can withstand fast water loss and gain and the second group that requires slow, controlled water loss and gain.

Although L_4s of *D. dipsaci* rehydrate very rapidly, there is a delay of several hours before the onset of locomotory activity (Fig. 2). Barrett (1982) termed this delay the 'lag phase' and considered that it may be necessary to restore membrane function. The

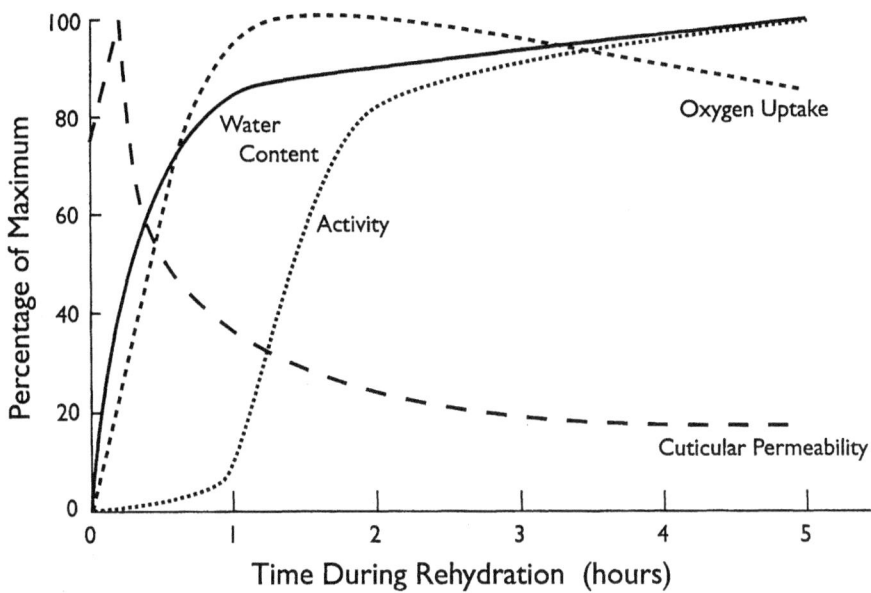

Fig. 2. Changes accompanying rehydration of L_4 of *Ditylenchus dipsaci* following placement of desiccated individuals in water at time zero. Nematode water content data were calculated from Perry (1977b). Cuticular permeability data were calculated from Wharton et al. (1988). Oxygen uptake and activity data were calculated from Barrett (1982). Activity is defined as the percentage of L_4s showing movement. The time difference between water uptake and activity is the 'lag phase' (Barrett, 1982). (Redrawn from Barrett, 1991.)

water permeability characteristics of the cuticles of *D. dipsaci* and *A. agrostis* are altered during desiccation and the permeability barrier associated with these species is restored during the lag phase (Wharton et al. 1985; Preston & Bird, 1987). The length of the lag phase in *D. dipsaci* increased as the severity of the desiccation stress during dehydration increased (Wharton & Aalders, 1999). A similar relationship between the length of the lag phase and the severity of desiccation has been demonstrated in *T. colubriformis* (Allan & Wharton, 1990). The restoration of the cuticular permeability barrier during rehydration can be prevented by inhibitors that block enzyme activity and post-transcriptional protein synthesis (Wharton et al. 1988), indicating an active repair mechanism. Leakage of inorganic ions during rehydration has been demonstrated in *Aphelenchus avenae* (Crowe, O'Dell & Armstrong, 1979) and in *Anguina tritici* (Womersley, 1981). The leakage ceases during the lag phase, indicating the repair of damaged membranes or the restoration of the permeability barrier due to a physical change associated with rehydration.

Morphological changes occur gradually throughout the lag phase. Muscle cells of L_4s of *D. dipsaci* increase in thickness and, in *D. dipsaci* and *A. agrostis*, small lipid droplets coalesce within the intestine to form large droplets (Wharton & Barrett, 1985; Wharton et al. 1985; Preston & Bird, 1987). There is a decrease in body length of *D. dipsaci* (Wharton et al. 1985) and *T. colubriformis* (Allan & Wharton, 1990) during the lag phase. This may indicate a contraction of the muscle cells as they recover and, in *T. colubriformis*, there is evidence of a change in the arrangements of muscle filaments in the contractile region of the muscle cells (Allan & Wharton, 1990).

Analyses of metabolic changes during rehydration have been confined almost entirely to L_4s of *D. dipsaci*. Barrett (1982) found that metabolism of L_4s, as measured by heat output, oxygen uptake or $^{14}CO_2$ production from labelled substrates, begins immediately after hydration. The metabolite profiles recover quickly during hydration with noticeable changes after 10 min and completion by 1 h. However, the ATP content does not recover as rapidly as those of the other metabolites; after 10 min there is little change and even after 1 h it is still low (Barrett, 1982). The slow trehalose depletion (up to 48 h to return to pre-desiccation levels) may be associated with the slow recovery of ATP levels. Mitochondria swell during rehydration before adopting a normal morphology (Wharton & Barrett, 1985); immediately after hydration, the mitochondria are essentially uncoupled and there is no oxidative phosphorylation (Barrett, 1982). Barrett (1982) suggested that during the dehydration-rehydration cycle, membrane function is disrupted and the lag phase reflects the time required to restore metabolic and ionic gradients.

There seems to be negligible protein synthesis during the first 2 h of rehydration and L_4s of *D. dipsaci* revive successfully in the presence of inhibitors of protein and RNA synthesis (Barrett, 1982). However, there is an increase in the activity of certain enzymes involved in prevention of cellular aging though free-radical scavenging reactions and negation of lipid peroxidation. For example, during rehydration of *A. avenae* an increase in superoxide dismutase activity occurs (Womersley, 1987) and catalase activity essentially triples during the first 4 h (Gresham & Womersley, 1991).

CONCLUSIONS

The urgent need for environmentally-acceptable methods to control pests has provided the impetus for research on the use of entomopathogenic nematodes as bioinsecticides and has renewed interest in studies on aspects of the survival of parasitic nematodes outside their hosts. In turn, this has generated further research on the morphological and biochemical adaptations associated with anhydrobiosis. It is clear, from this review, that only a limited number of nematode species have been used as the basis for detailed research and there is still much to be understood about the biochemical changes during desiccation and successful revival and the genetic control associated with induction of the dormant state. There is no 'model nematode' that can serve to provide the information about dormancy as species and stages of nematodes exhibit a variety of adaptations, and different combinations of these adaptations are associated with different nematodes to ensure anhydrobiotic survival.

ACKNOWLEDGEMENTS

IACR receives grant-aided support from the Biotechnology and Biological Sciences Research Council of the United Kingdom.

REFERENCES

ALLAN, G. S. & WHARTON, D. A. (1990). Anhydrobiosis in the infective juveniles of *Trichostrongylus colubriformis* (Nematoda: Trichostrongylidae). *International Journal for Parasitology* **20**, 183–192.

BARRETT, J. (1982). Metabolic responses to anabiosis in the fourth stage juveniles of *Ditylenchus dipsaci* (Nematoda). *Proceedings of the Royal Society of London B* **216**, 157–177.

BARRETT, J. (1991). Anhydrobiotic nematodes. *Agricultural Zoology Reviews* **4**, 161–176.

BARRETT, J. & BUTTERWORTH, P. E. (1985). DNA stability in the anabiotic fourth-stage juveniles of *Ditylenchus dipsaci* (Nematoda). *Annals of Applied Biology* **106**, 121–124.

BEHM, C. A. (1997). The role of trehalose in the physiology of nematodes. *International Journal for Parasitology* **27**, 215–229.

BIRD, A. F. (1983). Growth and moulting in nematodes: changes in the dimensions and morphology of *Rotylenchulus reniformis* from start to finish of moulting. *International Journal for Parasitology* **13**, 201–206.

BIRD, A. F. & BUTTROSE, M. S. (1974). Ultrastructural changes in the nematode *Anguina tritici* associated with anhydrobiosis. *Journal of Ultrastructural Research* **48**, 177–189.

BIRD, A. F. & SOEFFKY, A. (1972). Changes in the ultrastructure of the gelatinous matrix of *Meloidogyne javanica* during dehydration. *Journal of Nematology* **4**, 166–169.

BIRD, A. F. & ZUCKERMAN, B. M. (1989). Studies on the surface coat (glycocalyx) of the dauer larva of *Anguina agrostis*. *International Journal for Parasitology* **19**, 235–247.

BURNELL, A. M. & DOWDS, B. A. (1996). The genetic improvement of entomopathogenic nematodes and their symbiotic bacteria: phenotypic targets, genetic limitations and an assessment of possible hazards. *Biocontrol Science and Technology* **6**, 435–447.

CAMPBELL, L. R. & GAUGLER, R. (1991). Role of the sheath in desiccation tolerance of two entomopathogenic nematodes. *Nematologica* **37**, 324–332.

COSSINS, A. R. & BOWLER, K. (1987). *Temperature Biology of Animals*. London & New York, Chapman & Hall.

CROWE, J. H. & CROWE, L. M. (1999). Anhydrobiosis: the water replacement hypothesis. In *Survival of Entomopathogenic Nematodes* (ed. Glazer, P., Richardson, P., Boemare, N. & Coudert, F.), pp. 15–25. Luxembourg, Office for Official Publications of the European Communities.

CROWE, J. H., CARPENTER, J. F. & CROWE, L. M. (1998). The role of vitrification in anhydrobiosis. *Annual Review of Physiology* **60**, 73–103.

CROWE, J. H., HOEKSTRA, F. & CROWE, L. M. (1992). Anhydrobiosis. *Annual Review of Physiology* **54**, 579–599.

CROWE, J. H. & MADIN, K. A. (1975). Anhydrobiosis in nematodes: evaporative water loss and survival. *Journal of Experimental Zoology* **193**, 323–334.

CROWE, J. H., O'DELL, S. J. & ARMSTRONG, D. A. (1979). Anhydrobiosis in nematodes: permeability during rehydration. *Journal of Experimental Zoology* **207**, 431–438.

DEMEURE, Y. & FRECKMAN, D. W. (1981). Recent advances in the study of anhydrobiotic nematodes. In *Plant Parasitic Nematodes*, Vol 3 (ed. Zuckerman, B. M., Mai, W. F. & Rohde, R. F.), pp. 205–226. New York, Academic Press.

ELLENBY, C. (1946). Nature of the cyst wall of the potato-root eelworm *Heterodera rostochiensis* Wollenweber, and its permeability to water. *Nature* **157**, 302.

ELLENBY, C. (1968a). Desiccation survival in the plant parasitic nematodes, *Heterodera rostochiensis* Wollenweber and *Ditylenchus dipsaci* (Kuhn) Filipjev. *Proceedings of the Royal Society of London B* **169**, 203–213.

ELLENBY, C. (1968b). The survival of desiccated larvae of *Heterodera rostochiensis* and *H. schachtii*. *Nematologica* **14**, 544–548.

ELLENBY, C. (1968c). Desiccation survival of the infective larva of *Haemonchus contortus*. *Journal of Experimental Biology* **49**, 460–475.

ELLENBY, C. (1969). Dormancy and survival in nematodes. *Symposium of the Society for Experimental Biology* **23**, 83–97.

EVANS, A. A. F. (1987). Diapause in nematodes as a survival strategy. In *Vistas on Nematology* (ed. Veech, J. A. & Dickson, D. W.), pp. 180–187. Hyattsville, Society of Nematologists Inc.

EVANS, A. A. F. & PERRY, R. N. (1976). Survival strategies in nematodes. In *The Organisation of Nematodes* (ed. Croll, N. A.), pp. 383–424. London & New York, Academic Press.

GAUGLER, R., WILSON, M. & SHEARER, P. (1997). Field release and environmental fate of a transgenic entomopathogenic nematode. *Biological Control* **9**, 75–80.

GAUR, H. S. & PERRY, R. N. (1991a). The biology and control of the plant parasitic nematode *Rotylenchulus reniformis*. *Agricultural Zology Reviews* **4**, 177–212.

GAUR, H. S. & PERRY, R. N. (1991b). The role of the moulted cuticles in the desiccation survival of adults of *Rotylenchulus reniformis*. *Revue de Nématologie* **14**, 491–496.

GLAZER, I., KOZODOI, E., HASHMI, G. & GAUGLER, R. (1996). Biological characteristics of the entomopathogenic nematode *Heterorhabditis* sp.: a heat tolerant isolate from Israel. *Nematologica* **42**, 481–492.

GLAZER, I., RICHARDSON, P., BOEMARE, N. & COUDERT, F. (Eds) (1999). *Survival of Entomopathogenic Nematodes*. Luxembourg: Official Publications of the European Communities.

GRESHAM, A. & WOMERSLEY, C. Z. (1991). Modulation of catalase activity during the enforced induction of and revival from anhydrobiosis in nematodes. *FASEB Journal* **5A**, 682.

HIGA, L. M. & WOMERSLEY, C. Z. (1993). New insights into the anhydrobiotic phenomenon: the effects of trehalose content and differential rates of evaporative water loss on the survival of *Aphelenchus avenae*. *Journal of Experimental Zoology* **267**, 120–129.

HOOPER, D. J. (1971). Stem eelworm (*Ditylenchus dipsaci*), a seed and soil-borne pathogen of field beans (*Vicia faba*). *Plant Pathology* **20**, 25–27.

JONES, P. W., TYLKA, G. L. & PERRY, R. N. (1998). Hatching. In *The Physiology and Biochemistry of Free-living and Plant-parasitic Nematodes* (ed. Perry, R. N. & Wright, D. J.), pp. 181–212. Wallingford, CAB International.

KEILIN, D. (1959). The problem of anabiosis or latent life: history and current concepts. *Proceeding of the Royal Society B* **150**, 149–191.

KENYON, C. (1997). Environmental factors and gene activities that influence life span. In *C. elegans II* (ed. Riddle, D. L., Blumenthal, T., Meyer, B. J. & Priess, J. R.), pp. 791–813. Cold Spring Harbor Laboratory Press, New York.

LARSEN, M. N. & ROEPSTORFF, A. (1999). Seasonal variation in development and survival of *Ascaris suum* and *Trichuris suis* eggs on pastures. *Parasitology* **119**, 209–220.

LEE, D. L. (1972). Penetration of mammalian skin by the infective larva of *Nippostrongylus brasiliensis*. *Parasitology* **65**, 499–505.

LEVINE, H. & SLADE, L. (1992). Another view of trehalose for drying and stabilizing biological material. *BioPharm* **5**, 36–40.

LOOMIS, S. H., O'DELL, S. J. & CROWE, J. H. (1979). Anhydrobiosis in nematodes: inhibition of the browning reaction of reducing sugars with dry protein. *Journal of Experimental Zoology* **208**, 355–360.

MADIN, K. A. C. & CROWE, J. H. (1975). Anhydrobiosis in nematodes: carbohydrate and lipid metabolism during rehydration. *Journal of Experimental Zoology* **193**, 335–342.

MENTI, H., WRIGHT, D. J. & PERRY, R. N. (1997). Desiccation survival of populations of the entomopathogenic nematodes *Steinernema feltiae* and *Heterorhabditis megidis* from Greece and the UK. *Journal of Helminthology* **71**, 41–46.

MURRELL, K. D., GRAHAM, C. E. & MCGREEVY, M. (1983). *Strongyloides ratti* and *Trichinella spiralis*: net charge of epicuticle. *Experimental Parasitology* **55**, 331–339.

O'LEARY, S. A. & BURNELL, A. M. (1997). The isolation of mutants of *Heterorhabditis megidis* (strain UK211) with increased desiccation tolerance. *Fundamental and Applied Nematology* **20**, 197–205.

O'LEARY, S. A., BURNELL, A. M. & KUSEL, J. R. (1998). Biophysical properties of the surface of desiccation-tolerant mutants and parental strain of the entomopathogenic nematode *Heterorhabditis megidis* (strain UK211). *Parasitology* **117**, 337–345.

PATEL, M. N., PERRY, R. N. & WRIGHT, D. J. (1997). Desiccation survival and water contents of entomopathogenic nematodes, *Steinernema* spp. (Rhabditida: Steinernematidae). *International Journal for Parasitology* **27**, 61–70.

PERRY, R. N. (1977a). Desiccation survival of larval and adult stages of the plant parasitic nematodes *Ditylenchus dipsaci* and *D. myceliophagus*. *Parasitology* **74**, 139–148.

PERRY, R. N. (1977b). The water dynamics of stages of *Ditylenchus dipsaci* and *D. myceliophagus* during desiccation and rehydration. *Parasitology* **75**, 45–70.

PERRY, R. N. (1977c). The effect of previous desiccation on the ability of 4th-stage larvae of *Ditylenchus dipsaci* to control rate of water loss and to survive drying. *Parasitology* **75**, 215–231.

PERRY, R. N. (1983). The effect of potato root diffusate on the desiccation survival of unhatched *Globodera rostochiensis*. *Revue de Nématologie* **6**, 99–102.

PERRY, R. N. (1989). Dormancy and hatching of nematodes. *Parasitology Today* **5**, 377–383.

PERRY, R. N., CLARKE, A. J. & HENNESSY, J. (1980). The influence of osmotic pressure on the hatching of *Heterodera schachtii*. *Revue de Nématologie* **3**, 3–9.

PRECHT, H. (1958). Concepts of the temperature adaptation of unchanging reaction systems of cold-blooded animals. In *Physiological Adaptation* (ed. Prosser, C. L.), pp. 351–376. Washington, American Association for the Advancement of Science.

PRESTON, C. M. & BIRD, A. F. (1987). Physiological and morphological changes associated with recovery from anabiosis in the dauer larva of the nematode *Anguina agrostis*. *Parasitology* **44**, 125–133.

RIDDLE, D. L. & ALBERT, P. S. (1997). Genetic and environmental regulation of dauer larva development. In *C. elegans II* (ed. Riddle, D. L., Blumenthal, T., Meyer, B. J. & Priess, J. R.), pp. 739–768. Cold Spring Harbor Laboratory Press, New York.

ROBINSON, A. F., ORR, C. C. & HEINTZ, C. E. (1984). Some factors affecting survival of desiccation by infective juveniles of *Orrina phyllobia*. *Journal of Nematology* **16**, 86–91.

ROEPSTORFF, A. (1997). Helminth surveillance as a prerequisite for anthelmintic treatment in intensive sow herds. *Veterinary Parasitology* **73**, 139–151.

RÖSSNER, J. (1973). Anpassung wandernder Wurzelnematoden an Austrocknung im Boden. *Nematologica* **19**, 366–378.

RÖSSNER, J. & PERRY, R. N. (1975). Water loss and associated surface changes after desiccation in *Rotylenchus robustus*. *Nematologica* **21**, 438–442.

SOLOMON, A., PAPERNA, I. & GLAZER, I. (1999). Physiological and behavioural adaptation of *Steinernema feltiae* to desiccation stress. In *Survival of Entomopathogenic Nematodes* (ed. Glazer, P., Richardson, P., Boemare, N. & Coudert, F.), pp. 83–98. Luxembourg, Office for Official Publications of the European Communities.

STIRLING, G. R. (1991). *Biological Control of Plant Parasitic Nematodes*. Wallingford, CAB International.

TIMPER, P., KAYA, H. K. & JAFFEE, B. A. (1991). Survival of entomogenous nematodes in soil infested with the nematode-parasitic fungus *Hirsutella rhossiliensis* (Deuteromycotina: Hyphomycetes). *Biological Control* **1**, 42–50.

VELLAI, T., MOLNÁR, A., LAKATOS, L., BÁNFALVI, Z., FODOR, A. & SÁRINGER, G. (1999). Transgenic nematodes carrying a cloned stress resistance gene from yeast. In *Survival of Entomopathogenic Nematodes* (ed. Glazer, P., Richardson, P., Boemare, N. & Coudert, F.), pp. 105–119. Luxembourg, Office for Official Publications of the European Communities.

VINEY, M. (1996). A genetic analysis of reproduction in *Strongyloides ratti*. *Parasitology* **109**, 511–515.

WALLER, P. J. (1971). Structural differences in the egg envelope of *Haemonchus contortus* and *Trichostrongylus colubriformis* (Nematoda: Trichostrongylidae). *Parasitology* **62**, 157–160.

WALLER, P. J. & DONALD, A. D. (1970). The response to desiccation of eggs of *Trichostrongylus colubriformis* and *Haemonchus contortus* (Nematoda: Trichostrongylidae). *Parasitology* **61**, 195–204.

WHARTON, D. A. (1979). *Ascaris lumbricoides*: water loss during desiccation of embryonating eggs. *Experimental Parasitology* **48**, 398–406.

WHARTON, D. A. (1980). Studies on the function of the oxyurid egg-shell. *Parasitology* **81**, 103–113.

WHARTON, D. A. (1981). The initiation of coiling behaviour prior to desiccation in the infective larvae of *Trichostrongylus colubriformis*. *International Journal for Parasitology* **11**, 353–357.

WHARTON, D. A. (1996). Water loss and morphological changes during desiccation of the anhydrobiotic nematode *Ditylenchus dipsaci*. *The Journal of Experimental Biology* **199**, 1085–1093.

WHARTON, D. A. & AALDERS, O. (1999). Desiccation stress and recovery in the anhydrobiotic nematode *Ditylenchus dipsaci* (Nematoda: Anguinidae). *European Journal of Entomology* **96**, 199–203.

WHARTON, D. A. & BARRETT, J. (1985). Ultrastructural changes during recovery from anabiosis in the plant parasitic nematode, *Ditylenchus*. *Tissue and Cell* **17**, 79–96.

WHARTON, D. A., BARRETT, J. & PERRY, R. N. (1985). Water uptake and morphological changes during recovery from anabiosis in the plant parasitic nematode, *Ditylenchus dipsaci*. *Journal of Zoology* **206**, 391–402.

WHARTON, D. A. & LEMMON, J. (1998). Ultrastructural changes during desiccation of the anhydrobiotic nematode *Ditylenchus dipsaci*. *Tissue and Cell* **30**, 312–323.

WHARTON, D. A., PRESTON, C. M., BARRETT, J. & PERRY, R. N. (1988). Changes in cuticular permeability associated with recovery from anhydrobiosis in the plant parasitic nematode, *Ditylenchus dipsaci*. *Parasitology* **97**, 317–330.

WOMERSLEY, C. (1978). A comparison of the rate of drying of four nematode species using a liquid paraffin technique. *Annals of Applied Biology* **90**, 401–405.

WOMERSLEY, C. (1981). The effect of dehydration and rehydration on salt loss in the second-stage larvae of *Anguina tritici*. *Parasitology* **82**, 411–419.

WOMERSLEY, C. (1987). A reevaluation of strategies employed by nematode anhydrobiotes in relation to their natural environment. In *Vistas on Nematology* (ed. Veech, J. A. & Dickson, D. W.), pp. 165–173. Hyattsville, Society of Nematologists Inc.

WOMERSLEY, C. Z. (1990). Dehydration survival and anhydrobiotic survival. In *Entomopathogenic Nematodes in Biological Control* (ed. Gaugler, R. & Kaya, H. K.), pp. 117–137. Boca Raton, CRC Press.

WOMERSLEY, C. Z. & CHING, C. (1989). Natural dehydration regimes as a prerequisite for the successful induction of anhydrobiosis in the nematode *Rotylenchulus reniformis*. *Journal of Experimental Biology* **143**, 359–372

WOMERSLEY, C. Z. & HIGA, L. M. (1998). Trehalose: its role in the anhydrobiotic survival of *Ditylenchus myceliophagus*. *Nematologica* **44**, 269–291.

WOMERSLEY, C. & SMITH, L. (1981). Anhydrobiosis in nematodes. 1. The role of glycerol, myoinositol and trehalose during desiccation. *Comparative Biochemistry and Physiology* **70B**, 579–586.

WOMERSLEY, C., THOMPSON, S. N. & SMITH, L. (1982). Anhydrobiosis in nematodes. 2. Carbohydrate and lipid analysis in undesiccated and desiccated nematodes. *Journal of Nematology* **14**, 145–153.

WOMERSLEY, C. Z., WHARTON, D. A. & HIGA, L. M. (1998). Survival biology. In *The Physiology and Biochemistry of Free-living and Plant-parasitic Nematodes* (ed. Perry, R. N. & Wright, D. J.), pp. 271–302. Wallingford, CAB International.

Parasite adaptation to extreme conditions in a desert environment

R. C. TINSLEY*

School of Biological Sciences, University of Bristol, Bristol BS8 1UG, UK

SUMMARY

Deserts represent universally recognized extreme environments for animal life. This paper documents the highly specialized adaptations of *Pseudodiplorchis americanus*, a monogenean parasite of the desert toad, *Scaphiopus couchii*. Building on a long-term record of parasite population ecology (continuing since the early 1980s), field studies focus on the effects of severe drought in the Sonoran Desert, Arizona, in the mid 1990s. This provides a test of the ability of the host-parasite system to tolerate exceptional perturbation. The analysis provides new insight into parasite infection dynamics in a natural wildlife system through integration of host and parasite population age structure. The environmental check interrupted host recruitment in 1993-95 and parasite recruitment in 1995-97. This produced an imprint in age structure and infection levels recognizable over several years: parasite recruitment failure reduced transmission 2-3 years later. The host (maximum life span 17 years) tolerated the disruption but the impact was more serious for the parasite (life span 3 years) leading to extinction of some previously stable populations. Despite this demonstration of a rare event exacerbating external environmental constraints, experimental studies suggest that the internal (host) environment normally creates the most severe conditions affecting *P. americanus*. Only about 3% of parasites survive from invasion until first reproduction. Post-invasion factors including host immunity, characteristic of most parasite life cycles, constitute a greater constraint upon survival than external conditions, even in a desert environment.

Key words: Age analysis, population ecology, Monogenea, Polystomatidae, *Pseudodiplorchis*, life cycle adaptation.

INTRODUCTION

Deserts provide some of the harshest environmental conditions on Earth. Life in deserts requires adaptations to extremes of water deficit (a combination of low precipitation and high evaporation) and temperature. For most desert organisms there is also a requirement to withstand prolonged periods of starvation for which adaptations frequently involve dormancy. In most of the world's hot deserts, there are periodic weather patterns that, often very briefly, provide favourable conditions for life. As is well-documented, the passage of rainstorms across deserts may be followed by a sudden flourishing of organisms whose key attribute is the ability to wait, typically in a dormant state, until these conditions arise. Therefore, desert organisms show a suite of adaptations alongside specializations to tolerate or avoid drought, temperature fluctuations and periodic starvation. Characteristically, these include a rapid response to unpredictable opportunities and a lifestyle geared to rapid growth, reproduction and accumulation of reserves to enable survival through the next period of hostile conditions. By any ecological and physiological criteria, deserts represent extreme environments and animals that survive such conditions would be expected to have major specializations.

The extreme environment associated with hot deserts precludes a wide spectrum of animals that are unable to cope with the constraints. Amongst vertebrates, the class Amphibia should automatically be excluded because of limitations for terrestrial life including a highly permeable skin, inability to concentrate excretory products, and the requirement for breeding and early development to occur in water. Amongst parasites, the platyhelminth class Monogenea should equally be a 'non-starter'. These are typically ectoparasites of fishes, they possess a ciliated, swimming infective stage and lack a resting, resistant stage and tolerance of desiccation. Nevertheless, this study is based on a host-parasite association involving representatives of these two groups: their occurrence in deserts is a result of quite exceptional characteristics. The host is an anuran superbly adapted to the arid conditions in the southwestern deserts of North America, the spadefoot toad, *Scaphiopus couchii*. The parasite is *Pseudodiplorchis americanus*, a polystomatid monogenean. This system provides a special challenge to understanding the evolution of adaptations to extreme constraints: many of the components of the parasite's life cycle strategy have no known precedent within the platyhelminths and represent unique solutions to specific problems.

This paper has two aims. First, it presents a review of studies that document the specializations of *P. americanus* to the extreme conditions experienced during the life cycle. A previous account (Tinsley, 1995) has focused on the host-parasite interactions in this system concerned with pathology

* Tel: 0117 928 8660. Fax: 0117 925 7374. Email: r.c.tinsley@bristol.ac.uk

and regulation of disease. The first part of this paper represents a complementary overview of life cycle adaptations. The second aim is to explore the effects of extreme environmental perturbation on population biology, specifically the occurrence of severe drought in the Sonoran Desert study sites in 1993–1995. These conditions, associated with the lowest rainfall in a 34 year weather record for the area, provide a test of the ability of the host and parasite to tolerate environmental effects even more severe than those normally encountered. It would be predicted that there should have been major negative effects on host and parasite recruitment and, further, that extreme limitation of the host feeding season could have perturbed the interaction with its pathogenic, blood-feeding parasite, potentially contributing to more severe disease. This study aims to provide new insight into host-parasite ecology by applying techniques of age determination to both host and parasite in order to reconstruct the dynamic changes affecting the respective populations.

THE LIFE CYCLE ADAPTATIONS OF *PSEUDODIPLORCHIS AMERICANUS*

The Polystomatidae comprises an atypical family of monogeneans that has radiated within a diverse assemblage of tetrapod vertebrates, including lungfish, anuran and urodele amphibians, chelonians and one mammal (the hippopotamus) (Tinsley, 1990a). In the best known example, *Polystoma integerrimum*, there is an exact synchrony of host and parasite life cycles. The adult parasites, in the urinary bladder, produce eggs for about one week each year during the spawning of the European frog, *Rana temporaria*, and the target of invasion, when the eggs hatch some weeks later, is the tadpole. The same principle of reproductive synchrony applies to *Pseudodiplorchis americanus*. However, the timing is much more precise. The spadefoot toad hibernates below ground for at least 10 months each year; it has a brief activity season during the period of summer rainfall when it feeds intensively to accumulate energy reserves to survive the next long hibernation. Transmission involving the aquatic infective stage, the oncomiracidium, is focused on the period – normally a single night – when the toads enter temporary ponds to breed. The target of invasion is the adult host. *Pseudodiplorchis americanus* has a long uterus in which fully-developed oncomiracidia accumulate during host hibernation; these are discharged into water, hatch instantly, and invade via the host's nostrils. Juveniles migrate during the first week post-infection to the lungs where they develop for two weeks and then return to the oral cavity before migrating along the length of the alimentary tract to the urinary bladder. The worms reach sexual maturity about 1 month post-migration and then begin to accumulate offspring in preparation for the next opportunity for transmission during the following summer (Fig. 1).

The general pattern of this life cycle, particularly the synchrony of host and parasite oviposition, resembles that of *P. integerrimum* but the details differ. Indeed, there is a closer correspondence with other polystomatids that show adaptations for transmission in 'dynamic' environments including the fast-moving water of mountain torrents or the floodwaters of tropical forest. In both of these situations, polystomatids are specialized for 'instantaneous' host-to-host transmission (Tinsley, 1983, 1990a): this may be inferred from the anatomy of their reproductive systems but none of these life cycles has been investigated experimentally. Until the start of the present investigations, *P. americanus* was known only from a single brief taxonomic description (Rodgers & Kuntz, 1940). A series of studies has now provided comprehensive information on the biology of this parasite. These form the basis of the following review which examines critically the constraints of the desert environment and the corresponding parasite adaptations.

Adaptations to the constraints operating during host invasion: reproductive preparation

The environmental constraints that restrict the opportunity for transmission of *P. americanus* are rigidly defined. Larval invasion must occur in water and the appropriate circumstances are provided by the spawning of *S. couchii* in pools created by torrential summer rainfall. However, this allows a maximum timeframe of only 7 h since each episode is limited by the host's nocturnal habits. Instant host-to-host transfer is achieved by the production of larvae that complete development to the infective stage within the parent. Typically, the platyhelminth egg capsule is a relatively rigid structure designed to protect the embryo in the environments encountered after deposition. At the point of formation, the entire nutrient provision for larval development is packaged within the shell in the form of vitelline cells. With this reproductive pattern, output is constrained by parasite body size (of both the parent and the offspring) because of the requirement for storage space, and by the nutrient requirements of the developing larvae. Some platyhelminths adapted for ovoviviparity have modifications of this basic organisation (reviewed by Tinsley, 1983, 1990a), but the adaptations exhibited by *P. americanus* are without precedent amongst platyhelminths. Recently-formed embryos in the proximal uterus are surrounded by a thin capsule, about 100 μm in diameter, apparently with little potential for expansion. However, during passage along the uterus there is a progressive increase in egg capsule and embryo size. The fully-formed oncomiracidium is

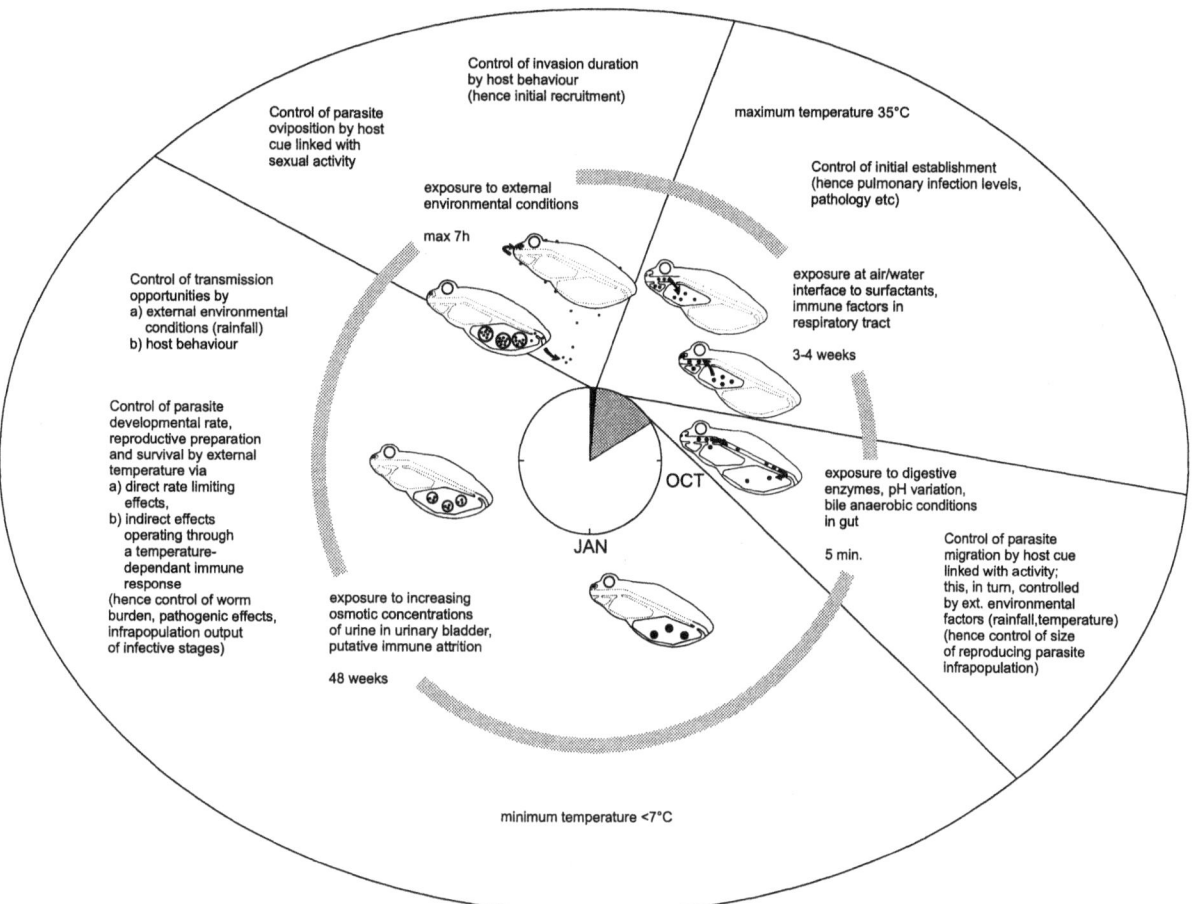

Fig. 1. Summary of the constraints operating during successive phases of the life cycle of *Pseudodiplorchis americanus*. Inner ring of annotations lists the variations in environmental conditions experienced by the parasite; outermost annotations identify the accompanying environmental controls on parasite biology. Central diagram correlates parasite life cycle events with the annual activity cycle of the host, *Scaphiopus couchii* (based on Tinsley & Jackson, 1986): transmission occurs during host spawning (▼); juvenile development and internal migration during host feeding (▽); maturation and accumulation of embryos *in utero* during host hibernation (▽). The cycle follows one cohort of parasites, but adult worms (☯) producing infective stages (·····) in one season's transmission may also survive to reproduce in the next year(s). Additionally, juveniles failing to migrate may remain in the respiratory tract throughout hibernation and then migrate to the bladder when the host again becomes active.

about 600 μm long and 250 μm wide. At this stage, the egg capsule forms a thin, loosely-fitting, transparent shroud wrapped around the oncomiracidium and, when distended, is over 800 μm in diameter. The highly flexible nature of this capsule, which is composed of elastin (Cable, Tocque & Tinsley, 1997), maximizes the numbers of oncomiracidia that can be stored *in utero*. The 'egg shell' is constructed with concentric layers of membranes. These are derived from the uterus wall and appear to become plastered onto the capsule surface. The resulting stacks of membranes can slide over one another, providing a mechanism by which the capsule can expand to accommodate the growth of the enclosed larva (Cable & Tinsley, 1991 a).

There is no provision of stored nutrients within the egg capsule; instead, the cytoplasmic lining of the capsule is modified to create a totally novel mechanism for the supply of nutrients. Cytoplasmic connections form placenta-like processes which 'plumb in' to the tegument of the developing larva. Transmission electron microscopy (TEM) shows that these are packed with glycogen rosettes and provide a route for the continuous transfer of nutrients from parent worm to offspring. The intestine of the adult parasite is highly branched, with diverticula passing between the folds of the uterus, ensuring close proximity between the sites of digestive uptake and the sites of nutrient delivery to the developing larvae (Tinsley, 1990 a; Cable & Tinsley, 1991 a).

Thus, the long uterus of *P. americanus*, which may accommodate up to 300 larvae, represents a remarkable adaptation (Fig. 2). In contrast to its presumed role in most platyhelminths as an inert storage chamber and passageway to the exterior, the uterus of *P. americanus* is a dynamic structure, responsible for the continuous production of new capsule membranes and for the transfer of nutrients to the accumulated embryos.

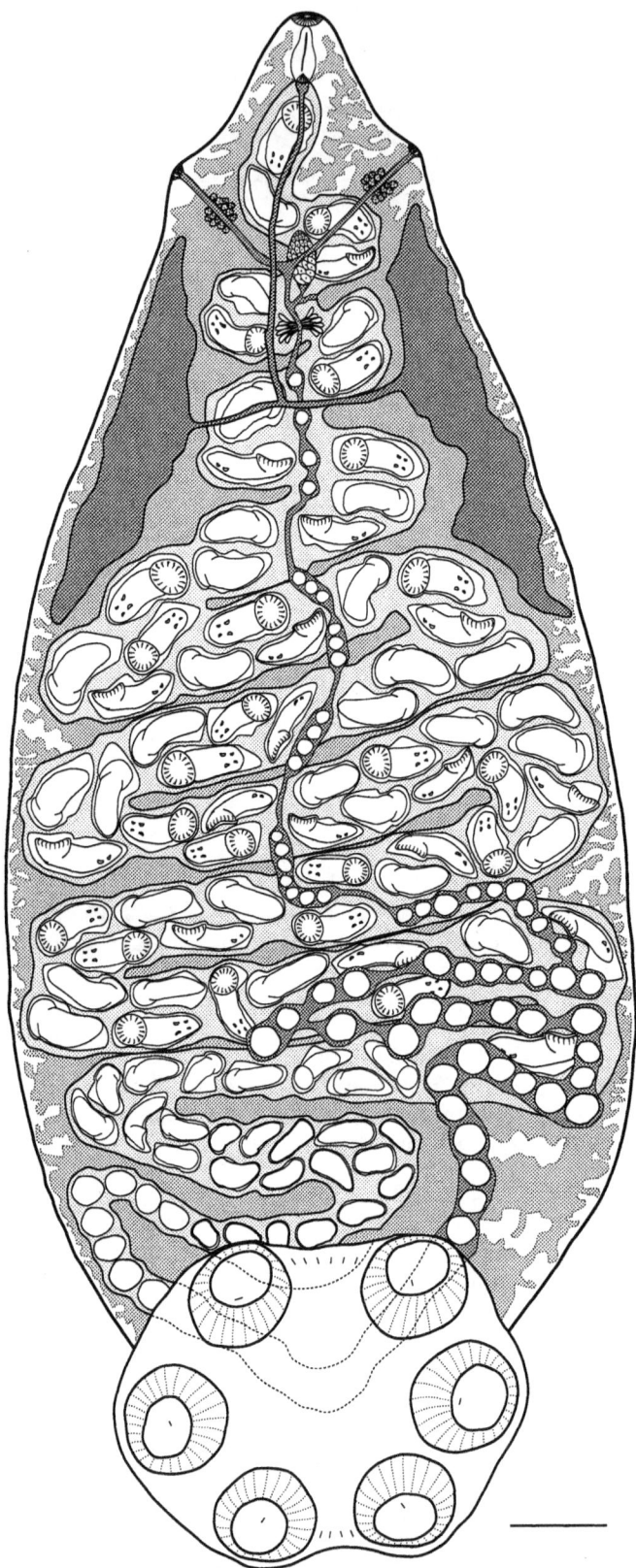

Fig. 2. *Pseudodiplorchis americanus*: diagrammatic representation of gravid adult showing extensive uterus with developing embryos in the descending limb and fully-developed encapsulated oncomiracidia in the ascending coils. Scale bar 0·5 mm.

Every aspect of this suite of adaptations complements the exacting demands of the life cycle. Parasites reach sexual maturity quickly, soon after migration to the urinary bladder, and begin to produce embryos while they are still relatively small (only 30% of the size of a one year old adult). Thus, accumulation of offspring proceeds in tandem with adult body growth. Larvae are produced continuously and numbers increase with time prior to transmission. The membranous egg capsule enables the close packing of maximum numbers of oncomiracidia within the uterus. Direct parent-to-offspring transfer of nutrients means that resources can be produced and supplied gradually over an extended period of time, instead of having to be stockpiled in the adult reproductive system before egg assembly can begin.

Probably the most important advantage of these reproductive adaptations relates to the unpredictable timing of transmission. Tocque & Tinsley (1991a) recorded data showing that, in successive years, the interval between transmission opportunities – and hence the period for production and storage of infective stages – varied from 11 to 13 months. The oncomiracidia of *P. americanus* must therefore be prepared for the earliest opportunity for transmission yet still remain in constant readiness if this is delayed. Continuous channelling of nutrients from the parent guarantees an adequate provision up to the moment of discharge and hatching. The process is efficient for the parent since transuterine nutrition requires investment of resources only for the precise period of need, and it also ensures that, when oncomiracidia are eventually released, they have maximum energy reserves for host invasion (Tinsley, 1990a; Cable & Tinsley, 1991a).

Although these adaptations for maintaining a prolonged state of readiness are remarkable, the stored infective stages of *P. americanus* do have a finite life-span. The data of Tocque & Tinsley (1991b), based on laboratory experimental infections, suggest that uterine larvae die about 2 months after they complete development to the infective stage. These disintegrate within the uterus and their nutrients may be recycled across the uterine wall or via the genito-intestinal canal to the gut (Cable & Tinsley, 1991a). In experimental infections (at 25 °C), Tocque & Tinsley (1991b) found that, after 6 months post-migration to the bladder, the complement of infective larvae achieved a steady state – about 100 oncomiracidia/parasite – with the fraction reaching the end of their storage life balanced by the numbers completing development to the infective stage. This corresponds with the output of 2 or 3 year old worms in the field, but output is also significantly affected by worm age and intraspecific competition (see below).

Adaptations to the constraints operating during host invasion: oviposition

The environmental events preceding transmission typically involve a long period of drought and steadily increasing temperatures during early summer. The 'monsoon' rains often arrive suddenly and spadefoot toads are aroused from hibernation below ground where they have remained inactive since the previous September. The stimulus for emergence is provided by the low frequency vibrations of rainfall drumming on the surface (Dimmitt & Ruibal, 1980) and the toads migrate after darkness to rapidly-forming pools. Assemblies of several hundred individuals gather in pools a few metres in diameter; the chorus of male mating calls, which may be heard by the human ear over distances of 1–2 km, attracts animals from a considerable area and spawning is rapid. At dawn, all the toads leave the water and bury themselves in shallow burrows in the desert soil to escape the daytime heat.

Given the preparation of immediately-infective larvae within the uterus of gravid parasites, the key requirement for exploiting this brief opportunity for transmission is a mechanism for synchronising discharge of oncomiracidia. Clearly this stimulus for parasite oviposition must be very precise and absolutely reliable: if gravid parasites discharge their larvae at any time outside the host's immersion in water the oncomiracidia will be lost. Environmental events accompanying entry of the toads into water could provide distinct signals for recognition by the parasites and Tinsley (1990a) discussed how various host physiological changes could represent the trigger, including the osmotic and chemical changes accompanying rehydration. However, the actual cue is associated with host sexual activity – a guarantee of the presence both of water, required for transmission by the ciliated infective stages, and of other potential hosts. Tinsley (1990a) summarized data recording the exact synchrony of larval discharge with amplexus by males and oviposition by females. Nevertheless, discharge is not linked with host gamete release. Indeed, it is important that the cue is disengaged from spawning *per se* because, with a male:female ratio of between 2:1 and 10:1 (Tinsley, 1990b), a majority of males in a given assembly may not actually mate. Instead, field and laboratory experiments demonstrate that the parasites (including those in chorusing unmated males) release their accumulated larvae in response to intense sexual excitement.

This general response can be confirmed by hormone treatment and behavioural manipulations (Tinsley 1990a, b), but the best evidence for the proximal mechanisms regulating oviposition in polystomatid monogeneans has been provided by experimental studies on the related *Polystoma nearcticum*, a urinary bladder parasite of the north American gray treefrog, *Hyla versicolor*. This system has the same requirement for synchrony of host and parasite oviposition, and the same exact link between host sexual excitement and the parasite's response. However, in this case, the stimulated parasite rapidly manufactures eggs from accumulated components (ova, sperm, vitelline cells and associated secretions) rather than the simple discharge of already-prepared larvae. Immunocytochemical studies by Armstrong et al. (1997) have monitored immunoreactivity of a FMRFamide-related peptide (FaRP), GYIRFamide, in the neurons associated with the ootype musculature (the chamber in which egg assembly occurs). Intense immunostaining in this nerve plexus occurred during periods of rapid parasite egg laying (coinciding with host sexual activity), whereas in hosts examined post-spawning, when parasite oviposition had ceased, there was little or no GYIRFamide immunoreactivity. These findings suggest that regulatory peptide expression at the egg assembly site is switched on and off to coincide with the short periods of *P. nearcticum* egg production. The FaRP may therefore serve as the trigger that coordinates the muscle contractions required for egg manufacture, and this control meets the ecological requirements for a very precise oviposition response by the parasite.

Survival in the external environment: characteristics of transmission

The problem of host-to-host transfer in a desert environment by a parasite with an aquatic infective stage appears to represent the most challenging demand for adaptation to extreme environmental constraints. Paradoxically, field evidence indicates that this segment of the *P. americanus* life cycle is accomplished easily because of two key factors. First, the characteristics of host behaviour provide a very specific opportunity and, second, the remarkable adaptations of the parasite's reproductive strategy achieve a finely-tuned response. The mating of anuran amphibians generates a mass assembly of animals which are otherwise solitary. For a short time each year, normally dispersed anuran populations become highly aggregated. The explosive breeding of *Scaphiopus*, triggered by intense rainfall and restricted to a single night after each storm, serves to produce maximum densities of potential hosts congregated in water which is essential for monogenean transmission (Tinsley, 1989, 1990b). Newly-formed temporary pools become foci for intense infection. However, whilst these characteristics serve to maximize transmission success, departure of the toads at dawn abruptly terminates invasion. Thus, each episode of infection is limited by the numbers of parasite larvae that can invade before this deadline. Since larvae are discharged at intervals throughout the night, especially because hosts must migrate varying distances to the newly-formed pools and may arrive relatively late within the total time-window, the exposure duration for many infective stages – and many potential hosts – is far less than the maximum 7 h. Males tend to remain in water from the variable time of their arrival until the assembly disperses, whereas each female remains only for the period of her own spawning, limiting involvement to around 4 h.

With these constraints restricting each transmission episode, larval characteristics promoting rapid host invasion should have selective advantage. The observations of Tinsley & Earle (1983) show that *P. americanus* has 'super-larvae', the largest recorded in the Monogenea, with a body length of 600 μm and a swimming life more than twice that typical of other monogeneans: oncomiracidia swim continuously for over 4 h at 25–27 °C. Uniquely, the ciliated stages are also resistant to desiccation. They survive drying on the skin of the host for up to 1 h (at 32 °C, r.h. 45%), and the unimpaired swimming ability of these larvae, if returned to water, demonstrates that the cilia are not affected. The tegument is tough and, whereas most monogenean oncomiracidia disintegrate rapidly after death, dead *P. americanus* larvae retain their integrity in water for several days (Tinsley & Earle, 1983; Cable & Tinsley, 1992a). The oncomiracidia possess four photoreceptors that are adapted for very low light intensity. Their fine structure, employing a quarter wavelength reflector, is found in some other polystomatids but in no other platyhelminths (Cable & Tinsley, 1991b). Tocque (1990) measured the energy reserves which contribute to the remarkable longevity and showed that glycogen content at hatching averages 0.216 μg/larva and declines exponentially during swimming life. At 22 °C, close to night-time water temperatures in the desert, 50% of the oncomiracidial population survives 29 h, with maximum longevity 96 h (Tinsley, unpublished). The survivorship curve is characterized by negligible mortality for the first 15 h of life, a period much greater than the entire transmission episode. Therefore, successful invasion is unlikely to be limited by larval survivorship. Rather, the exceptional attributes of the larva may reflect the energetic and other physiological demands of initial establishment in the host.

Ultimately, a parasite transmission stratagem inextricably linked with host reproduction is guaranteed to be successful – as long as the hosts do reproduce. The consequences for the parasite of host reproductive failure are considered below, under the heading host-parasite population ecology.

Survival in the external environment: transmission efficiency

Although the exposure is very brief, events occur so rapidly that dynamic changes can be recorded over a time scale of a few hours. Samples of mating populations of *S. couchii* taken at intervals during one night show a more or less exponential increase in infection levels during the 7 h window. Ten per cent of the night's total invasions occur in the 3 h until midnight, but more than 30% occur in the final hour before dawn (03.00–04.00 h) (Tinsley, 1989). Field studies have demonstrated the consistency of events. Infection levels of newly-invaded larvae are zero at

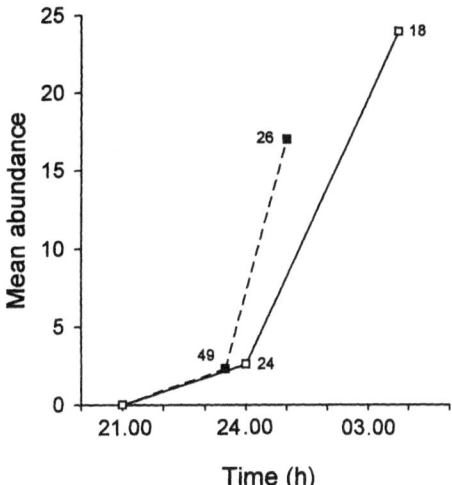

Fig. 3. *Pseudodiplorchis americanus*: short-term dynamics of invasion of *Scaphiopus couchii* illustrated by two different transmission episodes; (□—□) data from Tinsley & Jackson (1988), (■---■) data from Tinsley (1989, Fig. 2). Time (Standard Mountain Time, Arizona) represents the duration of the exposure from dusk until dawn. Sample sizes shown adjacent to respective points.

about 21·00 h on the first night of breeding (with animals having just emerged from hibernation). Data recorded by Tinsley (1989) show that mean abundance reached 2·3 worms/toad at 23·30 h, increasing to 17·0 worms/toad only 90 minutes later at 01·00 h. Comparable data were recorded by Tinsley & Jackson (1988) showing a mean abundance of 2·6 worms/toad at midnight, increasing to 23·9 worms/toad at 03·45 h (Fig. 3). Parasite population terms used in this account follow Bush et al. (1997).

In both these studies, a second torrential storm led to a repeat aggregation on the following night, producing a second input of invading larvae. In Tinsley & Jackson's (1988) study, the cumulative burdens acquired after the two exposures produced a mean abundance of 44 worms/toad. Exceptionally, in this season, there was a third storm and another transmission episode 11 days later. A record of cumulative recruitment for the population recorded 100 % prevalence and mean abundance 81 worms/toad, but this was taken at 03·00 h and would have underestimated final levels. Based on the relationship of recruitment with time in other studies, the last hour before dawn would be expected to have increased the final total to over 100 larvae/toad (Tinsley & Jackson, 1988). In the separate study of Tinsley (1989), a small sample taken at dawn showed that an equivalent final total was achieved by the end of the second night's exposure: a prevalence of 100 % and mean abundance 107 worms/toad.

These and other unpublished studies confirm a consistent, virtual saturation of the host population each season by invading *P. americanus* larvae. Prevalence in males is normally 100 % and around half of the individuals in populations acquire burdens exceeding 100 worms, up to a maximum of 400/host (Tinsley, 1989; Tinsley & Jackson, 1988). These field studies provide a detailed view of factors influencing individual invasion levels, including the effects of behavioural differences between males and females, and between mated and unmated males (see Tinsley, 1989, 1990 a, b). Other studies have included a more precise analysis of the output of infective stages from the reproducing parasite population. Tocque & Tinsley (1991 b) showed that the contribution of oncomiracidia by adult worms is age-dependent: 2nd year worms produce nearly 10 times the oncomiracidial output achieved by 1st year's; 3rd year's produce over twice the *per capita* output of 2nd year's; and, in this study, a rare 4 year old worm produced almost the same larval output as the combined total from the cohort of 73 1st year worms. Reproductive investment is also significantly affected by temperature and the duration of the host hibernation period (Tocque & Tinsley, 1991 a), and by competitive interactions that reduce larval production at high worm burdens (Tocque & Tinsley, 1991 b). Thus, larval output in transmission assemblies is influenced by the proportion of each age class present in the parasite suprapopulation, by environmental conditions preceding transmission, and by the distribution of infection levels within the host population and its effects on density-dependent parasite interactions.

A detailed analysis of *P. americanus* population dynamics is therefore considerably more complex than envisaged in the initial correlation of adult parasite population size with recruitment success (Tinsley & Jackson, 1988). Nevertheless, these field studies provide a useful estimate of invasion efficiency. Assuming a mean abundance of gravid parasites of 3·0 worms/toad, a mean output of 100 oncomiracidia/gravid worm, and a mean input of invading worms at the end of the season of around 100 larvae/toad, the probability of individual invasion success is about 0·3 (Tinsley & Jackson, 1988). This calculation may underestimate actual invasion probability. Comprehensive studies suggest that a mean of 100 larvae/gravid worm is achieved under only the most favourable conditions. In most seasons, output is considerably less. Tocque & Tinsley (1991 b) recorded an overall mean of 39 larvae/adult measured 11 months p.i., indicating a mean of about 60 at the time of transmission. On the other hand, previously-published records of the mean input of invasions/host also reflect best circumstances, based on years where there was repeated transmission. In many years the final burdens are lower. Tocque & Tinsley (1994 a) recorded invasion data for one population sample of mean abundance 59 worms/toad. Actually, therefore, these alternative (more conservative) data – mean output and input both around 60, with an adult suprapopulation of 3 worms/toad – produce exactly the same probability of invasion by individual larvae of around 0·3.

The transfer from external to internal environments: invasion

Uniquely, amongst monogenean life cycles so far described, the target for host invasion by the swimming oncomiracidium – the nostrils – is above the water surface. To reach this point of entry, larvae must break through the surface film of the water and migrate over exposed skin. Under normal conditions in mating aggregations, the skin is kept wet by movements of the toads in water, but larvae are nonetheless able to migrate over damp skin where only their ventral surface is in contact with moisture (Tinsley & Earle, 1983). Remarkably, as mentioned above, oncomiracidia can tolerate short-term drying on the skin. From the time of entry into the nostril chamber and sinuses, larvae live at an air/water interface. The ciliated cells are shed 1–2 h post-invasion (Cable & Tinsley, 1992a); all locomotion subsequent to initial host contact is by vigorous muscular crawling.

Internal environmental conditions: survival in the respiratory tract

As with other aspects of the biology of *P. americanus*, the environmental conditions experienced during initial establishment within the host have few precedents elsewhere in the Monogenea. Worms remain in the host nostrils and sinuses for 24 h after invasion, and during the ensuing days they migrate into the mouth and associated cavities (sub-lingual space, eustachian tubes, glottis and the vocal sac of males). They begin to feed on blood 24–48 h p.i. (Tinsley & Earle, 1983), but mucus and epithelial cells may be ingested by recently-invaded larvae. Progressively they migrate through the glottis and into the lungs and all occur in this site after 1 week p.i., remaining for 7–14 days. At approximately 3 weeks p.i., juvenile worms return to the buccal cavity where they accumulate prior to migration to the urinary bladder (Tinsley & Jackson, 1986).

The natural temperature regime experienced during this phase of parasite development is determined by host behaviour: this buffers the extreme characteristic of the desert habitats where soil surface temperatures can exceed 70 °C (Tocque & Tinsley, 1991a). During the day, the toads bury themselves about 5 cm below the surface (Ruibal, Tevis & Roig, 1969). Field records at this depth in July and August indicate that parasites experience a cycle with a mean daytime temperature of 34 °C for 7 h, falling at night to a mean of 22 °C for 5 h (Tocque & Tinsley, 1991a). On damp nights, when the toads emerge to forage, soil surface temperature varies between 18 and 24 °C.

Few monogeneans inhabit air-filled passageways where, even if the parasite's ventral surface is in contact with moist epithelia, the dorsal surface is at the air/water interface. Species of *Polystomoides* and *Neopolystoma* occur in the oral cavity and pharynx of chelonians (Tinsley & Earle, 1983), and *Gyrdicotylus gallieni* inhabits the oral membranes of *Xenopus laevis*. In this latter case, worms invade via the nostrils (as in *P. americanus*) but this portal is underwater at the time of entry (Harris & Tinsley, 1987; Jackson & Tinsley, 1994; Tinsley, 1996). *Pseudodiplorchis americanus* and the closely related *Neodiplorchis scaphiopodis* are unique amongst monogeneans in infecting vertebrate lungs (Tinsley & Earle, 1983). Here, the acid mucosubstance lining and phospholipid surfactant, together with the pulmonary immune defences, create a highly hostile environment for the juvenile worms. The environmental conditions parallel those experienced by pentastomid parasites in reptilian lungs. Riley & Henderson (see this volume) have described a remarkable defence mechanism based on massive membrane production and the secretion of lipids that apparently mimic host surfactant. Nothing is known of the responses of *P. americanus* in equivalent circumstances. Cable & Tinsley (1992a, b) found no obvious ultrastructural correlates of the hostile conditions in the respiratory tract: the tegument of juveniles is similar to that of oncomiracidia and there are no structural differences between the dorsal and ventral body surfaces. Membrane-bound tegumental vesicles probably contribute to the glycocalyx which, together with glandular secretions, may provide protection. However, in contrast to this lack of obvious adaptations to life in the respiratory tract, there are unique specializations that enable survival in the digestive tract.

Internal environmental conditions: migration through the alimentary tract

During the period when *P. americanus* occurs in the mouth, ingestion of prey by the host includes a majority of hard-bodied, abrasive desert invertebrates (beetles, ants, crickets) together with a remarkable range of noxious animals such as scorpions, solpugids, centipedes and pogonomyrmid ants (Tocque, Tinsley & Lamb, 1995). Exposure of parasites to a variety of toxic chemicals, especially acids, seems inevitable.

Migration through the stomach and intestine to the cloaca and urinary bladder is first possible at 4 weeks p.i. (at 25 °C), normally at the peak of the host's activity season when the gut is packed with digesting contents (see Tocque *et al.* 1995). Experimental trials indicate that movement is rapid, taking as little as 5 min for worms to traverse the 80–100 mm length of the alimentary tract. Tinsley & Jackson (1986) demonstrated that migration and survival in the potentially lethal conditions in the gut is dependent on a trigger factor: this both controls the moment of migration and activates the mechanisms providing protection. The cue is associated with host activity. In toads that enter dormancy before the time of migration, the parasites remain in the respiratory tract and undergo no more growth and development than that achieved prior to normal migration at 4 weeks p.i. Indeed, if hosts remain undisturbed during hibernation in the wild, or during laboratory maintenance simulating natural conditions, worms fail to migrate and remain in an arrested state for more than 1 year. These nevertheless retain the ability to migrate: when the host is aroused from dormancy, the delayed migrants are stimulated to undertake the rapid journey and then begin normal reproductive development leading to maturation around 1 month later.

Whilst the cue triggering migration has not been identified, it has emerged that anaesthetization of the host with tricaine methane sulphonate (MS222) provides a highly effective artificial stimulus facilitating experimental studies (Tinsley & Jackson, 1986). Worms transferred directly from the host respiratory tract to stomach and intestinal contents at varying stages of digestive activity all died within 2 min, and TEM showed that the tegument was stripped away. In starved toads, including those just emerged from hibernation, the mid-intestine contains a section of accumulated bile fluid and this is also lethal to transferred worms within 2 min. In complete contrast, worms that have begun migration, either naturally or in response to host anaesthetization with MS222, survive for 4–8 h in conditions that are lethal to non-migrants. Ultrastructural studies show that survival is conferred by the mass discharge of tegumental vesicles (Cable & Tinsley, 1992b). During development in the respiratory tract, the juvenile worms accumulate two types of electron-dense, membrane-bound vesicles, one PAS-positive and the other PAS-negative. These increase in density in the distal cytoplasm up to 28 d p.i. but then remain unaltered in numbers and appearance until migration, even if this is delayed for over 1 year. During transfer through the gut, there is a progressive discharge of vesicles from the surface layers and an initial compensatory transfer of vesicles from the perikarya into the distal cytoplasm. After depletion of the store of perikarya vesicles, the density in the surface layer also reduces. This dynamic flux suggests that mass release of vesicles contributes to rapid turnover of an outer coat that provides protection against digestion. If migrants are left for prolonged periods in gut contents (normally over 8 h), the store of tegumental vesicles is eventually exhausted and, at this point, the tegument begins to disintegrate and the worms die (Cable & Tinsley, 1992b; Cable, Harris & Tinsley, 1998).

Death of worms transferred to gut contents from the respiratory tract (without prior 'activation') occurs despite their store of tegumental vesicles, indicating that exposure to the hostile conditions in the gut does not trigger the protective secretions. Instead, vesicle discharge is stimulated in advance by a specific cue, presumably at the same point as activation of the migration response.

The arrested development of worms whose migration is delayed emphasizes the effect of environmental factors on this life cycle. Worms remaining in the respiratory tract are active and continue to feed (as indicated by the presence of haematin in the gut) but further growth and reproductive development is entirely inhibited. Worms from the same invading cohort that do migrate immediately begin rapid development. These developmental outcomes are unlikely to be determined by nutritional factors since parasites in the two sites have access to the same blood diet. The gut migration therefore has a central role in the life cycle, not only linking together the two phases of parasitization – of the respiratory and urinary tracts – but also initiating reproduction. Under experimental conditions at 25 °C, parasites reach sexual maturity one month post-migration and begin to accumulate embryos whilst their body growth continues. About 200 offspring are produced after 6 months, of which about half are fully-developed and infective (Tocque & Tinsley, 1991a, b). This appears to represent a very efficient course of preparation for the next transmission episode. However, information on development at the temperatures actually experienced in the desert indicates that the potential for reproduction during the first year post-infection is almost completely eliminated by this natural constraint.

Constraints of external environmental conditions: direct and indirect effects on parasite development and reproduction

Parallel field and laboratory studies have demonstrated that temperature exerts a powerful regulatory influence over parasite growth, development and reproductive output: fastest rates occur around 25 °C but are reduced by half at 20 °C and virtually halted at 16 °C. Based on the annual temperature cycle in the field, Tocque & Tinsley (1991a) showed that all parasite development is arrested from about October to April when temperatures at the soil depth occupied by hibernating toads are below 15 °C: this totally precludes growth and reproductive preparation for 6–7 months/year. Soil temperatures exceed 25 °C, providing optimum conditions for parasite development, for only about 3 months of the year. Temperatures exceed 20 °C for 4–5·5 months, i.e. during summer and early autumn following transmission in one year and late spring/early summer preceding transmission in the next year. For each cohort of invading parasites, juvenile development in the lungs occupies the first month of these favourable temperatures, leaving only 3–4.5 months for maturation and embryo production. Tocque & Tinsley (1991b) showed that 4 months at 25 °C is necessary for the first embryos developing *in utero* to reach the infective stage, and adult worms accumulated only about 30 infective larvae after 5 months post-migration. It follows that, as a result of the combined constraints of temperature and time, 1st year worms may make only a very minor contribution to transmission. Body growth continues throughout life but 2nd year worms, having already achieved initial development to maturity, benefit from the full period above 20 °C and make a larger reproductive contribution. This trend is maintained by 3rd year and very rare 4th year worms. The field records of Tocque & Tinsley (1991b) (based on a suprapopulation sampled 11 months after the previous year's transmission) showed means for 1st, 2nd and 3rd year adults of 5, 43 and 96 infective larvae/worm respectively, and a single 4th year worm with 326 larvae.

Because of this strict constraint of temperature regime, the period of reproductive development is also regulated by the timing of the 'monsoon' rains, and hence transmission, in adjacent years. Short 'years' occur when a late monsoon in one year is followed by early rains in the next. The study of Tocque & Tinsley (1991a) included one such 11 month inter-transmission period (first rains on 27 July and 1 July in successive years): field temperature records showed that parasites experienced only 17 weeks above 20 °C, precluding any contribution to transmission by 1st year adults. This was followed by a 13 month 'year' (first rains on 1 July and 29 July in the adjacent years), including 23 weeks above 20 °C, and gave the potential to produce around 30 infective larvae/1st year worm. Late invasion, resulting from host breeding in early August (see, for instance, Tocque, 1993), allows a very reduced time for development at optimum temperatures. Worms migrating in early September may not reach maturity before temperatures decline below 20 °C; these resume growth in May but remain small, with little reproductive development, at the start of the summer transmission. If host activity is precluded (by dry conditions and declining temperatures) before parasite migration, then the pre-migrants remain undeveloped in the respiratory tract throughout hibernation and migrate only when the hosts become active and enter breeding assemblies in the following summer (Tinsley & Jackson, 1986, 1988). These arrested stages, now 1 year p.i., are only 4 weeks more advanced in development than the new season's invading larvae.

Tinsley & Jackson (1988) recorded one suprapopulation of *P. americanus* in which the normal synchrony of the life cycle was highly disrupted, reflecting delayed migration and arrested development in the previous year. The host sample at the start of the transmission season (n = 92) contained a larger subset of parasites that had failed to complete reproductive preparation (mean abundance 3.37 worms/toad) than those prepared with infective stages *in utero* (mean abundance 2.95). About half of these 'failed' stages remained arrested in the respiratory tract, whilst the others that had migrated ranged from small immature worms to adults containing only developing embryos. Whilst these parasites had missed the once-per-year opportunity for transmission, it could be considered advantageous that they now had a full year to prepare an above average contribution for the next. However, given the normal 3 year life span, with an estimated 50% mortality between years 2 and 3 (Tocque & Tinsley, 1991b), inability to contribute to one of the years would be expected to have significant consequences for overall transmission. The actual effects of such disrupted life history schedules are explored in a case study of parasite population ecology later in this paper.

It seems surprising that there should be such a narrow 'margin of safety' in this life cycle. Year-to-year variations in temperature and rainfall have effects of such magnitude that a major fraction of the surviving parasite population may fail to contribute to annual transmission. Part of this vulnerability is attributable to the necessity for migration through the extreme environment represented by the alimentary tract. There are two consequences. On the one hand, the unique survival mechanism requires a cue from a host factor which is, in turn, dependent on external environmental conditions (rainfall and temperature); this creates the risk of disruption by chance factors. On the other hand, reproductive development is inhibited until migration has occurred. This dual effect reinforces the key importance of gut migration in the biology *P. americanus* (see above). Moreover, the vulnerability of the life cycle to disruption by environmental conditions is a direct consequence of the novel characteristic of a respiratory phase.

Constraints of internal (host) environmental conditions: effects on parasite survival

Following the manifest efficiency of the transmission process, it has emerged that the reproductive potential, represented by the annual input of invading worms, fails to have a commensurate effect on future transmission because of the powerful constraint of the natural temperature cycle in the desert. This reflects a direct rate-limiting effect of the environment on reproductive output. However, there is a second, far more important, constraint that limits realisation of the potential of the life cycle: a remarkably high mortality rate within established parasite infrapopulations. The major effects of this, too, are temperature dependent but, in this case, putatively operating through the host immune response.

Tinsley (1995) demonstrated that the virtual saturation of the mating host population by invading parasites is followed, each year, by a major pre-reproductive reduction in surviving worm numbers. The consistently high invasion levels, with prevalence 100% and mean intensities of 60–100 worms/host, fall in the period preceding the next transmission opportunity to prevalence 50%, mean intensity 5–6 worms/host. Maximum burdens following invasion, typically around 200 worms/host and sometimes 300–400, virtually never exceed 30 adults at the start of the next transmission season (Tinsley, 1995). The same effects can be demonstrated in laboratory experimental infections (Tinsley, 1989). Overall, 97% of the larvae that successfully invade the host population die before they can contribute to transmission. The basis of this estimate derives from the converse of the simple calculation (above) that, given relatively stable year-to-year population levels, where the mean abundance of invaded parasites is around 100 worms/toad, the mean abundance of adults surviving to reproduce one year later is around 3·0 worms/toad (Tinsley, 1995). This may actually be a significant overestimate: the adult reproducing populations are generally composed of 3 year classes of parasites, of which the 1st year adults – the survivors of the previous season's invasions – may comprise only 40% of the suprapopulation (Tocque & Tinsley, 1991b). On the other hand, the data cited by Tinsley (1989, 1995) are based on gravid adult parasite suprapopulations (worms with infective oncomiracidia *in utero*) and in many years there may also be a significant number of developing (not yet gravid) worms, the product of the same invasion. With these qualifications, therefore, the overall probability of survival from invasion to first transmission is likely to be close to 0·03 (as recorded by Tinsley, 1995).

Tinsley (1993) identified a succession of factors involved in parasite mortality, including losses attributable to host mortality (influenced by life history characteristics including longevity, and by stochastic factors), parasite-induced host mortality, parasite intra-specific competition. Other hazards include, in experimental studies, around 10% mortality during the course of gut migration (Cable & Tinsley, 1992b), and up to 10% attributed to the pathogenic effects of a microsporidean hyperparasite (Cable & Tinsley, 1992c).

However, even in combination, these mortality factors cannot contribute to more than a part of the total, almost overwhelming, parasite losses. Tinsley (1995) has reviewed evidence that regulation of the *P. americanus* populations strongly resembles the operation of a host immune response. In summary, a number of regularly-observed features of the population biology of *P. americanus* support this hypothesis (although there is not yet any experimental immunological evidence). First, studies at the same field sites in successive years have recorded a striking constancy in the surviving adult suprapopulations of about 3 worms/toad. This occurs despite year-to-year variations in the environmental conditions that influence invasion success. Apparently, differences in annual recruitment are overridden by post-invasion factors which exert a dominant control over surviving parasite numbers. The more-or-less constant annual 50% prevalence may represent the fraction of the host population that is susceptible to infection. Second, repeated annual infection by *P. americanus* should lead to progressively greater worm burdens during the host's life-time. However, infection levels rise to a plateau in toads aged about 6 years and then decline in the oldest age groups despite continued re-infection by parasites with a 3 year life span. This could provide evidence of acquired immunity. Third, experimental infection survived better (89% prevalence) in field-caught hosts that had pre-existing adult infections compared with those without a pre-existing infection (31% prevalence). This is suggestive of variation in susceptibility that predisposes to further infection (Tocque & Tinsley, 1994a). Fourth, there is comprehensive experimental evidence for an inverse relationship between parasite survival and temperature: this is consistent with the operation of an immune response that is temperature-dependent (as is characteristic of ectothermic vertebrates) (Tocque & Tinsley, 1994a). Finally, there is also indication that exceptionally high burdens of *P. americanus* in individual toads tend to co-occur with abnormally high burdens of other parasites (example cited in Tocque & Tinsley (1994a) and unpublished records). A behavioural influence is unlikely since the nematode, cestode and monogenean species involved have entirely different transmission routes. The findings are suggestive of pre-disposition to infection potentially reflecting impaired immune competence.

Evidence for the putative regulation of worm burdens is discussed in detail by Tocque & Tinsley (1994a) and Tinsley (1995). Based on experimental infections of *P. americanus* in wild-caught toads, Tocque & Tinsley (1994a) found that parasite survival was strongly temperature and time dependent. At 25 °C, there was no significant decline in the infections of 1st year worms for up to 5 months post-migration to the bladder (prevalence 94%, mean intensity 13–18 worms/host, maximum 30 worms). Subsequently, prevalence and intensity fell rapidly to 50%, maximum 19 worms after 6–8 months, and 25%, maximum 9 worms after 10 months. Only 2·1% of the initial experimental invasion survived to 10 months post-migration. Survival of pre-existing, older age classes (acquired in the field before experimental infection) followed a similar pattern. At 25 °C, infection levels were virtually identical after 3–4 months to those recorded at the time of host capture: prevalence 52%, intensity 6·2 worms/host, maximum 32 worms. However, prevalence declined to 27% after 7 months and 11% after 8–10 months with a corresponding decrease in intensity, and no pre-existing adult worms survived after 10 months. In both 1st year and pre-existing adult worm infections, the decline in infection levels was accompanied by a loss of the heaviest burdens with a corresponding decline in the variance/mean ratio.

In marked contrast, at 15–20 °C there was no decline in prevalence or intensity of either 1st year or pre-existing adult parasites for the duration of the laboratory study (14 months), and there was no change in the variance/mean ratio confirming the persistence of higher worm burdens.

These data suggest that the losses of *P. americanus* occur only during periods when the environmental temperatures (at the soil depth occupied by buried *S. couchii*) exceed 20 °C, a period comprising 4–5·5 months of each year (Tocque & Tinsley, 1991*a*).

In these laboratory studies, experimentally-infected hosts did not experience temperature manipulations during the period when the parasites occur in the host lungs. Potentially, this is the most hostile site for immune attack and respiratory infection coincides with the period of highest annual temperatures. TEM studies on the pathology of this phase have revealed evidence of immune attack where macrophage-like cells are associated with tegumental damage (Cable *et al.* Unpublished). Major parasite attrition might be expected during lung infection, but available quantitative evidence relates to total parasite losses rather than those occurring in this specific segment of the cycle.

These overall data, from a series of field studies and laboratory experiments, point consistently to the conclusion that the internal environment of the host presents by far the most extreme conditions for parasite survival. Indeed, if it were not for the amelioration of these hostile conditions by the period of low temperatures each year, the parasite would not survive in the Sonoran Desert from one transmission season to the next.

HOST-PARASITE POPULATION ECOLOGY

The field and laboratory studies reviewed above provide quantitative assessment of the effects of environmental conditions on the population biology of *Pseudodiplorchis americanus*. Most notably, there is an estimated 97% mortality of post-invasion parasites before the first opportunity for onward transmission. The studies also identify points where the life cycle is highly vulnerable to disruption, especially the indirect environmental control of gut migration and its effects on the future reproductive contribution of each year's invading cohort. The margin between the 'norm' for this life cycle and complete failure seems surprisingly narrow prompting the question whether natural circumstances could shift this apparently fragile balance towards extinction. The occurrence of a parasite in a given geographical area should provide unambiguous evidence that environmental conditions do not significantly threaten long-term survival. Population data for *P. americanus* at specific local field sites have been interpreted as demonstrating remarkable year-to-year stability of infection levels (Tinsley, 1995). This pattern has been maintained despite annual fluctuations in rainfall that should generate variation in larval recruitment. It may be concluded that parasite population biology is buffered against changes of this magnitude, particularly by the efficient transmission process that leads to massive invasion of target populations. Additionally, the 3 (very rarely 4) year life span of the parasite should provide a buffer against occasional recruitment failure. Analysis of weather records for the study area together with direct field observations from 1981 to 1992 indicate that relatively wetter and drier years have tended to alternate over this period (see below and Tinsley & Tocque, 1995) permitting the reversal of periodic environmental checks.

The second part of this account focuses on a 'test case' that reconstructs the effects of extreme environmental variation on the host-parasite interaction. The circumstances were provided by a period of severe summer drought in the study area, beginning in 1992 and intensifying in 1993 and 1994. The analysis considers the background to the environmental conditions and reviews data documenting the host and parasite populations from field studies conducted up to 1992. The after-effects of the drought conditions were recorded during summer fieldwork in 1995–1998 and the interpretation presented below provides a preliminary analysis of these results.

EFFECTS OF EXTREME ENVIRONMENTAL PERTURBATION ON THE HOST-PARASITE SYSTEM

The environment

Field study sites are located in the San Simon valley, Arizona, an area of desert scrub and desert grassland with mean annual rainfall of 223 mm/year (Tinsley & Earle, 1983). Long-term environmental records providing a general guide to conditions are available from two weather stations: Portal, covering the period 1914–1955, and the Southwestern Research Station, for 1965 to the present (http:www.wrcc.dri.edu/cgi-bin/cliMAIN.pl?azport).

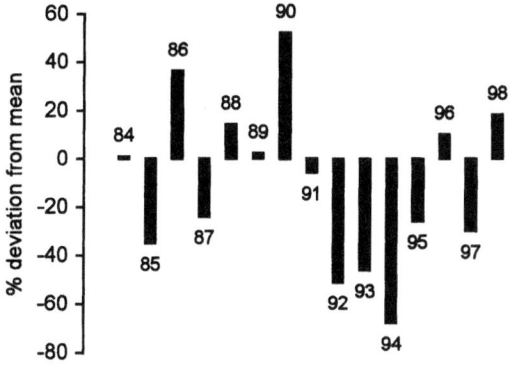

Fig. 4. Variations in July rainfall for the period 1984–1998. Zero represents the mean for this month based on the record of the Southwestern Research Station for 1965–1998. Bars for each year show the percentage deviation from this mean.

Both stations are at higher elevation than the desert field sites and have wetter and cooler conditions (mean for the 34 year record of the Research Station 541 mm/year, S.D. 120). Nevertheless, the annual profile of weather conditions is similar. Nearly half the annual rainfall occurs in July and August, facilitating a brief season of *Scaphiopus* activity until increasing aridity and declining temperatures bring surface activity to an end in September. Winter temperatures are cold and precipitation in November to April may fall as snow. Soil temperatures on the desert surface exhibit extreme variation: weekly minima are below freezing in November to March, but surface maxima are above 60 °C from April to September. In the buffered environment 15 cm beneath the soil surface, temperatures are below 10 °C for 4 months and below 15 °C for 6–7 months each year; at this depth, temperatures are above 25 °C for only 12 weeks each year (means of 3 years data from Tocque & Tinsley, 1991 a). An indication of rainfall variations influencing this study is provided by the records for July, normally the main period of surface activity for *Scaphiopus*. Fig. 4 compares July rainfall against the mean calculated from the 34 year data set (this is only a guide because conditions for host feeding and reproduction are also influenced by June and August rainfall). Other comparisons (below) employ 'percentage of the mean' using the average for the respective periods in the long-term weather record.

Assessment of environmental variation: rationale and interpretation

The seasonal restrictions imposed by the desert environment affect all aspects of *Scaphiopus* biology. Concentration of feeding into only a few weeks each year alternates with around 10 months of dormancy (and total starvation) and produces corresponding pulses of growth. In cross-section, bones show concentric growth rings which reflect age and annual growth rates. An initial study was based on sectioned bones from an overall sample of 694 animals (femurs from toads dissected to record parasite infection or digits from mark-released live toads) collected in 1986–1992 (for methods and interpretation see Tinsley & Tocque, 1995). The present study, in 1995–1998, examined sectioned bones from 650 *S. couchii* from populations sampled at 13 sites along a 110 km transect (including all the sites from the earlier study). These field sites are given the labels A–M in the following account. Data for each dissected *S. couchii* included age, body weight and length, organ weights including fatbodies, haematocrit, together with records of total parasite infection. Detailed analysis is in progress (including an extension to 1999).

Age analysis of all individuals in the *P. americanus* populations was undertaken in 1996–98, using the criteria of body size and reproductive development outlined by Tocque & Tinsley (1991 b).

Bone growth rings provide sensitive information which correlates *Scaphiopus* ecology with environmental conditions (Tinsley & Tocque, 1995). The width of each growth ring reflects feeding success in the given period. Population analysis, covering growth rings laid down in 1978–1991, showed a strong positive correlation between ring width and summer rainfall: this determines the abundance of invertebrate prey populations, frequency of feeding opportunities (damp nights when toads can emerge to forage), and duration of the activity season (and hence the overall period to accumulate nutrients). The rainfall record shows that there was an alternation of relatively wet and dry years over the 14 year period (wetter in 1979, 1981, 1983, 1984, 1986, 1988, 1990, 1991 and drier in 1978, 1980, 1982, 1985, 1987, 1989). About 50% of the animals sampled showed a corresponding alternation of thick and thin growth rings. The age structure of toads in breeding populations showed a series of dominant cohorts (a high frequency of a particular age class); the years in which these were born correlate with the higher rainfall years, reflecting improved opportunities for spawning, increased survival to metamorphosis in temporary pools, and favourable conditions for recent metamorphs to feed.

The overall weather record shows relatively wide variations in rainfall with the total in the wetter years more than twice that in the drier. In some years, at some local sites, dry conditions may have prevented breeding. More commonly in harsh years, breeding did occur but recruitment failed, either because the ponds dried before metamorphosis or because the feeding opportunities after metamorphosis were insufficient to allow adequate growth and lipid storage. These are the years characterized by a very low frequency of the corresponding age cohorts of toads within the total population. However, the weather record from the early 1980s to the early

Table 1. Résumé of rainfall patterns (hence general environmental conditions) and effects on host and parasite ecology, San Simon valley, Arizona 1993–98

	Summer rainfall	Host spawning	Host recruitment	Host feeding	Parasite invasion
1993	+	+	0	+	+ +
1994	+	+	0	+	+ + +
1995	+	0	0	+ +	0
1996	+ + + +	+ + + +	+ + +	+ + +	+
1997	+	+	+	+ +	+
1998	+ + +	+ + +	+ + +	+ +	+

Rainfall: below average in 1993, 1994, 1995 and 1997 (see Fig. 4), but extensive heavy July rains in 1996, 1998 with widespread flooding.

Host: general recruitment failure in 1993, 1994, 1995 producing a gap in maturing cohorts in 1995–97; first appearance of newly mature toads in mating assemblies in 1998.

Parasite: transmission successful in the 'drought years' 1993, 1994; invasion interrupted/poor in 1995 (local rainfall, little host spawning), 1996 (flooding), 1997 (local rainfall, little spawning) and 1998 (flooding); weak recruitment in some or all these years leading to reduced output of infective stages 2–3 years later; decline in infection levels in most populations but this was avoided at some local sites where transmission/recruitment was successful in 1996; otherwise some local populations reached point of extinction by 1998.

(negligible occurrence, 0; increasing extent of occurrence + → + + + +)

1990s, together with direct field observations, showed that conditions were sufficiently favourable, even in the worst years, for some breeding and foraging activity. The alternation of wet and dry summers led to the situation that, even if local conditions were exceptionally difficult in one year, better conditions in the next year would allow recovery (Fig. 4). Indeed, the evidence of this is provided in the alternation of wide and narrow growth rings in individuals and the regular succession of dominant age cohorts in the population age structure.

The adaptations of the parasite life cycle would predict a different scenario for the population biology of *P. americanus*. Recruitment into the parasite populations is likely to occur in most years as long as the toads assemble to spawn, regardless of the subsequent conditions determining host population recruitment. Studies at specific field sites during the period 1983–1990, when the rainfall pattern showed regular periodicity but never extreme fluctuations, showed that the parasite populations exhibited remarkable stability confirming relatively consistent recruitment (Tinsley, 1995).

A period of drier summers began in 1992 and continued through 1993 and 1994; effects were intensified by major reductions in spring rainfall, generally less than half of the mean, followed in each year by an exceptionally dry July (drier than all other years except one in the 34 year record of the Southwestern Research Station). August in 1993 was very wet but was followed by drought in September and October (combined rainfall only 16% of the mean), and any late breeding of *S. couchii* is unlikely to have led to successful recruitment. The 12 months influencing the 1994 activity season (September 1993–August 1994) received about half the mean annual rainfall, and the July 1994 rainfall was only 32% of the mean for this month. This was focused into one storm which stimulated emergence and breeding of *S. couchii* but the ponds quickly dried up (Fig. 4).

This perturbation provides the background for this analysis of environmental effects on the host-parasite system: a major impact on population biology would be predicted. Following the 3 dry years, including severe drought in 1993 and 1994, the 4 fieldwork seasons of this study comprised two relatively dry summers (1995, 1997) and two relatively wet (1996, 1998) (Fig. 4). The disruption of *S. couchii* recruitment continued in 1995 when spring rainfall was 40% below average and summer rainfall was 30% below average. The July and August rains failed to produce torrential downpours necessary to create successful breeding sites. Initially, 1996 continued this trend with rainfall in December 1995–May 1996 only 22% of the average. However, June was exceptionally wet – over 6 times the average – followed by a very wet July, causing extensive flooding of the desert habitats. This led to repeated spawning assemblies across the range of field sites and the prolonged damp conditions are likely to have favoured survival and development of the cohort of metamorphosed toads. Total summer rainfall in 1997 was close to average but storms were localized and there was little *S. couchii* breeding in many of the study sites. The final field season, in 1998, included torrential downpours in July, with

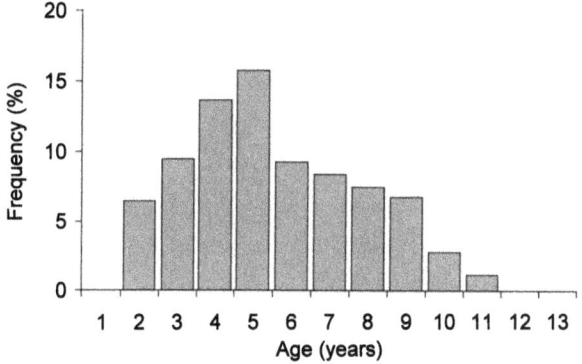

Fig. 5. Frequency distribution of age (numbers of growth rings) in male *Scaphiopus couchii* ($n = 359$) collected from breeding assemblies (data from Tinsley & Tocque, 1995). Combined sample of 7 years' records (1986–1992) from 4 study sites within a 7 km radius in the San Simon valley, Arizona.

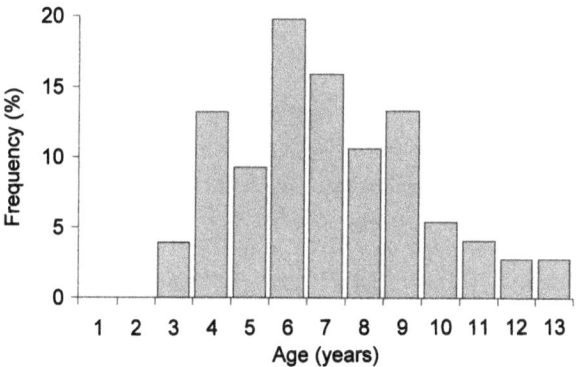

Fig. 6. *Scaphiopus couchii*: frequency distribution of age in the first post-drought year, 1995, based on a combined sample of 76 individuals from San Simon study sites, Arizona.

extensive flooding, and there was widespread *S. couchii* breeding. Key features are summarized in Fig. 4 and Table 1.

Host population ecology

Baseline information for the host populations at the main study sites is provided by Tinsley & Tocque (1995) derived from a combined sample of 7 years data, 1986–92 ($n = 518$). Counts of the annual growth rings in transverse sections of bone show that *S. couchii* is relatively long lived: most animals (65%) in breeding assemblies were at least 5 years old, about 33% were 7 years and older, and 5% were at least 10 years old. There are only minor differences between the survivorship of males and females: age at first appearance in breeding populations is 2 years for males and 3 years for females but greater numbers reach maturity in the following 1 or 2 years, contributing to an increasing frequency of representation of age classes up to 5 years; maximum longevity is slightly greater in females (13 years) than males (11 years) (Fig. 5 shows records for males).

However, these data do not constitute a general life table for *S. couchii* because the 7 years data set showed very different characteristics in each annual 'snap shot' influenced particularly by a succession of dominant cohorts. These reflect the outcome of successful recruitment originating in years particularly favourable for spawning and juvenile survival (relatively wetter years). Each dominant cohort formed a peak in the age distribution shifting by one year in each successive annual sample until it was replaced as the dominant cohort by the next successful age class (see Tinsley & Tocque, 1995). Because of this dynamic population structure, interpretation of life history characteristics is complicated. Nevertheless, the general patterns, including the record of recruitment (representation of the youngest age classes) and of longevity, provide important evidence of the effect of the period of severe drought on host ecology.

In 1995, the first "post-drought" field season, low intensity rainfall produced few waterbodies in which spawning was successful. However, conditions were ideal for feeding. Records of gut contents showed food intake equivalent to that documented by Tocque *et al.* (1995) which would have restored energy reserves and gonadal investment after the stress of the preceding drought years. Age analysis of population samples in 1995 suggests a slight shift to the right (to older age classes) in the sample of 76 *S. couchii* (Fig. 6). There were few animals in breeding assemblies with less than 4 annual growth rings: 2 year olds were missing and 3 year olds comprised only 4% of the population. This reflects the failure of recruitment in the preceding years, both from drying up of ponds before tadpole metamorphosis and high mortality of any post-metamorphs. For established adults, however, there is no evidence that the harsh conditions resulted in selective mortality. Indeed, modal age spans the 6th and 7th year classes in 1995 in comparison with the 4th and 5th years in the 1986–92 records and there is a greater representation of the oldest age classes (28% of toads are 10 years and older compared with 5% in the 7 year data set). Detailed comparison is complicated by cohort interpretation and the aggregation of data from discrete sites; however, there is no evidence from population age structure that adult toads experienced significant negative age-specific effects from the successive seasons of severe drought.

The following years' data demonstrate progressively ageing populations. At site H, one of the key sites followed since 1983, age structure showed a mean of 7 years in 1995, with 11% ($n = 19$) aged 10 years or older. In 1996 and 1997 ($n = 54$), over two-thirds (69%) were 7 and over, and in 1998 ($n = 31$), 42% of toads were aged 10 years and over (maximum 15 years). Nevertheless, this site received a trickle of young recruits (aged 2–4 years) (which depressed the mean age). At other sites, these younger age groups

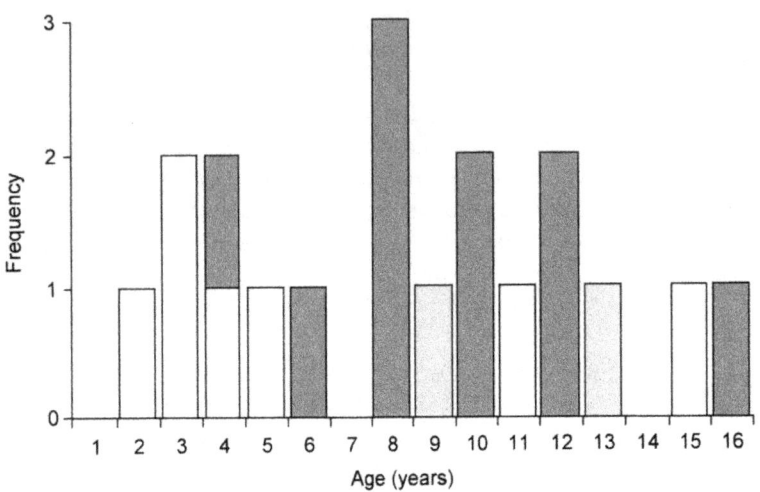

Fig. 7. Frequency distribution of age of *Scaphiopus couchii* at Site F, 1998 ($n = 19$). Shading represents *Pseudodiplorchis americanus* infection status: (□) uninfected (principally youngest individuals); (▨) infected by adults whose uterine larvae are retarded in development (not yet infective); (■) infected by gravid adult parasites. Wide host age range promotes transfer of infection between older and younger cohorts, especially to new recruits breeding for the first time and previously uninfected.

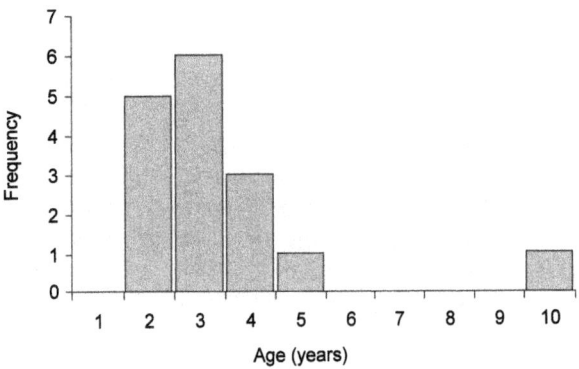

Fig. 8. Frequency distribution of age of *Scaphiopus couchii* at Site D, 1998 ($n = 16$), showing influx of young individuals breeding for the first time.

were lacking: thus, at site B in 1995, 1996 and 1997, mean age increased from 6·3 to 7·8 and 8·2 years respectively, and no toads in any year were aged < 6 years ($n = 29$). This confirms the repeated failure of recruitment. At site F, the 1995 sample indicated a relatively young population: mean 4·9 years, max 7·0 years, 78% aged 5 years or less. The next years illustrate negligible recruitment to balance increasing age: in 1996, mean 7·0 years, 36% aged 5 years or less ($n = 22$); in 1997, mean 7·6 years, 8% aged 5 years or less ($n = 13$).

In 1998, significant recruitment into breeding populations was recorded at some sites. At site F, an influx of younger toads occurred alongside the ageing individuals referred to above: in a sample of 19 toads virtually all year classes between 2 and 16 were represented, with 42% aged 10 years and older (exactly as at site H), but 32% aged 5 years and younger, and an overall mean of 8·4 years (Fig. 7). This appearance in the breeding assemblies of new recruits, 2 year olds from the wet summer of 1996, provides evidence of the first successful breeding since the 1993–1994 drought years. Confirmation is given by successive samples from site D that show a major reversal in age structure. In 1997, the sample showed a mean age of 7·1 years with no individuals under 5 years; in 1998, mean age was 3·4 years: one toad had 10 annual growth rings, one had 5 rings, and all others (88%) had 2–4 growth rings (Fig. 8). The 2 ring animals were attributable unambiguously to the good spawning season in 1996 and probably represent the start of a new dominant cohort. The animals with 3–5 rings were atypical of their presumed age classes: comparison with Tinsley & Tocque's (1995) data set indicates that these were around 10 mm (up to 20%) shorter and 10–15 g (up to 50%) lighter than expected at age 3–5 years, falling instead within the size range of the 2 year olds. No toads with less than 5 annual rings occurred in samples from spawning aggregations at this site in the preceding years and it is likely that 15 of the 16 toads in the 1998 sample were first-time breeders. This provides evidence that, where *Scaphiopus* breeding and metamorphosis did occur in the drought years, 1993–95, the surviving recruits carry a signal of both stunted growth and delayed maturation.

In summary, this end-point in the series of field samples demonstrates that the *S. couchii* populations did experience a major check in recruitment for up to 3 successive years (from 1992), but the populations were sustained by ageing individuals whose longevity represents a very important survival characteristic. However, the eventual success of recruitment (from spawning in 1996) produced a major shift in population age structure and the new cohorts of younger age classes set in place the future survival of the breeding populations.

Environmental constraints in a desert

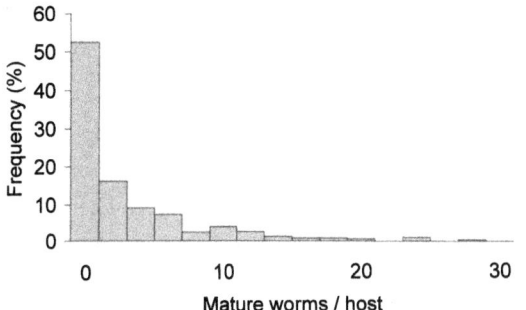

Fig. 9. *Pseudodiplorchis americanus*: frequency distribution of infection levels of gravid adult parasites in male *Scaphiopus couchii* ($n = 297$) preceding the period of extreme drought (data from Tinsley, 1993).

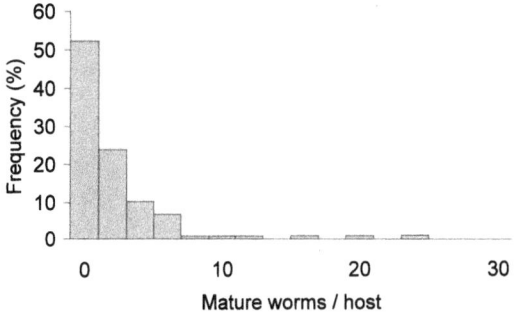

Fig. 10. *Pseudodiplorchis americanus*: frequency distribution of infection levels of gravid adult parasites in 1995, following the drought period ($n = 88$ *S. couchii*) (based on the same samples contributing to host age distribution in Fig. 6).

Parasite population ecology

Baseline data for the parasite suprapopulations is provided by the records of Tinsley (1993, 1995). Amongst male *S. couchii* sampled over several years in the 1980s ($n = 297$), the prevalence of adult (gravid) parasites was 48% and mean intensity was 5·9 worms/host. The distribution of infection levels was over-dispersed: 25% of toads carried small burdens of 1–4 worms/host and 9% carried over 9 worms each up to a maximum of 27 (although this minority of the host population actually contains nearly half (48%) of the total parasite population) (Fig. 9). This pattern of infection has been found to be remarkably stable in records from the same field area between 1983 and 1990, with prevalence consistently close to 50% and an overall mean abundance across the 8 year data set of 3·1 worms/host (based on adult parasites with infective larvae *in utero*) (Tinsley, 1995).

At the end of the 'drought years', the records of gravid parasites in the same field area in 1995 provide a guide to effects on overall infection levels. In a combined sample of 88 toads, prevalence was 48%, mean intensity 4·5 worms/host and mean abundance 2·4 worms/toad. Figure 10 shows a slightly higher representation of the lowest worm burdens; nevertheless, the pattern shown by this smaller sample is comparable with the 'pre-drought'

data, with 34% of toads with burdens of 1–4 worms and 6% with over 9 worms/host, maximum 23. Detailed comparison is influenced by the aggregation of data from various sites and the differences in sample size. Nevertheless, there is no suggestion of truncation of the distribution pattern attributable to density-dependent parasite-induced host mortality. Correlation with data on host population age structure for the same samples (Figs 5 and 6) also provides no indication of selective host mortality: as noted above, the age profile shows a slight shift towards older age classes and a decrease in representation of the youngest ages probably reflecting reduced recruitment. This youngest fraction of the host population is not infected until first entry into the breeding assemblies so its poorer representation cannot reflect potential negative effects of infection. Mean and maximum ages are slightly increased and there is no reduction in the 'middle' year classes (6–8 years) where worm burdens tend to be heaviest (Tinsley, 1995).

More precise interpretation of the outcome of parasite transmission was undertaken in 1996–98 with age analysis of all parasite individuals using the criteria defined by Tocque & Tinsley (1991*b*). At site E in 1996 (Table 2), the overall reproducing parasite population (prevalence 58%, mean abundance 2·0 worms/toad) is attributable almost entirely to a single large cohort of worms, those invading in 1994 (prevalence 54%, mean abundance 1·9). There are no parasites representing the 1993 transmission season and input in 1995 was very weak (prevalence 13%, mean abundance 0·2). This age distribution is likely to have had a negative effect on transmission in 1996: according to Tocque & Tinsley (1991*b*), the output of oncomiracidia from 3 year old worms normally contributes 40% of total output from the suprapopulation (because of the much greater *per capita* production of offspring by these older worms). This contribution was entirely lacking in 1996. The second important feature of these data concerns future transmission. The large cohort of 2nd year worms present in 1996 is likely to be much reduced in 1997: Tocque & Tinsley's analysis indicated a 50% decrease in worm numbers between years 2 and 3. Moreover, the very small 1995 cohort would have made a limited contribution to transmission in 1997 and 1998 when the age-specific contribution to larval output should be greatest. This age analysis shows that, although overall infections appear relatively robust, the strength lies in a single age class and this identifies a future weakness in transmission potential regardless of invasion success in 1996.

The same effects are evident at other breeding/transmission sites in the 1996 season. At site I (20 km south of site E), overall infection levels appear moderately high in comparison with the long-term, baseline data (prevalence 41%, mean abundance 3·1) but, as at site E, a single year class

Table 2. Characteristics of *Pseudodiplorchis americanus* infection in *Scaphiopus couchii* (n = 24) at Site E, 1996

Parasite age/state	Cohort origin	Prevalence	Intensity	Mean abundance
Delayed/non-gravid	1995	0	0	0
1st year gravid	1995	12·5	1·3	0·17
2nd year gravid	1994	54·2	3·5	1·88
3rd year gravid	1993	0	0	0
Total gravid	1993/5	58·3	3·43	2·00

Table 3. Characteristics of *Pseudodiplorchis americanus* infection in *Scaphiopus couchii* (n = 17) at Site I, 1996

Parasite age/state	Cohort origin	Prevalence	Intensity	Mean abundance
Overall 1st year	1995	11·8	3·0	0·35
Delayed/non-gravid	1995	5·9	3·0	0·18
1st year gravid	1995	5·9	3·0	0·18
2nd year gravid	1994	35·3	6·5	2·29
3rd year gravid	1993	5·9	11·0	0·65
Total gravid	1993/5	41·2	7·57	3·12

Table 4. Characteristics of *Pseudodiplorchis americanus* infection in *Scaphiopus couchii* (n = 21) at Site J, 1996

Parasite age/state	Cohort origin	Prevalence	Intensity	Mean abundance
Overall 1st year	1995	47·6	4·9	2·33
Delayed/non-gravid	1995	28·6	4·5	1·29
1st year gravid	1995	19·0	5·5	1·05
2nd year gravid	1994	66·7	5·7	3·81
3rd year gravid	1993	4·7	2·0	0·10
Total gravid	1993/5	66·7	7·21	4·81

Table 5. Characteristics of *Pseudodiplorchis americanus* infection in *Scaphiopus couchii* (n = 13) at Site F, 1997

Parasite age/state	Cohort origin	Prevalence	Intensity	Mean abundance
Overall 1st year	1996	61·5	4·4	2·69
Delayed/non-gravid	1996	30·8	3·8	1·15
1st year gravid	1996	46·2	3·3	1·54
2nd year gravid	1995	23·1	2·3	0·54
3rd year gravid	1994	7·7	2·0	0·15
Total gravid	1994/6	69·2	3·22	2·23

(1994) is responsible for most of the suprapopulation (Table 3). Poor invasion success in 1995 is reflected by infections in only 2 out of 17 toads, carrying only 15% of the total worm burden present in the 1994 cohort; only a single animal carried infection from 1993. Again, the very small 1st year class has important future significance for reduced larval output over the following two transmission seasons. The 1995 cohort also shows the retarded development of some parasites and their failure to produce infective stages in 1996: this probably reflects invasion occurring relatively late in the host's activity season (see above). At site J (18 km south of site I), there was effective transmission in 1995 (prevalence 48%, mean abundance 2·3) but over half of the surviving worms failed to complete development within the year, probably because of late transmission and delayed migration (Table 4). This site confirms the other indications that 1994, a year characterized by severe drought, was highly successful in terms of parasite invasion (prevalence 67%, mean abundance 3·8 of two year old worms) with about 40% more worms than in the 1st year class. Again, the 1993 year class is very weakly represented, contributing little to overall larval output.

It is striking how consistent are the general trends in population dynamics from these and other widely separated sites across the desert. In summary, the age analysis revealed that production of infective stages by the parasite suprapopulation in 1996 depended disproportionately on only 1 of the 3 parasite year classes (invading in 1994). A significant impact on transmission success would be predicted. In subsequent years, the negative influence of the weak 1993 cohort on total larval output would be displaced, but the small 1995 cohort could cause a ripple effect on transmission for 3 years with the potential to prejudice future survival. Two alternative outcomes became evident over these subsequent years: recovery at some sites and extinction at others.

The outcome of environmental perturbation: recovery

In 1997, at site F (Table 5), the parasite suprapopulation shows the 'tail-end' of the widely successful 1994 cohort, now reduced to 3rd year

Table 6. Characteristics of *Pseudodiplorchis americanus* infection in *Scaphiopus couchii* ($n = 19$) at Site F, 1998

Parasite age/state	Cohort origin	Prevalence	Intensity	Mean abundance
Overall 1st year	1997	36·8	2·3	0·84
Delayed/ non-gravid	1997	10·5	4·0	0·42
1st year gravid	1997	26·3	1·6	0·42
2nd year gravid	1996	36·8	4·4	1·63
3rd year gravid	1995	21·1	2·3	0·47
Total gravid	1995/7	53·0	4·80	2·53

worms (prevalence 8%, mean abundance 0·15). This is accompanied by a weak representation of 2nd year worms (from the 1995 season) (prevalence 23%, mean abundance 0·54). Thus, the two year classes that should, according to the analysis of Tocque & Tinsley (1991b), make a combined contribution of over 80% of the infective stages released into transmission sites, are represented by a tiny population (combined mean abundance only 0·7 worms/toad). This confirms the predicted effect on future transmission. However, the 1997 sample demonstrates that, at this site, the 1996 invasion was relatively successful. Overall prevalence of 1st year parasites in the 1997 host mating assembly was 62%, mean abundance 2·7. Their potential contribution to transmission was greatly reduced because 43% of the worms had failed to complete reproductive development. Nevertheless, this 1996 cohort provides a basis for the recovery of the parasite suprapopulation in future years.

Confirmation of this recovery is provided by data for this same site in 1998 (Table 6). The "good" 1994 age class has now disappeared and the 1996 cohort has a dominant role. In this sample, the numbers of surviving 2 year old worms represent only 60% of the total recorded in the previous year originating from the same invasion. Nevertheless, this cohort contains more parasites than the 1st and 3rd year classes combined. The 1995 cohort is represented by only 9 parasites amongst the 19 toads. The 1997 cohort reflects poor transmission and development was also disrupted since 50% of the parasites failed to produce infective offspring during the year. However, the contribution from all 3 years combines to represent a transmission potential (prevalence 53%, mean abundance 2·5 gravid worms/toad) close to the original baseline levels (prevalence 50%, mean abundance 3).

Despite this superficial similarity to 1983–1990 infection levels, the 1998 sample at this site is too heterogeneous for detailed comparison. The age profile of the host population comprises 2 distinct subsets (Fig. 7). Sixty three per cent of the sample is aged 8 years and older (up to 16 years), and 83% of these carry *P. americanus* infections (gravid and non-gravid) from infections in 1995–1997. These older toads represent a major influence on transmission. The remainder of the sample comprises young toads, including a majority of 2–4 year olds (probably first-time breeders and all uninfected). This influx of naïve toads reduces prevalence and abundance. However, the overlap in time and space between these older and younger host generations would allow the transfer and establishment of a subset of reproducing parasites in the toads that will contribute to future breeding (and transmission) assemblies.

The outcome of environmental perturbation: extinction

An alternative outcome of the ripple-effect, attributable to poor transmission success beginning in 1995, is evident in other 1998 field samples. At a series of sites where, in the 1983–90 record, infection levels had been consistently relatively high, the 1995–97 samples showed a year-on-year decline in worm burdens. These were sites where the 1996 season, characterized by extensive flooding, did not result in the recruitment required to reverse the downward trend. In the 4th year of the series, this led finally to the disappearance of *P. americanus*. Thus, at site D, a sample of 16 male *S. couchii* carried no established parasite infections in 1998. The age structure data for the host population sample makes an important contribution to interpretation of this outcome (Fig. 8). Since *Scaphiopus* is not infected before breeding, the mass appearance in the mating assembly of toads that had not previously bred has the direct effect that *none* of these animals (94% of the sample, 2 year olds together with stunted 3–5 year olds) would have had pre-existing (reproducing) *P. americanus* infection. As outlined above, this major shift in host age structure – to young age classes – is important for the future survival of the *Scaphiopus* populations. However, it has a major negative effect on parasite prevalence. Indeed, since the single older toad in the sample (10 years old) was also uninfected, all contribution to transmission by this fraction of the overall host population was abolished (Fig. 8).

Since parasite transmission is almost instantaneous, samples of the hosts aggregated in breeding assemblies provide two distinct assessments of the parasite suprapopulation. First, the animals examined provide a direct representation of pre-existing infection in this subset of the host population, i.e. parasites established for at least one year since the last exposure to infection. Second, these same animals contain evidence of larval invasion, derived in the few hours before collection, that provides an indirect reflection of adult parasite occurrence in the wider host population. Amongst the sample of 16 toads at Site D, one carried a single post-oncomiracidium in the nostrils. This could not have been derived from any individuals in the population sample; instead, it indicates that the breeding assembly (of about 50 toads at the time of sampling, at 02·00 h) did contain at least one gravid *P. americanus* that contributed to transmission. However, from this measure of wider invasion success within the host population (mean abundance 0·06 worms/toad), and the estimates that fewer than 3% of invading larvae survive to maturity (see above), it may be predicted that successful recruitment into the reproducing adult parasite population in 1999 is highly unlikely. At this site, therefore, the 1998 data suggest that the parasite population is at the point of extinction.

DISCUSSION

Extreme environmental perturbation

This study provides new insight into parasite population dynamics based on the application of techniques to determine age in both parasite and host individuals. Specifically, the fieldwork area in Arizona experienced a period of exceptional drought, with 3 successive years of the lowest summer rainfall in a 34 year weather record. The data, in combination with long-term records of the parasite and host populations at specific sites in the desert, have revealed effects operating over a time-course of several years after this major environmental perturbation.

Effects on the host populations

Analysis of age structure of the *S. couchii* populations provided no evidence for major mortality of already established adult populations; instead, these were found to have aged progressively until, in 1997, nearly half of some populations were aged 10 years or more (maximum 17 years). Survival can be attributed to an exceptional ability to feed intensively, even with limited opportunities, and to store lipid reserves for prolonged hibernation. Field studies show that some *S. couchii* contain sufficient fat to survive two years of dormancy and total starvation (Seymour, 1973; Tocque, 1993; Tinsley, 1995). Whilst the population age structure data provided no indication of unusual adult mortality, the effect of several successive years of failed recruitment at some sites was clearly demonstrated by a complete absence of the youngest cohorts in the age profile. Baseline data for the same *S. couchii* populations sampled in 1986–92 indicated that, on average, 35% of individuals in breeding assemblies were aged 2–4 years representing recent recruits, including first-time breeders. By 1997, when the succession of recruitment failures had gained maximum impact, there were no individuals aged under 7 years at some breeding sites. Most of the offspring born in the drought years had probably died, although the 1998 data demonstrated that there were some survivors characterized by stunted growth and delayed maturation. The ability of *S. couchii* populations to tolerate a succession of years when recruitment is severely reduced can be attributed to the relatively long life span of this species. Toads aged up to 16 years (female) and 17 years (male) spawned successfully in mating assemblies. Host longevity enables the populations to 'ride out' an environmental check until recruitment can be resumed in favourable years. In this study, 1996 provided exactly these conditions for recovery with twice the mean June and July rainfall, the highest in the 34 year Southwestern Research Station record. This enabled widespread successful breeding across flooded desert habitats with metamorphosis sufficiently early in the season to allow good feeding in preparation for hibernation. The outcome was reflected in the mass appearance of 2 year old adults entering breeding assemblies for the first time in 1998 and likely to represent a dominant cohort in succeeding years.

At the start of this study on the effects of the severe drought, it was not known that adult toads could survive such restriction in their activity season (limiting opportunities to feed and accumulate lipid reserves for hibernation), nor that age structure would shift into such old age classes. Now, given this knowledge, the effects of the successive drought years appear relatively straightforward: a failure to recruit for 3 years in succession led to a progressively ageing population which subsequently reproduced very successfully with the return of favourable conditions. For the parasite populations, the effects were not as straightforward.

Effects on the parasite populations

The parasite population data provide conclusive evidence that the 1994 season, characterized by exceptionally low rainfall and almost complete failure of host recruitment, resulted in highly effective transmission. This outcome created a signal recognizable in widespread high prevalence and intensity of *P. americanus* infection, specifically of 2 year old parasites in 1996 and 3 year-olds in 1997. A single major storm occurred in July 1994 which prompted mass spawning, but the newly-created ponds quickly dried up and most *S. couchii* progeny died. In contrast, parasite recruitment is geared to short-term episodes of transmission (each a maximum of 7 h), so the single period of spawning – although subsequently unsuccessful for the host – will have permitted mass parasite invasion.

The link between parasite transmission and host spawning guaranteed successful invasion in 1994 despite the brief exposure. However, the same principle was responsible for very weak transmission in 1995: here, limited rainfall in the form of low intensity showers prompted *S. couchii* emergence and provided good feeding but there was insufficient runoff to create breeding ponds. In this year, therefore, there was little host spawning and, significantly, little parasite invasion. At widely separated sites across the desert, this outcome was recorded in very low infection levels of 1st year worms in 1996, 2nd years in 1997 and 3rd years (at some sites) in 1998.

Paradoxically, the very heavy rainfall of 1996 that was responsible for major host breeding had a variable effect on parasite transmission. In many areas, discrete breeding ponds that usually form in shallow depressions, including ditches, were transformed into large expanses of moving floodwater. Spawning toads, that typically occur in high-density aggregations, were dispersed, and parasite infective stages, usually released into confined bodies of standing water, would have been washed away. At some study sites, a cohort of parasites confirmed that transmission did occur; however, at most others, this year class was missing.

The summer of 1997 rainfall pattern closely resembled that in 1995 (almost identical June and July combined rainfall in these 2 years) and produced a third successive year in which invasion levels were much reduced.

Local variations in rainfall intensity are likely to have produced conditions at some sites favourable for parasite transmission, irrespective of the dominant environmental influence recorded at a majority of other sites. Thus, in rare cases there was a good 1995 year class recorded in 1996 (e.g. site J, see above), and, as mentioned, a good 1996 year class in 1997 (e.g. site F). Nevertheless, despite these local exceptions, the rainfall patterns during the period 1995–1997 were sufficiently pronounced to produce a strong negative effect on the annual succession of parasite age cohorts. At some sites, 3 successive years of poor recruitment could no longer be tolerated and the parasite suprapopulations were virtually extinct in 1998.

Factors contributing to extinction

The present analysis provides detailed evidence for the sequence of events leading to local extinction. Three factors contributed. First, the succession of years (1993, 1994 and 1995) in which *S. couchii* recruitment was precluded led to a shift in host population age structure that had a knock-on effect on the parasite populations. The lack of input of the youngest age classes into host mating aggregations, especially those breeding for the first time, removed the normal 'diluting effect' of uninfected (previously unexposed) immigrants leading to an increase in parasite prevalence. The increase in the proportion of older toads increased the representation of age classes that typically have lower infection levels, particularly the absence of the highest worm burdens. This effect would, to some extent, counteract the increase in prevalence caused by the lack of the diluting influence of the youngest cohorts but, more significantly, it would be predicted to result in a decrease in intensity.

Second, the parasite populations also experienced reduced recruitment over several years (1995, 1996 and 1997). Because parasite transmission depends heavily on worms that are 2 and 3 years old, missed invasion has a delayed effect on future reproduction: it produces a ripple in the succession of parasite age classes so that there is greatly reduced output of infective stages 2 and 3 years after the poor recruitment. In some of the present study sites, the trough in transmission potential, attributable to reduced invasion beginning in 1995, exerted its maximum effect in 1997 and 1998.

Third, successful host breeding in 1996 led to the mass appearance of newly matured recruits in the mating sites in 1998. At the same time, older toads (with 3–5 growth rings) that represented the stunted survivors of recruitment in the drought years also entered the mating sites as delayed first-time breeders. The effect on host population age structure was a sudden massive switch from older age classes (at some sites 40% of toads were > 10 years and none were < 7 years) to the youngest (70% or more of toads aged 2–3 years). Significantly, the overall population was swamped with first-time breeders that had never previously been exposed to *P. americanus* invasion (94% of the sample collected at Site D) coinciding with minimum population size of reproducing parasites in the remainder of the host population. Thus, parasite extinction was determined by the combination of three independent factors involving the delayed effects of events affecting the host populations 2–5 years previously and the parasite populations 1–3 years previously.

The longer-term data set reveals that each of these events may be intermittent features of the host-parasite interaction. Records of *S. couchii* population dynamics by Tinsley & Tocque (1995) showed a series of years – 1986, 1987 and 1988 – in which modal age was influenced by a dominant cohort of toads aged 5, 6 and 7 years in the successive annual samples (originating from a very wet summer, favourable for breeding and juvenile survival, in 1981). These host age-classes would be expected to have maximum infection levels with a succession of parasite age classes contributing to transmission. In contrast, in 1989, modal age was 3 years (comprising about 30% of the total population) originating from a wet summer in 1986 and including a majority of first-time breeders (uninfected by *P. americanus*) that would have greatly reduced prevalence and transmission. Similarly, in 1990 there was a major influx of 2 year-old toads (originating in 1988) that would also have made no contribution to transmission. However, in each of these data sets there was a spectrum of other age classes which would have maintained the reproducing parasite population and served to infect each annual influx of new recruits. In the present study, this reservoir of older infected toads was at a minimum and the continuity between older and younger host generations, important for parasite population stability, was broken.

Factors involved in recovery

Although local parasite extinction was documented at a series of sites, parasite population recovery occurred at others. Two key features distinguish the sites, D and F described above, where these contrasting outcomes were recorded. First, recovery was attributable to local conditions that promoted successful invasion in 1996 and interrupted the succession of poor recruitment seasons in 1995–97. Second, where recovery occurred, there was a spatial and temporal overlap in the two discrete subsets of the host population comprising older toads with reproducing parasite infections and young, naïve, recruits without pre-existing infections. These circumstances promoted transfer of infection between the host cohorts at Site F (Fig. 7). At this and other sites analysed in 1998, there is an apparent trend towards lower worm burdens. Interpretation is affected by the statistical chance of locating relatively rare high intensities in small samples (combined total $n = 108$ in 1998). If the worm burdens encountered are representative of the metapopulations of the wider study area, then loss of the highest burdens could have important effects on the host-parasite interaction. It would remove the infections in which the greatest pathogenic effects are evident. Tocque & Tinsley (1992) quantified the parasite-induced removal of host blood, and Tocque (1993) and Tocque & Tinsley (1994b) demonstrated a density-dependent effect of infection on host blood composition and fat reserves. A trend towards lower worm burdens could produce the outcome, at a population level, that *P. americanus* becomes more 'benign'. It would also remove those infrapopulations that make the greatest contribution to reproduction: Tinsley (1993) showed that 9% of toads, those with the highest burdens, carry 48% of the total parasite population. Their elimination would reduce the intensity of transmission and hence future recruitment into the reproducing parasite population. Coincidentally, a reduction in the highest worm burdens would also reduce the competitive interactions demonstrated by Tocque & Tinsley (1991b) and hence intraspecific negative effects on transmission. These trends would be predicted from the experimental studies carried out on this system but confirmation awaits further data analysis and examination of samples taken in 1999 (in progress).

Loss of the heaviest worm burdens could be attributed to density-dependent, parasite-induced host mortality. However, the present comprehensive data provide no evidence for this. The host population age profile shows no indication of selective mortality of age-classes that have the highest worm burdens (typically 6–8 year olds) nor of the oldest ages that might have accumulated the effects of repeated challenge. Indeed, in the absence of effective recruitment, mean host age increased smoothly across the years of this study. Instead, a loss of high burdens would be better attributed to an increase in host resistance (larger numbers of older hosts that typically have lower infection intensity) and a reduction in invasion success.

Life cycle characteristics

Given the superb adaptations of *P. americanus* to prepare during the long period of host inactivity for the moment of transmission, the very brief episode of host-to-host transfer seems to be accomplished with great ease. Indeed, this may be quantified by the calculation that the probability of individual invasion success is about 0·3. Where equivalent data exist for helminths that produce hundreds, thousands or even hundreds of thousands of offspring/parasite/day (see for instance Tinsley, 1990a), comparison suggests that *P. americanus* has one of the highest transmission efficiencies amongst helminth parasites. For the few hours each year of host-to-host transfer, *P. americanus* operates in "ancestral mode", equivalent to a fish parasite; significantly, however, this efficiency is maximized because all the difficulties of host availability and host location, inherent in most parasite life cycles, are reduced to a minimum.

Insight into the optimum circumstances for transmission is provided by the contrasting outcomes of the 1994 and 1996 seasons. Severe drought in 1994 (July rainfall only one-third of the mean) produced intense transmission because this was focused into the few, restricted host breeding sites. The very wet conditions in 1996 (beginning early with June rainfall 6 times the mean) produced poor parasite recruitment because of extensive flooding. Thus, the life cycle is most efficient when conditions for host-to-host transfer are, by comparison with other parasites employing aquatic invasion, most extreme.

For this parasite, adapted for transmission in a desert, the most hostile environmental conditions begin once it is within the host. No other monogeneans experience such diverse physiological conditions as *P. americanus*. Developing stages in the respiratory tract live at the air/water interface and, whilst in the lungs, are exposed to surfactants. Migration through the digestive tract involves exposure to extremes of pH, digestive enzymes, bile and to anaerobic conditions. Within the urinary bladder, the worms experience wide variations in urine composition: little or no urination occurs during the 10 months of host hibernation each year and there is a progressive increase in the osmolarity of the urine, up to 200 osmol/l, as urea accumulates and water is withdrawn by the host (Ruibal *et al.* 1969) (Fig. 1).

Migration of *P. americanus* through the digestive tract involves exposure to environmental conditions that are, by all criteria, extreme. Of course, the other major helminth groups have all adapted to survive conditions in the gut but the challenge to *P. americanus* must be assessed alongside the capabilities of the class Monogenea in which the overwhelming majority of the c. 1500 species occur on the external surfaces (skin and gills) of fishes and therefore experience an aquatic environment. The severity of the conditions within the digestive tract for *P. americanus* is clearly demonstrated by the rapid death (in 1–2 min) of pre-migrants transferred experimentally to stomach or intestinal contents (Tinsley & Jackson, 1986). That migrating worms can tolerate these conditions for 4–8 h is a measure of the effectiveness of their tegumental adaptations.

Whilst these adaptations for gut migration represent a unique response to hostile conditions, this component of the life cycle also creates a weak link. Because migration requires a host cue that is linked to external environmental conditions (promoting host activity), the course of parasite development is prone to disruption by chance variations in rainfall and temperature. The present age-specific data highlight the population consequences of this life cycle requirement. The disturbed weather patterns occurring during this study were probably responsible for the outcome, noted in many local populations cited above, that up to half of each invading cohort failed to complete reproductive development, precluding a contribution to transmission one year p.i. This vulnerability of the life cycle design can result in a loss of one year of reproductive output out of a normal maximum of three.

CONCLUSIONS

The environmental perturbations recorded in this analysis proved to have more serious effects on the parasite than its host. In both cases, recruitment was interrupted (to a greater or lesser extent) for 3 successive years: 1993–95 for the host and 1995–97 for the parasite. In both, the effect was recognizable in terms of missing year classes in the succession of annual age cohorts. The different outcomes are attributable to differences in longevity. The relatively long-lived host can tolerate a succession of years when recruitment fails until population numbers are restored in years favourable for offspring survival. For the parasite, however, which lives for up to 3 (very rarely 4) years, interruption to the succession of annual cohorts has serious consequences for future transmission. This study emphasizes the relatively long time course of these perturbations, particularly that the major effects of failed parasite recruitment occur 2 and 3 years afterwards when these year classes would normally make their greatest contribution to further transmission.

The overview of the population biology of *P. americanus* has implications at three levels for assessing environmental constraints on parasite survival. First, the desert environment, characterized by extremes of physical conditions that are hostile to organisms without specific adaptations, normally has only relatively minor effects on the survival of *P. americanus*. Clearly there are major constraints on both the host and its parasite which restrict 'external' activity to relatively brief episodes: feeding, breeding and other desert surface activities for the toad, and host-to-host transmission for the parasite. For both partners, the major part of the year is spent protected from the harsh external environment: the host buried in deep burrows beneath the soil surface, and the parasite buried within the host. A major part of the adaptation to the desert environment involves preparation to exploit the briefly-favourable conditions and, particularly, the precise timing of the response. Clearly, the remarkable and in many cases unique specializations of *P. americanus* are absolutely essential to allow a member of a parasite group so apparently ill-suited to desert conditions to exploit its elusive host. Thus, it would be difficult to envisage how a parasite could cope with the unpredictable period between transmission episodes and yet release infective larvae instantly, when the opportunity arises, without a very special mechanism of *in utero* nutrition of embryos and, then, an immediate response to a precise oviposition trigger (see above).

Counterintuitively, the most important constraint imposed on this system by the desert environment concerns low temperatures rather than high. The winter temperatures in the Sonoran Desert create a powerful negative effect on the rates of parasite development and reproduction. The temperature threshold – around 15 °C – determines that development and reproduction can normally proceed only between May and October. This regulates the ability of the parasite to complete reproductive development between transmission seasons and the numbers of infective stages available for infection. From a critical viewpoint therefore, the external environment in the Sonoran Desert does represent a significant constraint on the life cycle of *P. americanus*, but this is not a function of any its *extreme* characteristics. In comparison with this specialized desert environment, exactly the same environmental check on parasite population biology occurs in the equable temperate habitats of *Discocotyle sagittata*, *Fasciola hepatica* and *Ascaris lumbricoides*. In these unrelated helminths, development of stages in the external environment (especially eggs) is halted below 10 °C for the same period, over half of each year, between October/November and May (see Gannicott & Tinsley, 1998 a, b). On the other hand, paradoxically, the low temperature check in the *P. americanus* life cycle has its most important effect to the benefit of the parasite, increasing parasite survival (see below).

Second, the greatest attrition of the parasite populations occurs within the host. Whilst the probability of successful host-to-host transfer is at least 0·3, the probability of subsequent survival until the first opportunity for onward transmission is < 0·03. A series of factors are involved, including the hazards of migration along the alimentary tract. However, the major fraction of parasite mortality appears best interpreted as a function of a temperature-dependent host immune response (Tinsley, 1995 and references therein). The conclusion, that the environment created by host defences represents the most hostile part of the parasite life cycle, is not novel. However, this *Scaphiopus–Pseudodiplorchis* system provides comprehensive data for quantitative comparisons. Annual infections disappear in 50% of the host population (perhaps the resistant fraction of the population) but worm burdens are reduced in all toads. Field observations are mirrored by laboratory experimental findings confirming a progressive decline in worm burdens with time. The margin between parasite survival and elimination is remarkably narrow: 'normally' only about 3% survive from invasion to the first contribution to reproduction.

For a host-parasite system involving an ectothermic vertebrate host, this second environmental constraint on parasite biology – immunity – clearly interacts with the first, the seasonal variations in temperature. Indeed, laboratory experiments show that if environmental temperatures remain high throughout the year then *P. americanus* would not survive from one transmission season to the next. Temperature therefore has a dual effect on parasite population dynamics: first, direct temperature-dependent effects influence the rates of parasite life history processes, and second, indirect effects operating through host physiology influence the ability of the immune system to control the parasite. As outlined by Tinsley (1995), significant departures from the temperature cycles normally experienced by the *Scaphiopus–Pseudodiplorchis* system could result in parasite extinction and hence may exert a major influence on geographical distribution (unless there are geographical variants with different temperature thresholds).

The third level of environmental constraint documented in this study is distinguished by its ability to cause local parasite extinction. The severe drought that triggered this effect constitutes a rare event (by the evidence of the lowest summer rainfall in the existing records for the area); the fundamental challenge to parasite survival was created by the series of consecutive years of recruitment failure attributable both to summers that were too dry and too wet. A 3 year parasite life span actually provides considerable buffering against occasional episodes precluding transmission. However, the inability of *P. americanus* to cope with a succession of such constraints could be predicted from knowledge of this life span. Tocque & Tinsley (1991b) noted the rare occurrence of a parasite (1 out of 178 worms in their study) that survived to 4 years. The relevance of this trait includes both an ability to bridge a more extended interruption to recruitment and a marked age-dependent increase in reproductive output. Thus, this single 4 year old worm contained a complement of infective stages equal to that in the total of 73 first year worms in the same supra-population sample. In areas where repeated recruitment failures are frequent, such differences in parasite longevity, if heritable, would have important selective advantage and would provide the basis for future evolution by *P. americanus* in its adaptation to extreme environmental constraints. The present data emphasize that the environmental effects leading to parasite extinction were also intimately linked with the succession of drought-induced recruitment failures of the host: this broke the overlap between older infected and younger naïve cohorts within the *S. couchii* mating populations.

Each of the environmental constraints encountered in the life cycle of *P. americanus* is met by superb adaptations. Many of the specializations are unique and their combined effect is that the external environment normally exerts only a moderating influence on parasite population biology. In contrast, the severity of the conditions imposed within the internal host environment creates an almost overwhelming constraint. The attrition of parasite stages in the lungs and bladder has the hallmarks of an immune response, although evidence is so far only circumstantial (see Tinsley, 1995): it may make a major contribution to the pre-reproductive mortality of at least 97% of each invading parasite cohort. The value of this assessment of parasite survival in extreme desert conditions is that it puts into perspective the relative difficulties of the external and internal environments. With appropriate adaptations, this monogenean parasite experiences rate-limiting regulation by the external factors, but these are principally temperature-dependent effects common to most environments. Many aspects of the life cycle require special adaptation to other constraints occurring within the host (such as the extreme conditions that are a natural feature of the gut). However, the conclusion emerges that the interaction with the living reactive host environment – and its response directed specifically at the parasite – creates a far greater constraint on parasite survival. Thus, whilst this paper has focused on the desert environment to illustrate the evolution of remarkable parasitic adaptations, the analysis serves additionally to reinforce the conclusion that the most severe conditions in this case study may actually be found in every parasite life cycle – those created by the host.

ACKNOWLEDGEMENTS

The initial fieldwork, on which this review was based, was supported by NERC grants GR3/5903 and 7314. I am very grateful for assistance during the 1995–98 fieldwork from Jo Cable, Abi Gannicott, Anna Henderson, Jeanne Pimenta, Emily Scott and Matthew Tinsley. In subsequent laboratory work, I am greatly indebted to Joe Jackson for very extensive histological analysis and other help and to Matthew Tinsley for data analysis. I thank Tim Colborn for preparation of Figs 1 and 2. I also thank the staff and volunteers of the Southwestern Research Station of the American Museum of Natural History, and the Natural Environment Research Council for support (grant GR3/10108).

REFERENCES

ARMSTRONG, E. P., HALTON, D. W., TINSLEY, R. C., CABLE, J., JOHNSTON, R. N., JOHNSTON, C. F. & SHAW, C. (1997). Immunocytochemical evidence for the involvement of an FMRFamide-related peptide in egg production in the flatworm parasite *Polystoma nearcticum*. *Journal of Comparative Neurology* **377**, 41–8.

BUSH, A. O., LAFFERTY, K. D., LOTZ, J. M. & SHOSTAK, A. W. (1997). Parasitology meets ecology on its own terms: Margolis *et al.* revisited. *Journal of Parasitology* **83**, 575–83.

CABLE, J., HARRIS, P. D. & TINSLEY, R. C. (1998). Life history specializations of monogenean flatworms: a review of experimental and microscopical studies. *Microscopy Research and Technique* **42**, 186–99.

CABLE, J. & TINSLEY, R. C. (1991a). Intra-uterine larval development in the polystomatid monogeneans *Pseudodiplorchis americanus* and *Neodiplorchis scaphiopodis*. *Parasitology* **103**, 253–66.

CABLE, J. & TINSLEY, R. C. (1991b). The ultrastructure of photoreceptors in *Pseudodiplorchis americanus* and *Neodiplorchis scaphiopodis* (Monogenea: Polystomatidae). *International Journal for Parasitology* **21**, 81–90.

CABLE, J. & TINSLEY, R. C. (1992a). Tegumental ultrastructure of *Pseudodiplorchis americanus* larvae (Monogenea: Polystomatidae). *International Journal for Parasitology* **22**, 819–29.

CABLE, J. & TINSLEY, R. C. (1992b). Unique ultrastructural adaptations of *Pseudodiplorchis americanus* (Polystomatidae: Monogenea) to a sequence of hostile conditions following host infection. *Parasitology* **105**, 229–41.

CABLE, J. & TINSLEY, R. C. (1992c). Microsporidean hyperparasites and bacteria associated with *Pseudodiplorchis americanus* (Monogenea: Polystomatidae). *Canadian Journal of Zoology* **70**, 523–29.

CABLE, J., TOCQUE, K. & TINSLEY, R. C. (1997). Histological analysis of the egg capsule of the ovoviviparous polystomatid monogenean, *Pseudodiplorchis americanus*. *International Journal for Parasitology* **27**, 1075–80.

DIMMITT, M. A. & RUIBAL, R. (1980). Environmental correlates of emergence in spadefoot toads (*Scaphiopus*). *Journal of Herpetology* **14**, 21–9.

GANNICOTT, A. M. & TINSLEY, R. C. (1998a). Larval survival characteristics and behaviour of the gill monogenean *Discocotyle sagittata*. *Parasitology* **117**, 491–8.

GANNICOTT, A. M. & TINSLEY, R. C. (1998b). Environmental control of transmission of *Discocotyle sagittata* (Monogenea): egg production and development. *Parasitology* **117**, 499–504.

HARRIS, P. D. & TINSLEY, R. C. (1987). The biology of *Gyrdicotylus gallieni* (Gyrodactylidea), an unusual viviparous monogenean from the African clawed toad, *Xenopus laevis*. *Journal of Zoology (London)* **212**, 325–46.

JACKSON, J. A. & TINSLEY, R. C. (1994). Infrapopulation dynamics of *Gyrdicotylus gallieni* (Monogenea: Gyrodactylidae). *Parasitology* **108**, 447–52.

RODGERS, L. O. & KUNTZ, R. E. (1940). A new polystomatid monogenean fluke from a spadefoot. *The Wasmann Collector* **4**, 37–40.

RUIBAL, R., TEVIS, L. & ROIG, V. (1969). The terrestrial ecology of the spadefoot toad *Scaphiopus hammondii*. *Copeia* **1969**, 571–84.

SEYMOUR, R. S. (1973). Energy metabolism of dormant spadefoot toads. *Copeia* **1973**, 435–45.

TINSLEY, R. C. (1983). Ovoviviparity in platyhelminth life cycles. *Parasitology* **86**, 161–196.

TINSLEY, R. C. (1989). Effects of host sex on transmission success. *Parasitology Today* **5**, 190–5.

TINSLEY, R. C. (1990a). Host behaviour and opportunism in parasite life cycles. In *Parasitism and host behaviour*, (ed. Barnard, C. J. & Behnke, J. M.), pp. 158–92. London: Taylor & Francis.

TINSLEY, R. C. (1990b). The influence of parasite infection on mating success in Spadefoot toads, *Scaphiopus couchii*. *American Zoologist* **30**, 313–24.

TINSLEY, R. C. (1993). The population biology of polystomatid monogeneans. *Bulletin Français de la Pêche et de la Pisciculture*, **328**, 120–36.

TINSLEY, R. C. (1995). Parasitic disease in amphibians: control by the regulation of worm burdens. *Parasitology* **111**, S153–78.

TINSLEY, R. C. (1996). Parasites of *Xenopus*. In *The Biology of Xenopus*, (ed. Tinsley, R. C. & Kobel, H. R.), pp. 233–61. Oxford: Oxford University Press.

TINSLEY, R. C. & EARLE, C. M. (1983). Invasion of vertebrate lungs by the polystomatid monogeneans *Pseudodiplorchis americanus* and *Neodiplorchis scaphiopodis*. *Parasitology* **86**, 501–17.

TINSLEY, R. C. & JACKSON, H. C. (1986). Intestinal migration in the life cycle of *Pseudodiplorchis americanus* (Monogenea). *Parasitology* **93**, 451–69.

TINSLEY, R. C. & JACKSON, H. C. (1988). Pulsed transmission of *Pseudodiplorchis americanus* (Monogenea) between desert hosts (*Scaphiopus couchii*). *Parasitology* **97**, 437–52.

TINSLEY, R. C. & TOCQUE, K. (1995). The population dynamics of a desert anuran, *Scaphiopus couchii*. *Australian Journal of Ecology* **20**, 376–84.

TOCQUE, K. (1990). The reproductive strategy of a monogenean parasite in a desert environment. PhD. thesis. London University.

TOCQUE, K. (1993). The relationship between parasite burden and host resources in the desert toad (*Scaphiopus couchii*), under natural environmental conditions. *Journal of Animal Ecology* **62**, 683–93.

TOCQUE, K. & TINSLEY, R. C. (1991a). The influence of desert temperature cycles on the reproductive biology of *Pseudodiplorchis americanus* (Monogenea). *Parasitology* **103**, 111–20.

TOCQUE, K. & TINSLEY, R. C. (1991b). Asymmetric reproductive output by the monogenean *Pseudodiplorchis americanus*. *Parasitology* **102**, 213–30.

TOCQUE, K. & TINSLEY, R. C. (1992). Ingestion of host blood by the monogenean *Pseudodiplorchis americanus*: a quantitative analysis. *Parasitology* **104**, 283–9.

TOCQUE, K. & TINSLEY, R. C. (1994a). Survival of *Pseudodiplorchis americanus* (Monogenea) under controlled environmental conditions. *Parasitology* **108**, 185–94.

TOCQUE, K. & TINSLEY, R. C. (1994b). The relationship between *Pseudodiplorchis americanus* (Monogenea) density and host resources under controlled environmental conditions. *Parasitology* **108**, 175–83.

TOCQUE, K., TINSLEY, R. C. & LAMB, T. (1995). Ecological constraints on feeding and growth of *Scaphiopus couchii*. *Herpetological Journal* **5**, 257–65.

The survival of monogenean (platyhelminth) parasites on fish skin

G. C. KEARN*

School of Biological Sciences, University of East Anglia, Norwich NR4 7TJ, UK

SUMMARY

This review deals with the problems faced by those monogenean (platyhelminth) parasites that attach themselves to fish skin. The structure of the skin and the ways in which the posterior hook-bearing haptor achieves virtually permanent attachment to the skin are considered. Small marginal hooklets are specialized for attachment to superficial host epidermal cells, finding anchorage in the terminal web of keratinous tonofilaments, while large hooks (hamuli) may penetrate into and lodge in the collagenous dermis. The complementary roles of suction and sticky secretions in haptor attachment and the role of the pharynx in temporary attachment during feeding are also considered. During leech-like locomotion the haptor is briefly detached and, at this critical time, the anterior end is strongly fixed to the wet, current-swept and possibly slimy skin by a sticky secretion. This secretion is deployed on paired pads or discs, the latter sometimes backed up by suction. After attachment by the haptor is re-established, the special tegument covering the anterior adhesive areas may be instrumental in their instant release. The role of fish skin in the phenomenon of host specificity and in the generation of a defensive response against monogeneans is considered and site-specificity of parasites on the host's body is discussed. Possible selection pressures exerted by predatory 'cleaner' organisms are briefly evaluated.

Key words: Monogeneans, fish skin, attachment, host-specificity, immunity, site-specificity, predation.

INTRODUCTION

The surfaces of aquatic animals are exposed to the attentions of epizoic and epiphytic organisms. Settling organisms may be destroyed by organelles such as the nematocysts of cnidarians and the pedicellariae of echinoderms or deterred by the mucous layer on the skin surface of fishes. The effectiveness of fish mucus in this respect is supported by the observation that fishes that permit epiphytic growth, such as sea horses (*Hippocampus*), do so only on highly specialized so-called flame cone cells at the epidermal surface (Bereiter-Hahn et al. 1980). Parasites pose a greater threat to aquatic organisms and, without a substantial protective cuticle or shell, the external surfaces of fishes seem particularly vulnerable. In fact, a range of protozoans, leeches, crustaceans and monogeneans (platyhelminths) habitually feed on the exposed cellular epidermis or on blood from accessible blood vessels deeper in the skin. Some of these skin parasites make only brief visits, but others, in particular many monopisthocotylean monogeneans, spend most of their lives attached to fishes, where they feed on epidermis (see Kearn, 1998).

Intimate contact between monogenean skin parasites and the host's skin serves three functions and is achieved by three quite different organs. The modified posterior body region or haptor (Fig. 1A, h) provides for prolonged, virtually permanent attachment to the host's skin. At the opposite end of the body, most monogeneans have paired, ventrally directed, anterior adhesive areas (Fig. 1A, a) which are used for brief attachment (rarely more than 1 s), permitting the parasite to undergo locomotion by detaching the haptor without completely losing contact with the host. A monogenean moves like a leech (Fig. 1B). First the body is extended (Fig. 1B, 1, 2) and the anterior adhesive areas are attached to the substrate. Then the haptor is detached (Fig. 1B, 3), drawn forward (Fig. 1B, 4) and reattached close to the head. Finally the head is released (Fig. 1B, 5). During feeding, the head region is attached independently of the anterior adhesive areas by the centrally located everted pharynx (Fig. 1A, p), which in medium-sized monogeneans typically maintains contact with the host's skin for a few minutes. In terms of attachment effectiveness, there is less demand on the pharynx during feeding than on the anterior adhesive areas during locomotion because the haptor remains attached during feeding.

Attachment to fish skin, whether temporary or permanent, is not as straightforward as it may seem. Fish surfaces are wet, notoriously slippery and, even in the most sedentary of fishes, swept at times by powerful water currents. In addition, the very features of the surface that appear to equate with host vulnerability to parasites, namely exposed living cells and proximity of blood vessels, may also be associated with powerful host immune responses which are potentially capable of deterring parasite settlement, preventing feeding or even rejecting established parasites. Because they are exposed on

* Tel: +01603 592251. Fax: +01603 592250. E-mail: g.kearn@uea.ac.uk

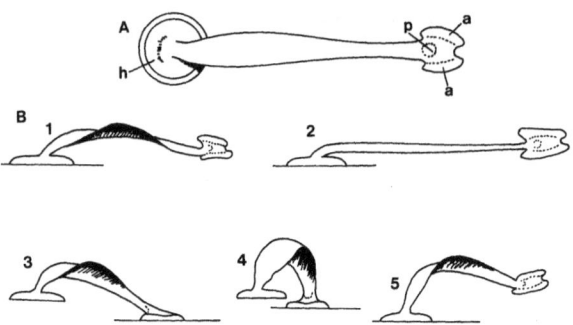

Fig. 1. Sketches of the monogenean *Nitzschia sturionis* undergoing locomotion. (A) Parasite in dorsal view attached by the haptor (h) with body extended. (B) 1–5, Successive stages in a locomotory 'step'. Parasite in lateral view. a, Anterior adhesive area; p, pharynx. Redrawn from Bychowsky (1957).

the host's surface, skin parasites may also fall victim to predators ('cleaner' organisms). This range of problems faced by monogenean skin parasites and some of the solutions adopted will be explored in this review.

FISH SKIN

Structure

The skin of vertebrates consists of two major layers, an outer epidermis derived from the ectoderm and an inner dermis derived from the mesoderm (Fig. 2).

The epidermis. There is a general similarity between cyclostomes, bony fishes and cartilaginous fishes (chondrichthyans) in the structure of the epidermis and in the cytology of its components (Fig. 3; Whitear, 1986a; Whitear & Moate, 1998). In fishes, the epidermis is a stratified squamous epithelium, ranging in thickness from two layers or tiers of cells in larvae to ten or more layers (Whitear, 1986a). The cells of the basal layer are cuboidal or columnar, but do not constitute a germinal layer in fishes because mitotic activity occurs throughout the epidermis, although most common in the deeper layers (Bullock, Marks & Roberts, 1978a). The cells of the epidermis are connected by desmosomes, which are specialized regions of strong attachment between cells (Whitear, 1986a). The major differentiation product of the epidermal cells of all living vertebrates is the α-keratin filament or tonofilament (Skerrow, 1986; Matoltsy & Bereiter-Hahn, 1986), hence the terms 'filament-containing cell' and 'keratocyte'. The tonofilament is an intermediate filament with a diameter of about 8 nm. Bundles of unbranched tonofilaments course through the cytoplasm forming an irregular interwoven mesh. Some of these bundles are intimately associated with desmosomes. According to Fuchs & Hanukoglu (1986), the longitudinal component of this network makes a greater contribution to the strength and special characteristics of these cells than any other cytoplasmic component. The epidermal cells are also called Malpighian cells, but according to Whitear (1986a) the classical and correct term is 'epithelial cell'.

The cytoarchitecture of the exposed epithelial cells of the fish epidermis differs from that of deeper cells. In these superficial exposed cells the tonofilaments are arranged to form an extensive terminal web between the outermost cell membrane and the perinuclear organelle complex (Fig. 4). The exposed surfaces of the superficial epithelial cells are characterized by the presence of elliptical whorls of microridges, resembling fingerprints (Schliwa, 1975). The superficial epithelial cells are secretory (Whitear, 1970) and Whitear (1986a) regarded these cells as the first of four sources of secretion in fish skin. In the central perinuclear zone of the superficial epithelial cells of *Lebistes reticulatus*, Schliwa (1975) found ovoid or spherical, membrane-bound

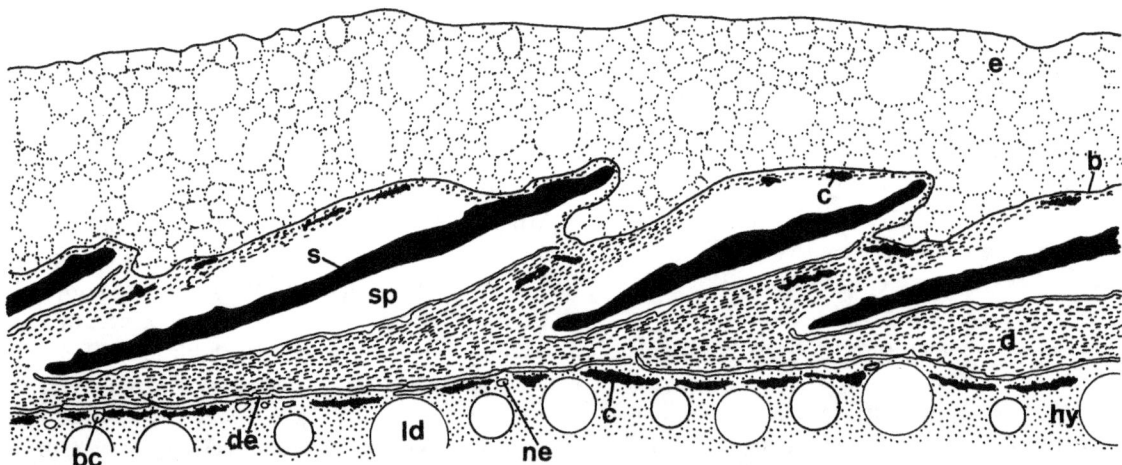

Fig. 2. Diagrammatic longitudinal section through the skin of a bony fish. b, Basement membrane of epidermis; bc, blood capillary; c, chromatophore; d, dermis; de, dermal endothelium; e, epidermis; hy, hypodermis; ld, lipid droplet; ne, nerve; s, scale; sp, scale pocket. Based on a drawing of *Phoxinus phoxinus* skin by Whitear, Mittal & Lane (1980).

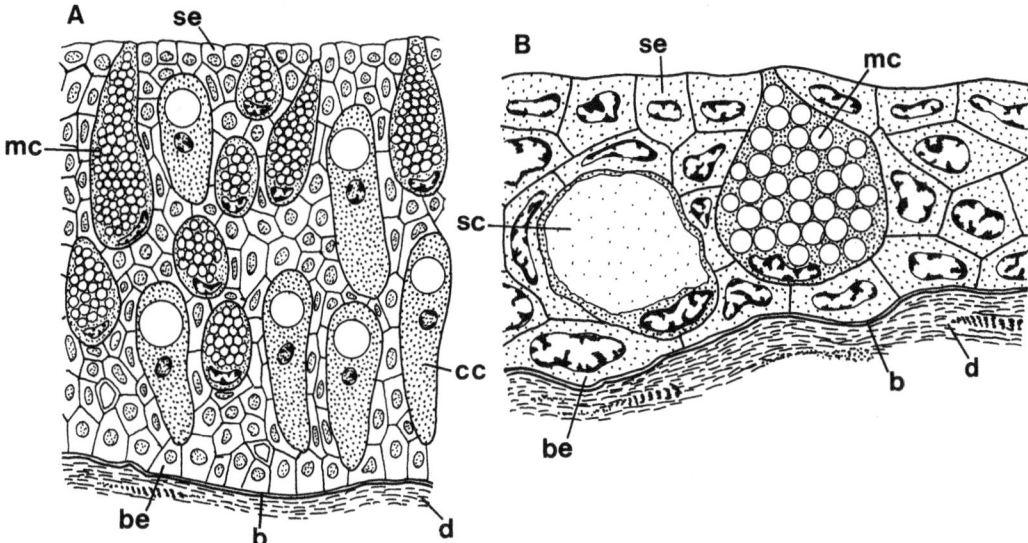

Fig. 3. Diagrammatic sections through the epidermis of (A) a bony fish, the eel *Anguilla anguilla* and (B) a cartilaginous fish, *Raja* sp. b, Basement membrane of epidermis; be, basal epithelial cell; cc, club cell; d, dermis; mc, mucous (goblet) cell; sc, sacciform cell; se, superficial epithelial cell. Redrawn from Whitear (1986a).

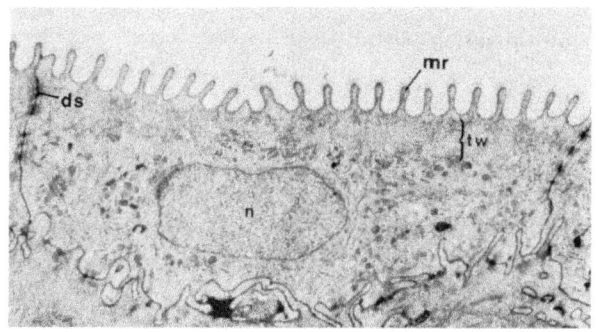

Fig. 4. Transmission electron microscope photograph of a section through a surface epithelial cell in the epidermis of *Lebistes reticulatus*. ds, Desmosome; mr, surface microridge; n, nucleus; tw, terminal web. Reproduced with permission of Academic Press from Schliwa (1975).

secretory vesicles about 0·4 μm in diameter in association with rough endoplasmic reticulum and Golgi systems (Fig. 4). According to Schliwa, these vesicles are the only cell organelles present in the terminal web zone. There is evidence that the contents of these vesicles contribute to the fuzzy coat (cuticle or glycocalyx) attached to the outer surface of the cell and that their boundary membranes are incorporated into the cell's surface membrane (Whitear, 1970; Schliwa, 1975).

Whitear (1986a) identified three other sources of secretion in fish skin. First, mucous cells or goblet cells (Fig. 3) are exocrine unicellular glands producing a mucoid secretion in the form of membrane-bounded globules, which, according to Fletcher, Jones & Reid (1976), contain glycoproteins. These globules normally remain intact until shortly before discharge and are commonly electron-lucent and finely granular in transmission electron microscope sections. According to Roberts & Bullock (1980), there may be regional differences in the density of mucous cells on the surface of a fish and temporal differences in the production of mucus by the skin. Similar cells producing a proteinaceous rather than a mucoid secretion were termed serous goblet cells by Whitear (1986a). Secondly, sacciform cells (Fig. 3B) accumulate their secretion in a large membrane-bound vacuole, typically fed by smaller vacuoles or channels. Thirdly, the club cell (Fig. 3A) is a large, sometimes binuclear cell in the middle layers of the epidermis. Some club cells have a secretory vacuole but others do not. However, a peculiarity common to all is the presence of coiled filaments rather than typical bundles of tonofilaments in association with their desmosomes. Club cells are the source of the pheromone that releases a fright reaction in ostariophysans but not in other fishes (see Whitear, 1986a). Lymphocytes, macrophages and various types of granulocytes have been found in the epidermis (Whitear, 1986a). Leucocytes are most numerous between and above the basal layer cells, but may occur at any level in the epidermis and may be exposed at the surface if the skin is damaged.

A specific junctional structure, the basement membrane, separates the epidermis from the dermis beneath (Figs 2, 3). This membrane has an important part to play in the regulatory influence exerted on the epidermis by the dermis (Krieg & Timpl, 1986). The basement membrane consists of two layers, an outer *lamina lucida* and an inner *lamina densa* (Krieg & Timpl, 1986). The *lamina lucida* is attached to the basal epidermal cells by hemidesmosomes. Fine anchoring filaments run from the regions of these hemidesmosomes, cross the *lamina lucida* and connect the plasma membrane of the basal epithelial cell to the *lamina densa*. A major component of the *lamina densa* is a minor collagen (type IV) which does not form broad-banded fibres. Other anchoring fibrils connect the *lamina densa* to the dermis.

The dermis. The dermis (Fig. 2) exerts a dominant and instructive influence on the overlying epidermis, providing a suitable substrate and nutrient supply for its maintenance, ordered proliferation and stratification (Sengel, 1986). The dermis is essentially a connective tissue layer of which the major fibrillar component is collagen, contributing 70–80% of its dry weight (Uitto, 1986). The major collagens providing tensile strength to the dermis are the interstitial collagens, which, unlike type IV collagen present in the basement membrane, form broad-banded extracellular fibres. Chromatophores, blood capillaries, lymphatic vessels and nerves occur within the dermis and it is also the site of development of the bony scales of teleosts and the denticles of cartilaginous fishes. A bony scale develops in an oblique pocket (Fig. 2) and consists of a plate of collagenous tissue with superficial calcification, surrounded by scleroblasts and fibroblasts (see Whitear, 1986b). The scale is anchored in the pocket by bundles of collagen fibres attached to the calcified layer. In some teleosts, the posterior edge of the obliquely orientated scale protrudes above the surface plane of the skin, but the protruding scale usually retains a thin covering of epidermis and overlaps the epidermis covering the scale behind. The flow of water over the swimming fish may be influenced by projecting scales, especially if they have ctenoid (spiny) margins, and mucus released from beneath overlapping scales may also be hydrodynamically important (Whitear, 1986b). The placoid scales or denticles of cartilaginous fishes are isolated structures consisting of dermal dentine on a base of bone and covered by a cap of hard enameloid of epidermal origin. Unlike most scales, these denticles penetrate the epidermis and are partly exposed above it (Spearman, 1973, Fig. 21; Whitear, 1986b).

A thin single sheet of cells joined by desmosomes called the dermal endothelium provides the inner boundary of the dermis (Fig. 2). This endothelium has a basement membrane on one or both sides and a conspicuous feature is the presence of *caveolae intracellulares*, which sometimes join to form a channel across the cell (Whitear, 1986b). The tissue below the dermal endothelium is the hypodermis containing deep chromatophores and cells containing fat interspersed with connective tissue. Blood vessels and nerve bundles approaching the skin are also present (Fig. 2; Whitear, 1986b).

Function

Fish epidermis and its secretions have been implicated in osmoregulation, in locomotion (by virtue of the lubricating properties of mucus) and in pheromonal communication (Fletcher, 1978; Pickering & Richards, 1980), and the dermis provides mechanical strength mainly through the dermal collagen and scales (Fletcher, 1978). The importance of maintaining the integrity of the skin is emphasized by the speed at which the epidermis repairs trauma to the skin surface (Bullock, Marks & Roberts, 1978b). This enhanced repair capability is also associated with the importance of the skin in defending the fish against parasites, pathogens and pollutants.

Increased secretion of skin mucus and invasion of the dermis and epidermis by leucocytes have been observed following exposure of fishes (rainbow trout, *Oncorhynchus mykiss*) to polluted water (Iger, Jenner & Wendelaar Bonga, 1994). Complement has been identified in skin mucus (Sakai, 1992). Peroxidase and catalase have been detected in skin and in skin mucus and may be responsible for the bactericidal properties of the latter, together with lysozyme and immunoglobulins (see Fletcher, 1978). Fletcher & Grant (1969) detected antibody activity in surface mucus of plaice, *Pleuronectes platessa*, after immunization with human erythrocytes.

There are indications that fishes possess the ability to produce secretory antibodies independently of systemic production of immunoglobulins (Ig) circulating in blood serum. Fletcher & White (1973) suggested that the mucous surfaces (intestine lining, body surface) of plaice may be protected by antibody synthesized locally as well as by antibody derived from the serum (transudated antibody), the dominant mechanism being determined by the route by which the antigen enters the fish. Thus fish may have a mechanism comparable to the production of mammalian secretory Ig, but, in fishes, encompassing the mucus-secreting epidermis as well as internal sites such as the intestinal lining and liver. Lobb & Clem (1981) found evidence that immunoglobulins in cutaneous mucus and bile of the sheepshead, *Archosargus probatocephalus*, have insufficient specific activities to account for their presence by a transudative or active transport process from the serum. Their data strongly suggest that Ig in cutaneous mucus and in bile is locally synthesized. Rombout *et al.* (1993) working with carp, *Cyprinus carpio*, also found evidence that mucus and serum Ig are structurally and functionally different and that a specific mucosal defence system is important. Using fluorescent antibody techniques, St. Louis-Cormier, Osterland & Anderson (1984) identified dermal foci of Ig-producing plasma cells in the rainbow trout. Positive fluorescent staining between cells of the epidermis suggested intercellular transport of Ig from the dermal plasma cells to the external skin surface and they also reported finding some plasma cells in the surface mucus.

The question is whether the independent secretory immune system of the skin of fishes or any other

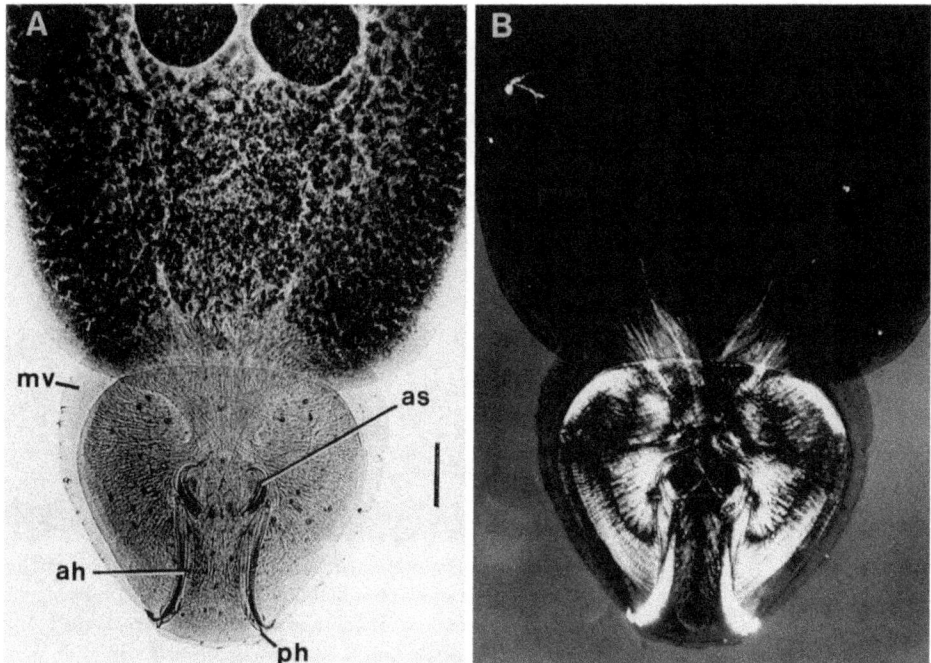

Fig. 5. The posterior region of the body of the same specimen of *Entobdella soleae* seen in (A) bright field illumination and (B) polarized light. Note strongly birefringent muscle concentrated in the haptor in B. The hamuli are also birefringent. ah, Anterior hamulus; as, accessory sclerite; mv, marginal valve; ph, posterior hamulus. Scale bar, 250 µm.

properties of the skin have any influence on multicellular skin parasites such as monogeneans.

PERMANENT ATTACHMENT

The haptor

A posterior haptor is a feature of all monogeneans. The haptors of many skin parasites (e.g. capsalids, gyrodactylids) are packed with muscle, so that in polarized light images they show a high level of birefringence compared with the rest of the animal (Fig. 5). As the well-developed musculature indicates, suction is commonly employed for attachment to fish skin and there are some monogeneans that secrete an adhesive, but most monogeneans rely at least partly on hard hook-shaped sclerites to pin themselves to the host. These hooks come in two sizes: small so-called 'marginal' hooklets and larger hamuli.

Marginal hooklets. Each marginal hooklet has a curved hook or 'blade' which protrudes through the tegument and a 'handle' and 'guard' embedded in the haptor tissue (Fig. 6A). There is a separate sclerite ('domus') shaped like an elongated cup associated with each hooklet (Fig. 6H). In some monogeneans there is a hinge between the blade/guard and the handle. The blade/guard regions are surprisingly uniform in shape and in size, in gill-parasitic as well as in skin-parasitic monogeneans (Fig. 6). The blade/guard region usually falls between 4 and 10 µm in length. The overall length of marginal hooklets ranges from 7 or 8 µm to over 40 µm, but much of this variation is attributable to the varied length of the handles (Fig. 6). Contraction of tiny muscles attached to the blade, guard, handle and domus produce gaffing movements of the blade.

In terms of size, each marginal hooklet is highly suitable for pinning a small parasite to a host epidermal cell. Fig. 7 is reproduced from Cone & Wiles (1989) and shows the blade of a marginal hooklet of *Gyrodactylus colemanensis* attached to an epidermal cell of its host (*Salmo gairdneri*), the traction exerted by the hooklet lifting the surface of the cell. It is remarkable that the cell, although perforated, remains attached to its neighbours, presumably by cell junctions (although Cone & Wiles observed none). Cone & Wiles observed *G. colemanensis* attached to the same site for 12 h, so the damaged epithelial cell seems capable of providing secure anchorage for the hooklet for at least this long.

It seems likely that monogenean marginal hooklets have evolved in response to the special challenges of attachment to fish epidermal cells. The uniform shapes and limited size range of marginal hooklets have probably been imposed by the unique and conservative cytoarchitecture of the superficial epithelial cells, in particular by the tangled bed of α-keratin tonofilaments (terminal web) just below the cell surface. The tiny hooklets are firmly rooted in this terminal web, which is sufficiently strong to resist tearing of the cell, so that mechanical damage is restricted to the small puncture where the hook

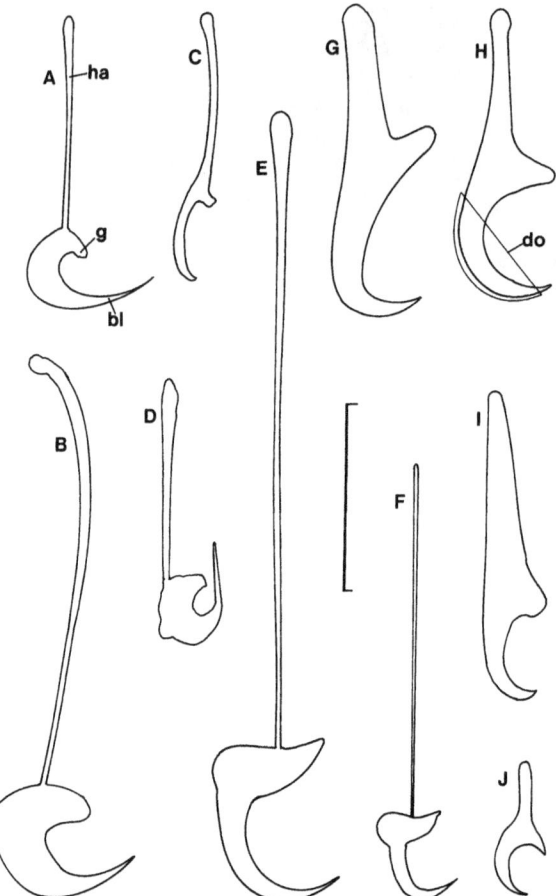

Fig. 6. A selection of marginal hooklets of some skin-parasitic monogeneans, illustrating the range of shape and size. (A, B) Acanthocotylids; (C) a calceostomatid; (D) *Enoplocotyle*; (E, F) gyrodactylids; (G–J) capsalids (H, *Entobdella soleae*). Domus (do) omitted in all except H. bl, Blade; g, guard; ha, handle. Redrawn from various sources to approximately the same scale; scale bar, 10 μm.

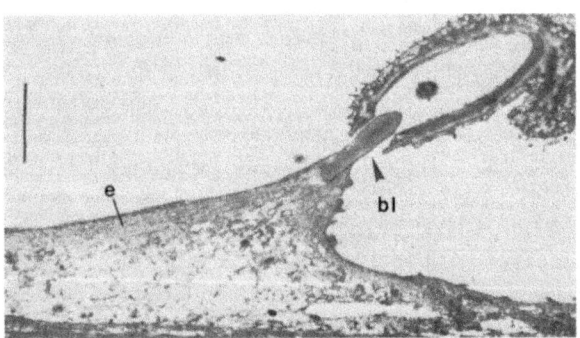

Fig. 7. Transmission electron microscope photograph of a section through a marginal hooklet of *Gyrodactylus colemanensis* attached to a surface epithelial cell on the caudal fin of *Salmo gairdneri*. bl, Blade; e, host epidermal cell. Scale bar, 2·5 μm. Reproduced with permission of the *Journal of the Helminthological Society of Washington* from Cone & Wiles (1989).

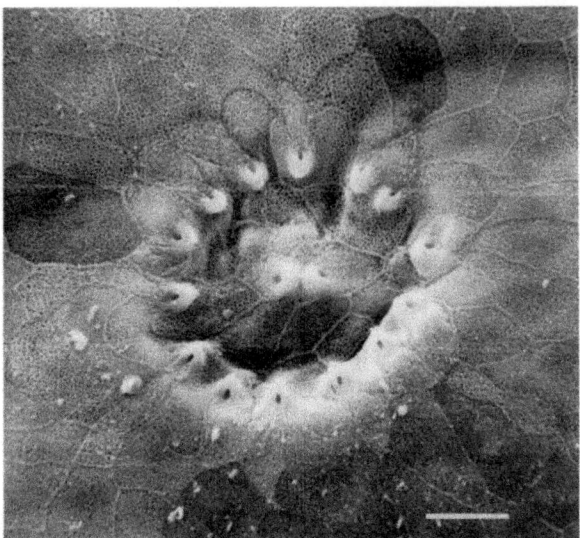

Fig. 8. Scanning electron microscope photograph of the skin surface of the hagfish *Eptatretus stoutii*, showing the site of attachment of the haptor of *Myxinidocotyle californica*. Scale bar, 25 μm. Reproduced with permission of Springer-Verlag from Malmberg & Fernholm (1991). Photograph by B. Fernholm. © by Springer-Verlag, 1991.

point enters. It may be a coincidence that marginal hooklets contain significant amounts of sulphur (Lyons, 1966; Shinn, Gibson & Sommerville, 1995) and may themselves be composed of a keratin-like protein, like the tonofilaments of the host in which they are embedded. On the other hand, molecular similarity between intimately associated parasite and host structures may be less provocative to host tissue.

Spreading the load by employing many hooklets to provide multiple points of attachment to the epidermis has proved to be a highly successful means of pinning a relatively small multicellular parasite to fish skin (Fig. 8). Most larvae (oncomiracidia) of skin-parasitic monogeneans do not exceed 300 μm in length and rely on marginal hooklets for attachment (Fig. 9). Typically, 14 or 16 of these hooklets appear during embryological development as tiny intracellular spicules (see Lyons, 1966). These later move to the surface of the developing disc-shaped haptor where they may be arranged radially with the blades exposed and close to the disc margin (hence 'marginal' hooklets). However, in many monogeneans two of the hooklets occupy a central location on the disc (Fig. 10B). This arrangement may provide a more stable platform for the parasite and Malmberg & Fernholm (1991) found that the two central hooklets of the acanthocotylid *Myxinidocotyle californica* oppose each other, like a pair of pincers (Fig. 8). Thus, at the time of hatching most oncomiracidia are already equipped with an array of fully formed and fully functional hooklets for attachment to the host's exposed epidermal surface.

The oncomiracidia of some gill-parasitic monogeneans have been recovered from fish skin, e.g. *Neodactylogyrus crucifer* (Fig. 9C; Kearn, 1968) and *Urocleidus adspectus* (see Cone & Burt, 1981). Over

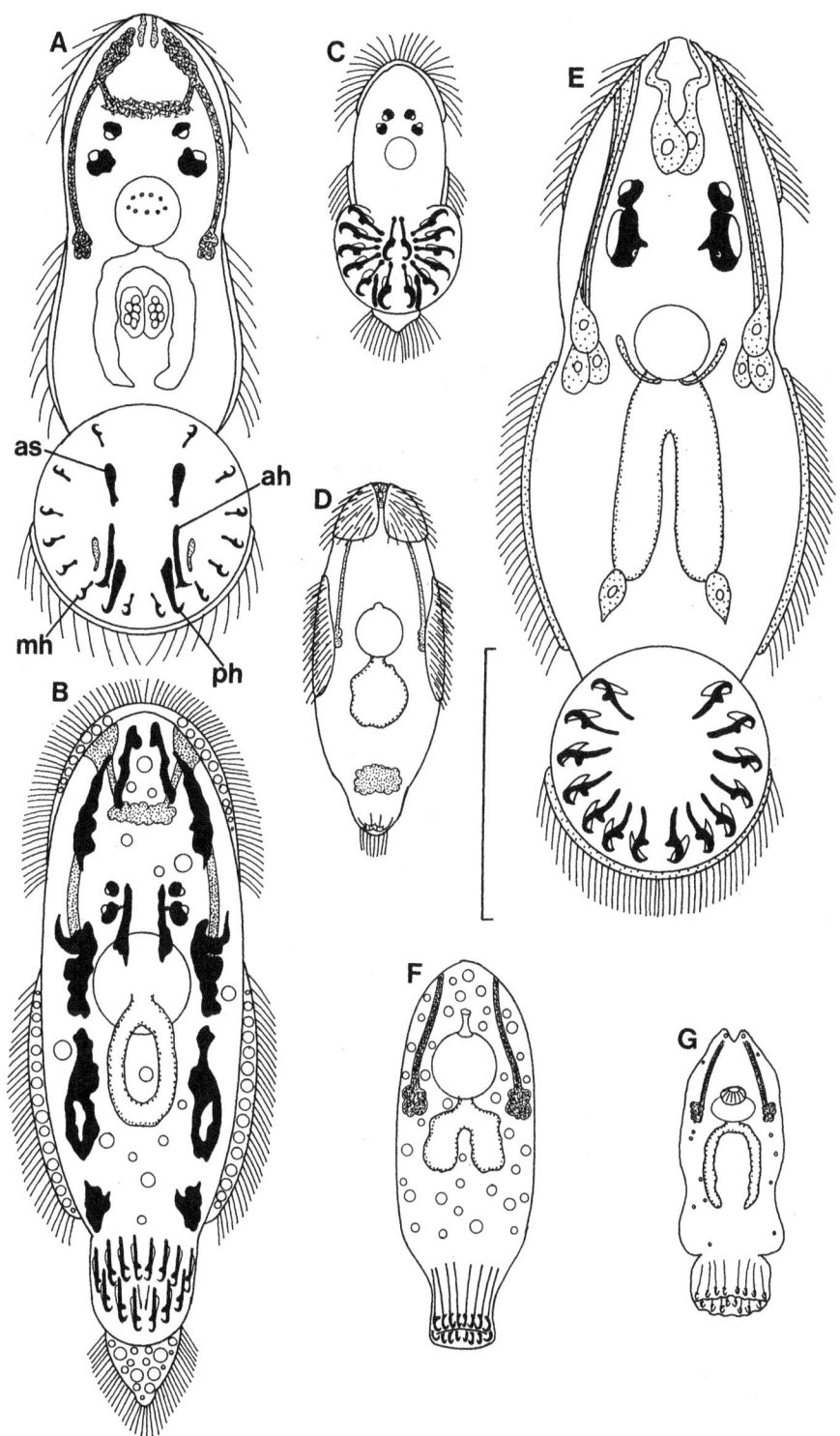

Fig. 9. Some oncomiracidia that establish themselves on fish skin. (A) *Benedenia lutjani*; (B) *Capsala martinieri*; (C) *Neodactylogyrus crucifer*; (D) *Leptocotyle minor*; (E) *Dendromonocotyle ardea*; (F) *Acanthocotyle lobianchi*; (G) *Enoplocotyle kidakoi*. Drawn to approximately the same scale; scale bar, 100 µm. ah, Anterior hamulus; as, accessory sclerite; mh, marginal hooklet; ph, posterior hamulus. Redrawn from the following: (A) Whittington & Kearn (1993); (B) Kearn (1963a); (C) Kearn (1968); (D) Kearn (1965); (E) Chisholm & Whittington (1995); (F) Kearn (1967a); (G) Kearn (1993).

an 8 h period, Cone & Burt (1981) exposed young perch, *Perca flavescens*, measuring 2·5–4 cm in length, to large numbers of oncomiracidia of *U. adspectus*. One day later they found that relatively few parasites were attached to the gills (8–15%), the rest being distributed uniformly over the body surface along the full length of the host. During the next 4 days this surface distribution changed, skin parasites disappearing progressively from the body surface in an anterior direction and increasing in

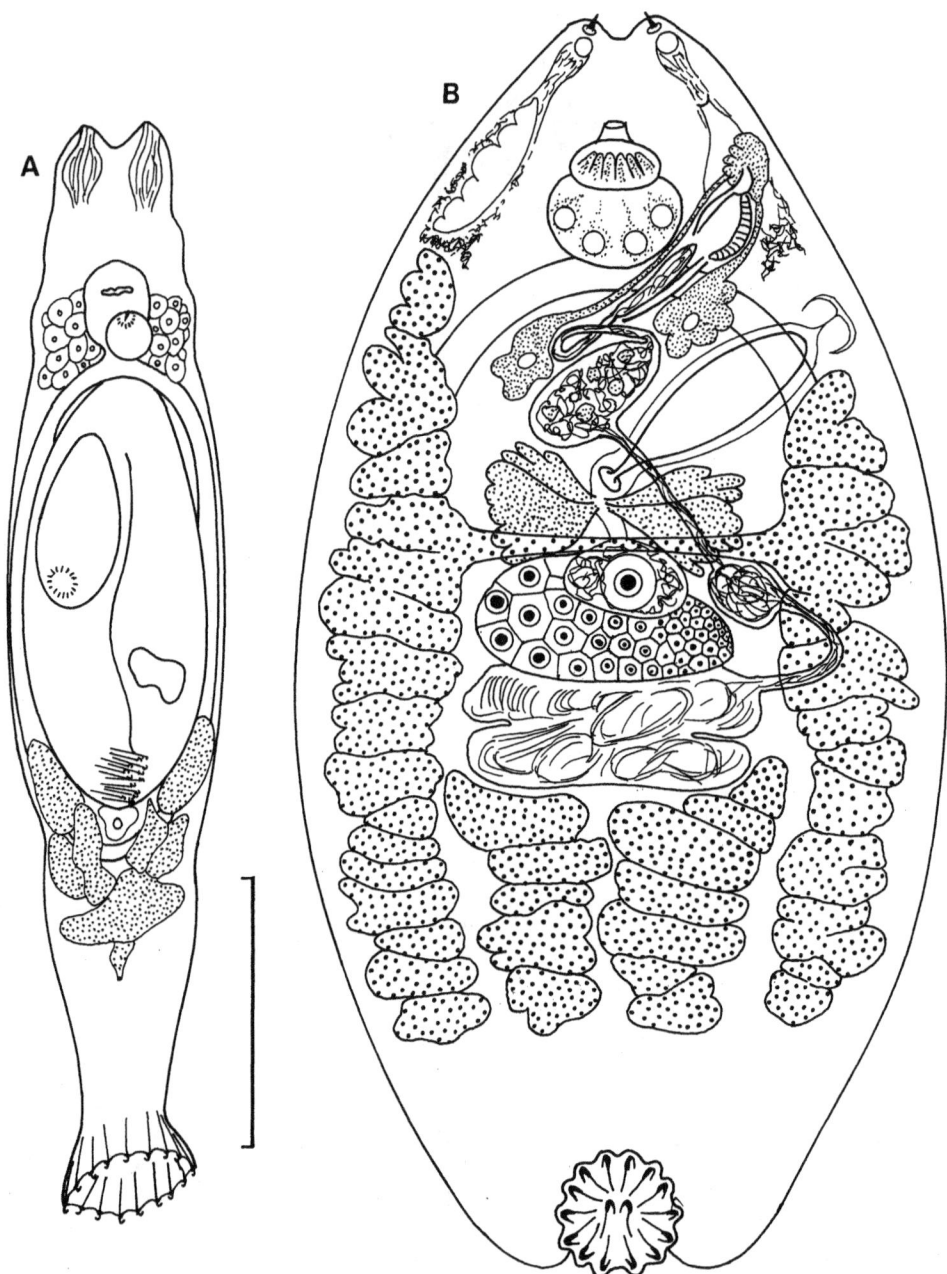

Fig. 10. Two skin-parasitic monogeneans that use only marginal hooklets for attachment of the adult. (A) *Anacanthocotyle anacanthocotyle*; (B) *Enoplocotyle kidakoi*. Redrawn from the following: (A) Kritsky & Fritts (1970); (B) Kearn (1993). Scale bar, 100 μm.

numbers on the gills. Since there was no decrease in intensity of infection during this period, significant mortality of skin parasites seems unlikely and *Urocleidus* appears to be capable of finding its way forwards on the host, ultimately reaching the gills.

There are a few adult monogeneans that rely entirely on marginal hooklets for attachment to fish skin, but these parasites are relatively small. *Anacanthocotyle anacanthocotyle* (Fig. 10A) is a viviparous gyrodactylid monogenean found on the skin of a freshwater fish *Astyanax fasciatus* (see Kritsky & Fritts, 1970). Its haptor is armed with 16 radially-arranged marginal hooklets and supports a mature parasite with an average length of just over 300 μm and a width of about 80 μm. The load on the haptor will fluctuate depending on the stage of development of the contained embryo, which is not born until it is virtually as large as its parent. Probably the largest monogenean to depend entirely on marginal hooklets for attachment is *Enoplocotyle kidakoi*, an oviparous parasite with 16 marginal hooklets (14 peripheral and 2 central) (Fig. 10B). The body of the adult parasite is significantly larger than that of *Anacanthocotyle*, measuring as much as 634 μm in length and 253 μm in breadth (Kearn, 1993). However, *Enoplocotyle* spp. parasitize moray eels which are inactive fishes inhabiting crevices and caves (Hobson, 1974). Because of the sluggish habits

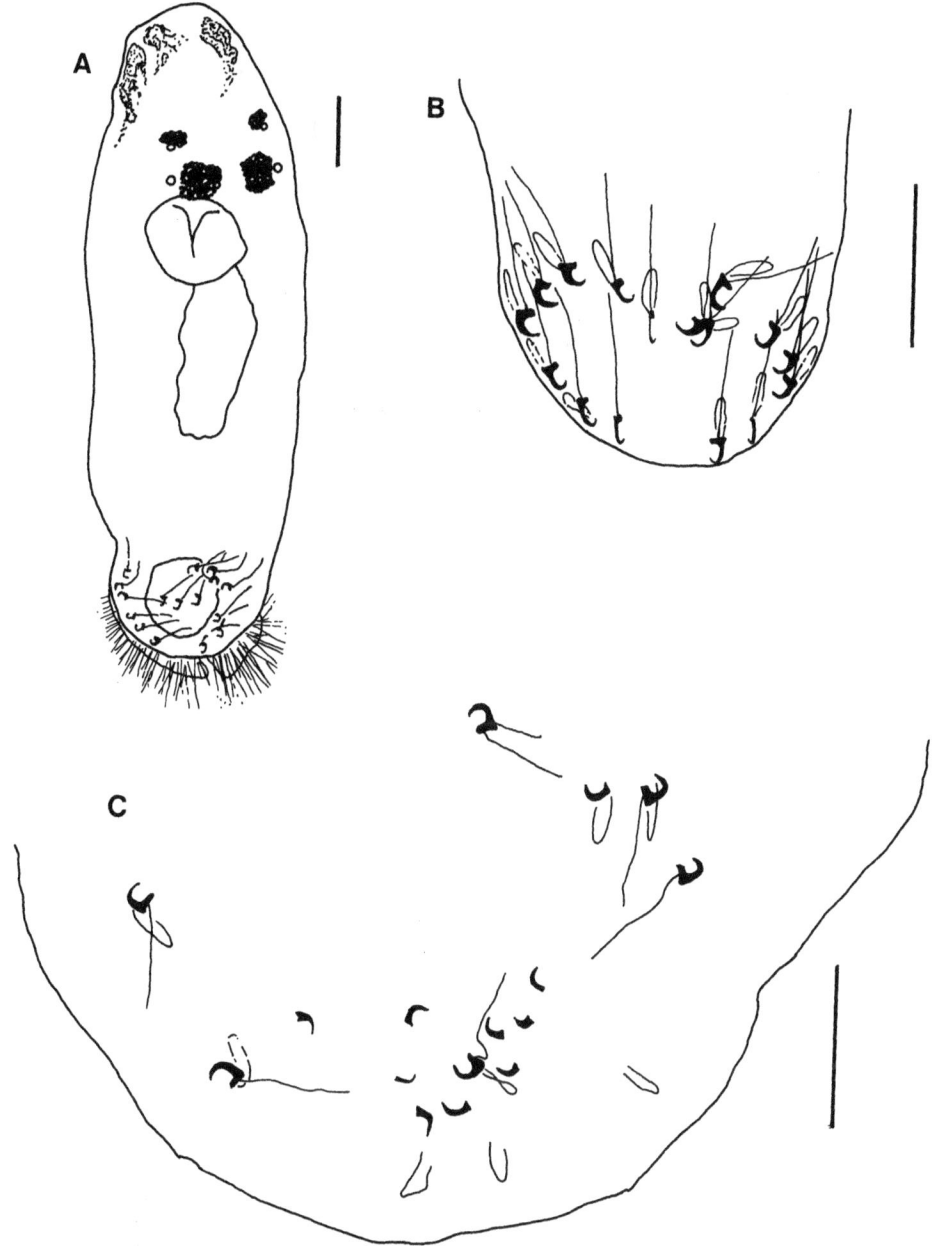

Fig. 11. Marginal hooklets of *Anoplodiscus spari*. (A) Whole oncomiracidium; (B) haptor of oncomiracidium enlarged; (C) haptor of juvenile parasite about 350 μm in total length. Scale bars, 20 μm. Reproduced from Ogawa & Egusa (1981) with kind permission of Kluwer Academic Publishers.

of these eels their skin parasites may be subjected to relatively weak water currents compared with parasites of active hosts and marginal hooklets may be adequate to resist dislodgement.

Marginal hooklets persist without further growth in most adult skin parasites, and in many adults they probably contribute little to attachment. However, in the disc-shaped haptor of the capsalid *Entobdella soleae*, the persistent peripheral hooklets probably have a minor but significant role as 'tent pegs' preventing inefficient inward movement of the edge of the disc when suction is generated (see below). In addition, the two centrally placed marginal hooklets of capsalids continue to grow and, as in *E. soleae*, become spatially and functionally associated, as accessory sclerites, with the anterior hamuli (Fig. 5; see below).

In *Anoplodiscus spari* from the skin of the sea bream *Acanthopagrus schlegeli*, the sixteen marginal hooklets of the oncomiracidium (Fig. 11 A, B) disintegrate and are absorbed during post-oncomiracidial development (Fig. 11 C). The six marginal hooklets of the microbothriid *Leptocotyle minor* from the skin of the dogfish *Scyliorhinus canicula*, never progress beyond tiny spicules in the oncomiracidium (Fig. 9 D). Possible reasons for these features in *A. spari* and *L. minor* will be discussed below.

Hamuli. Marginal hooklets are essentially larval structures, specifically for attachment to fish epi-

Fig. 12. A selection of hamuli of some skin-parasitic monogeneans, illustrating the range of shape and size. (A–D) Capsalids (A, *Entobdella soleae*); (E, F) gyrodactylids; (G) a calceostomatid; (H) *Dendromonocotyle*. Scale bars, 50 μm. Note scale difference between A, B, C, E (central scale bar) and D, F, G, H (scale bar at lower right). Redrawn from various sources.

dermal cells. If the ontogenetic sequence in monogeneans is a faithful reflection of evolutionary progress within the group, then the first monogeneans may have attached themselves to the host's exposed epidermal surface by marginal hooklets and may have been similar in size to oncomiracidia. Any increase in size of these ancestral monogeneans or an increase in activity or swimming speeds of their fish hosts could not be supported by marginal hooklets only. Many monogeneans have solved this problem by acquiring one or two pairs of larger hooks or hamuli (acanthocotylids have a different solution – see below). These are not enlarged marginal hooklets but are independent sclerites (Fig. 12). Like marginal hooklets they originate as intracellular spicules (Lyons, 1966) and contain substantial amounts of sulphur (Lyons, 1966; Kayton, 1983; Shinn *et al.* 1995). Hamuli may first appear in the embryo (as in many capsalids; Fig. 9A), but their growth usually continues in post-oncomiracidia and may proceed throughout adult life (as in the anterior hamuli of *E. soleae*; see Kearn, 1990). Since most marginal hooklets complete their growth before hatching, hamuli soon outstrip them in size and many are sufficiently large (Fig. 12 A, B, E, F) to penetrate through the epidermis and provide firm anchorage in the tough collagenous dermis. In some capsalids with two pairs of hamuli, such as *E. soleae*, the posterior hamuli are smaller than the anterior hamuli (Fig. 5) and make their major contribution to attachment in the oncomiracidium and early post-oncomiracidium. The hooked regions of these small posterior hamuli may not reach the dermis.

The adult haptor and the role of sclerites. Most gyrodactylids possess a single pair of hamuli and 16 peripherally situated marginal hooklets (Fig. 13). Lester (1972) observed that *G. alexanderi* was unable to attach its haptor to glass unless sticky secretions from the anterior adhesive areas had already been deposited at the site. The implication is that the parasite does not employ suction or haptor secretion to achieve attachment.

In the oviparous gyrodactylids the hamuli are already present as a pair of spicules in the freshly hatched oncomiracidium, but are well developed in the new-born daughters of viviparous gyrodactylids. Most species of the viviparous genus *Gyrodactylus* (Fig. 13A) range in length from 0·4 to 0·8 mm and other monogeneans of similar size lack hamuli, relying only on marginal hooklets for attachment. Kearn (1994, 1998) has offered a possible explanation for this anomaly. He pointed out that autoinfection in gyrodactylids leads to large infrapopulations, the feeding activities of which are potentially damaging, if not lethal, to the host. The impact of such large populations would be less if the individual parasites were small and selection may have favoured a reduction in overall size, rendering the hamuli redundant. In spite of this, hamuli have not yet been lost, except in isancistrines such as *Anacanthocotyle* (see above; Fig. 10A), but there have been reports that the hamuli do not penetrate host skin (Lester, 1972; Harris, 1982; Cone & Odense, 1984; Cone & Wiles, 1989; Buchmann, 1999). However, Lester (1972) claimed that the hamuli of *Gyrodactylus alexanderi* are used to anchor the parasite when strong shearing forces are exerted upon it, e.g. by fish movement. Thus some gyrodactylids may only use their hamuli to supplement their marginal hooklets at times of extra stress, such as when carrying a full-term embryo or when the host swims vigorously. According to Harris (1982), skin parasites such as *G. bullatarudis* and *G. gasterostei* are unable to use their hamuli to pierce the skin because the points of these sclerites are covered with tegument.

The hamuli of *Gyrodactylus* spp. are supported by dorsal and ventral bars. In the related genera *Macrogyrodactylus*, *Swingleus*, *Polyclithrum* and *Accessorius*, there are additional, often elaborate supporting sclerites (Fig. 13). There is also a tendency in all four genera for separation of the marginal hooklets into a fan-shaped posteriorly directed array of 8, 10 or 14 hooklets, while the remaining hooklets are

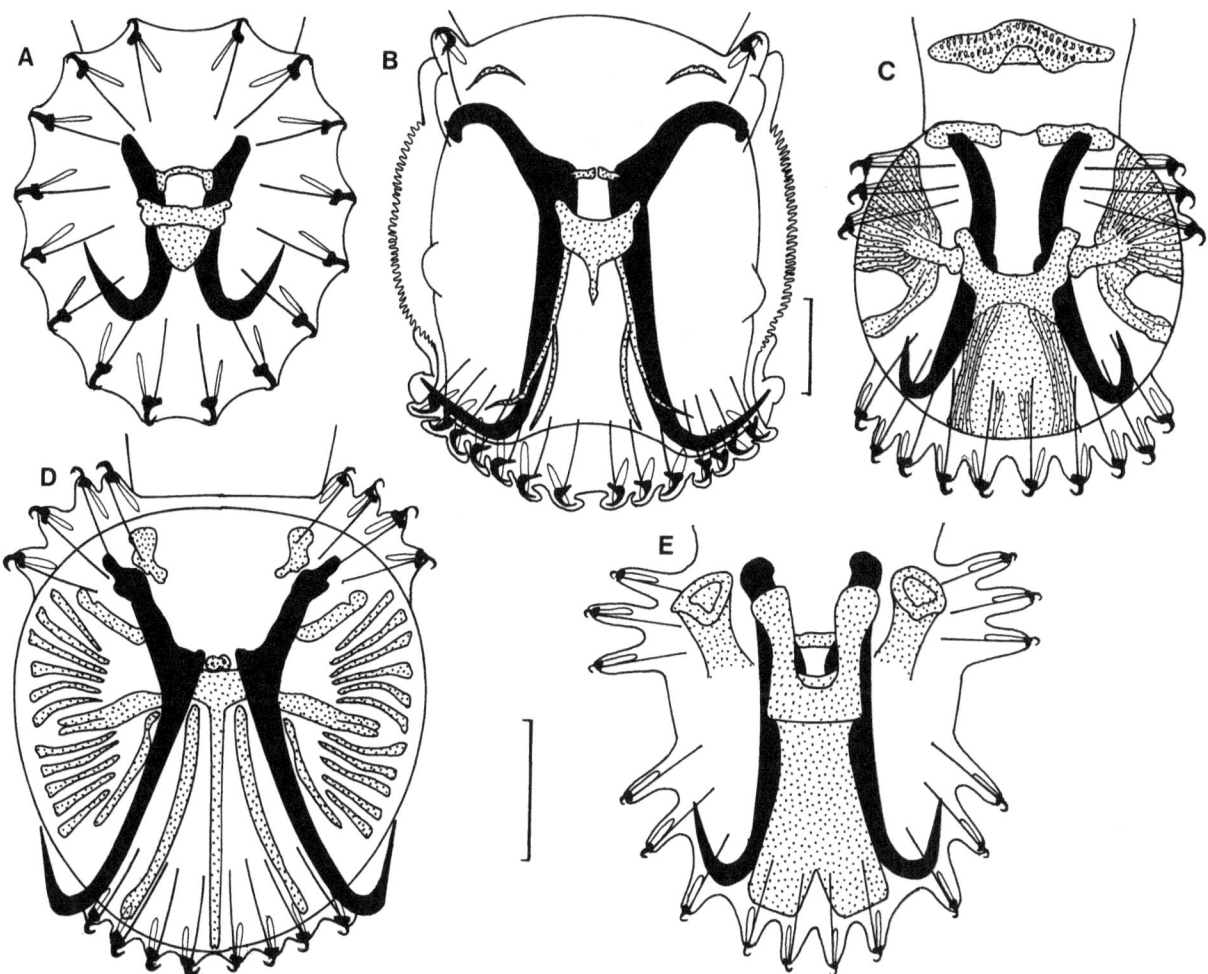

Fig. 13. Elaboration of the haptor in some gyrodactylid monogeneans. (A) *Gyrodactylus*; (B) *Macrogyrodactylus*; (C) *Swingleus*; (D) *Polyclithrum*; (E) *Accessorius*. Marginal hooklets and hamuli shown black, supporting sclerites stippled. A, C–E, drawn to approximately the same scale; lower scale bar, 50 μm; upper scale bar for B, 100 μm. Redrawn from the following: (B) El-Naggar & Serag (1987); (C) Rogers (1968); (D) Rogers (1967); (E) Jara, An & Cone (1991).

positioned in two anterolateral groups, each of 4, 3 or 1 hooklet respectively (Fig. 13). The significance of these embellishments and hook rearrangements is unknown.

Many capsalids have two pairs of hamuli. In *Entobdella soleae* the persistently growing central pair of marginal hooklets (= accessory sclerites) (see above; Fig. 5) keep pace with the growing anterior hamuli (Kearn, 1963b), but they differ from them in orientation. The accessory sclerites project from the ventral surface of the central region of the disc so that they push down into the host's skin (Kearn, 1964). On each side a tendon originates from an extrinsic muscle in the posterior region of the body, passes through a notch at the proximal end of the corresponding prop-like accessory sclerite and attaches to the anterior end of the anterior hamulus (Kearn, 1964). The way these muscles, tendons, accessory sclerites and anterior hamuli interact to generate suction between the haptor disc of *E. soleae* and the host's skin has been described elsewhere (Kearn, 1964, 1994, 1998). In simple terms, the anterior hamuli are embedded in the disc and when their anterior ends are raised by the muscle/tendon system, the centre of the disc is lifted and suction is produced. Inward movement of the edge of the disc and influx of seawater beneath the disc are prevented by 'tent pegs' (peripheral marginal hooklets) and by a flexible marginal valve. In addition, the efficiency of the suctorial haptor is enhanced by a major feature of the skin of the host, *Solea solea*. This is the presence of dermal scales, which resist the tendency for the accessory sclerites to sink into the skin when the roof of the disc is lifted. In other capsalids, such as *Benedenia*, haptor anatomy is similar, but the tendons are attached directly to the ventral surface of the disc, not to the hamuli (Fig. 14A; Kearn, 1994).

Observations made on locomotion in *E. soleae* have revealed that the hooks are not always used to pin the adult haptor to the skin. After a locomotory 'step' on the host, the attached haptor was observed to glide slowly backwards until lodged securely beneath the projecting edge of a host scale (Kearn, 1988a). It is therefore possible that *E. soleae* and its

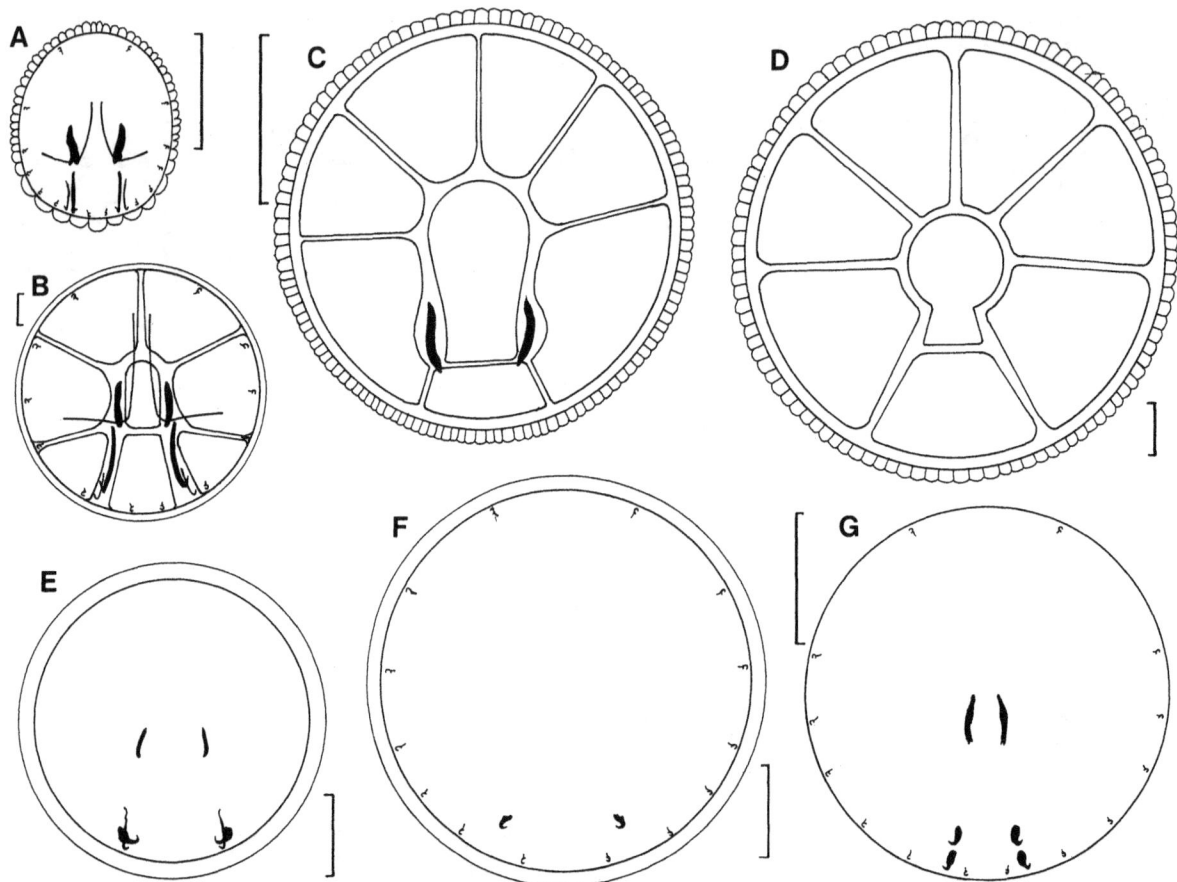

Fig. 14. The haptors of some capsalid skin parasites. (A) *Benedenia lutjani*; (B) *Pseudobenedenia nototheniae*; (C) *Capsala pricei*; (D) *Capsala martinieri*; (E) *Nitzschia sturionis*; (F) *Calicobenedenia polyprioni*; (G) *Trimusculotrema uarnaki*. Scale bars for A–C, F, G, 250 μm; scale bars for D, E, 1 mm. Redrawn from the following: (A) Whittington & Kearn (1993); (B) Dollfus & Euzet (1964); (C) Hidalgo Escalante (1958); (D) Lamothe-Argumedo (1997); (E) original; (F) Kritsky & Fennessy (1999); (G) Whittington & Barton (1990).

relatives have a way of generating suction that is independent of sclerite participation. The haptor of *E. soleae* does, in fact, contain abundant intrinsic muscles (Fig. 5B), including dorsoventral fibres, which are potentially capable of creating a drop in pressure between the haptor and the host's skin.

If there are two ways of generating suction their relative contributions to the day-to-day life of the parasite are of obvious interest. Perhaps these capsalids employ energy expensive modes of attachment only at times when shear forces are high, such as when the swimming host accelerates. A similar suggestion has been made for the employment of hamuli in some gyrodactylids (see above).

A microscopist might be forgiven for concluding that *E. soleae* has no glands associated with the haptor, but intensive study of living adult parasites has revealed the presence of narrow, secretion-filled ducts opening at sites on the concave ventral surface of the haptor (Kearn, 1974). These sites are symmetrically arranged with two located in anterolateral positions, lateral to the accessory sclerites, and two in posterolateral positions close to the hooked regions of the anterior hamuli. These are probably the same gland ducts that have been identified in the haptor of the oncomiracidium, each duct arising from a single gland cell located in the posterior region of the body (Kearn, 1974). Similar ducts occur in the adult haptor of *Benedenia seriolae* (see Kearn, Ogawa & Maeno, 1992). In *E. soleae*, these glands resemble other unicellular glands opening at various sites on the body surface, especially in the adult (El-Naggar & Kearn, 1983), and they may all have the same, currently unknown, function. The body glands are able to release their secretion into the water, but the only opportunity for the haptor glands to do so would be during the brief interval when the haptor is detached during locomotion. There is no evidence that the haptor of *E. soleae* has adhesive properties (Kearn, 1974).

Haptors with reduced sclerites. Many capsalids have greatly reduced their median sclerites (anterior and posterior hamuli and accessory sclerites) with concomitant loss of the facility to generate suction using extrinsic muscles. This trend is associated with two different parallel developments. The first of these concerns the promotion of suction generated by intrinsic haptor muscles. The second involves the development of adhesive secretions.

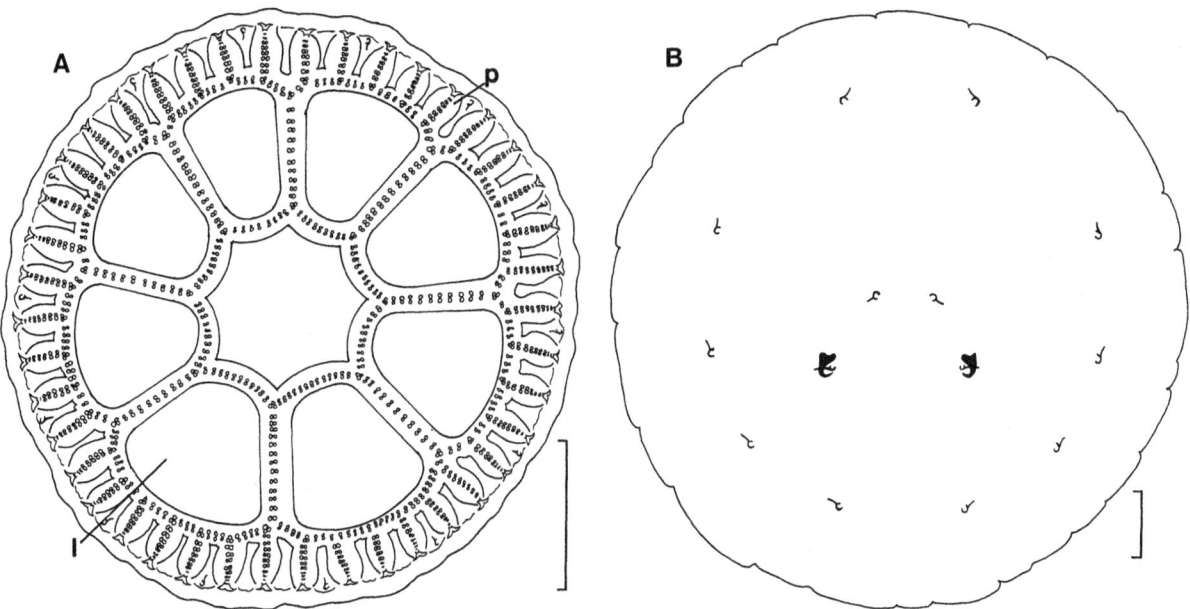

Fig. 15. The haptors of (A) *Dendromonocotyle taeniurae* and (B) *Neocalceostomoides brisbanensis*. Scale bar for A, 250 μm; scale bar for B, 50 μm. l, Loculus; p, papilla. Redrawn from the following: (A) Euzet & Maillard (1967); (B) Whittington & Kearn (1995).

In *Pseudobenedenia nototheniae* from the skin of *Notothenia neglecta*, the suctorial haptor is divided into separate muscular compartments or loculi (Fig. 14B), each presumably capable of generating suction independently of the others. However, the median sclerites of *P. nototheniae* are well developed and the parasite seems likely to retain the ability to generate suction using its extrinsic muscles and tendons. *Capsala pricei* from the skin of *Makaira mitzukurii* has haptor loculi but there is only one pair of central sclerites (probably the accessory sclerites) (Fig. 14C). This progression culminates in *C. martinieri* from the skin of the sunfish *Mola mola*, in which there is no trace of the median sclerites in the adult (Fig. 14D) (the oncomiracidium has one central pair; Fig. 9B). *C. martinieri* relies entirely on loculi and their intrinsic muscles to generate suction.

Attachment by loculi seems to be a very effective means of attachment to fish skin because it is able to support very large parasites. *C. pricei* and *C. martinieri* have circular bodies with diameters of about 13 mm and 2 cm, respectively and circular haptors with diameters of about 6 mm and 8 mm, respectively (see Hidalgo Escalante, 1958; Kearn, 1994, respectively).

Subdivision of the haptor to provide smaller suctorial units has not taken place in all large capsalids. *Nitzschia sturionis* (Fig. 1) may exceed 1 cm in length and inhabits the relatively flat, skin-covered lining of the buccal cavity of the sturgeon *Huso huso*. *Nitzschia* has a non-loculate circular haptor measuring up to 3 mm in diameter, with substantially reduced median sclerites (Fig. 14E). The haptor lacks conspicuous glands and has prominent dorsoventral and circular muscles (G. C. Kearn, unpublished observations). Bychowsky (1957) believed that these muscles create suction. He stated that 'if a very thin capillary, which allows passage for the water from the outside, is placed under the disc of the attached worm, there is little difficulty in removing the worm, and in most cases it will fall off itself'. He also observed that 'it is easier to tear the worm in two than to pull it from its place of attachment'. Bychowsky's suggestion is supported by the presence of a thin projecting flap that may act as a marginal valve, around the haptor perimeter (Fig. 14E).

Another relatively large capsalid, *Calicobenedenia polyprioni* from the skin of *Polyprion americanus* has a non-loculate haptor possessing only one pair of greatly reduced hamuli (Fig. 14F). The presence of what appears to be a marginal valve around the disc perimeter (see Kritsky & Fennessy, 1999) suggests that this haptor is also suctorial in function.

Monocotylid monogeneans are parasites of cartilaginous fishes and most of them inhabit internal sites, including gills, cloaca, nasal fossae and body cavity. Monocotylids have loculate haptors, a single pair of hamuli and retain in the adult 14 peripheral marginal hooklets. Species of the genus *Dendromonocotyle* (Fig. 15A) are the only monocotylids recorded so far from the skin of their hosts (rays of the families Dasyatididae, Myliobatididae and Urolophidae) (Chisholm & Whittington, 1995), although some monocotylids attach themselves to flat non-respiratory skin surfaces inside the gill chamber (see Kearn, 1994). Some *Dendromonocotyle* species have substantially reduced their hamuli (Fig. 12H) and in others these are lost (Chisholm & Whittington, 1995). It is not clear whether *Dendro-*

monocotyle is a survivor of early skin-parasitic monocotylids that also gave rise to the endoparasites or whether *Dendromonocotyle* evolved at a later stage from an endoparasitic ancestor. Whichever evolutionary route has led to these skin parasites, there is a remarkable convergence with *Capsala* spp., extending to their loculate haptors and reduction of their median sclerites (Fig. 15A). An added refinement in *Dendromonocotyle* is that all the septa between the loculi have acquired rows of tiny projecting sclerites (Fig. 15A). These may counteract the tendency for the septa edges to slide inwards when suction is generated within the loculus, thereby leading to an overall improvement in suction efficiency. Radially-arranged papillae, armed with similar but morphologically distinct sclerites, reinforce the margins of the disc where slipping may be more significant.

The second development associated with reduction of the median sclerites concerns the adoption of cement for attachment to the host's skin. Adhesive secretions are widespread in platyhelminths, in non-parasitic aquatic and terrestrial species as well as in parasites. It is somewhat surprising that the adhesive secretions produced by platyhelminths function in water and even more surprising that they are able to glue parasites to slimy fish skin sufficiently firmly to resist dislodgement by water currents. Temporary attachment of the heads of monogeneans during locomotion is commonly achieved by means of adhesive secretions (see below), but there are few monogeneans that employ adhesives for permanent attachment to skin.

Species of the capsalid genus *Trimusculotrema* attach themselves to the skin of stingrays by means of a circular haptor, lacking loculi and with greatly reduced median sclerites (Fig. 14G). Kearn & Whittington (in Kearn, 1994) were able to study a living specimen of an undescribed species attached to glass. It was found that the haptor resembled adhesive tape in the way that it resisted removal from the substrate and this indicates attachment by cement rather than by suction. Consistent with this interpretation is the absence of a marginal valve and, in this particular individual, the presence of a haptor deformity likely to prevent establishment of a marginal seal (see Kearn, 1994). A marginal valve is also absent in *T. uarnaki*, but Whittington & Barton (1990) noted that the haptor is rapidly detached and reattached during locomotion. So, if *T. uarnaki* attaches its haptor by means of cement, it is able to detach it rapidly (cf. the microbothriid *Leptocotyle minor*, see below).

A parallel development seems to have taken place in calceostomatine monogeneans. *Neocalceostomatoides brisbanensis* inhabits the buccal and gill chambers of its host, *Arius graeffei*, where its thin and flexible circular haptor (Fig. 15B) is attached to relatively flat surfaces (Kearn, Whittington & Evans-Gowing, 1995). The haptor was found to have similar adhesive properties to that of *Trimusculotrema* spp. and large quantities of secretion from gland cells in and near the haptor are directed into the tegument covering the ventral surface of the haptor. It is assumed that this secretion reaches the surface of the haptor by exocytosis and is responsible for the adhesive properties of the disc. Thus, the cement appears to be a highly specialized glycocalyx. The haptor sclerites and the musculature of the haptor are reduced (Fig. 15B).

In the parasites described above, hooks are retained, although they are reduced in size. In microbothriid monogeneans such as *Leptocotyle minor* from the dogfish *Scyliorhinus canicula*, hooks are absent in the adult parasite, although spicules present in the haptor of the oncomiracidium (Fig. 9D) are thought to be vestigial marginal hooklets (Kearn & Gowing, 1990). The reason for this is immediately apparent if the site of attachment of *L. minor* is examined. The relatively small hookless haptor (Fig. 16A) is attached to a single host denticle. Hooks are unable to function since the exposed surface of a denticle is a hard enameloid substance lacking a covering of skin.

Kearn (1965) has shown that *L. minor* uses cement to attach itself to this hard surface, but, in spite of being cemented permanently to a naked denticle, *L. minor* is able to move (Kearn, 1965). However, after attachment of the head to a new denticle, removal of the haptor is a slow stripping process that takes between 5 and 10 s to complete, so that a locomotory step takes significantly longer than in a monogenean like *E. soleae*.

There is another group of parasites that, like the microbothriids, attach themselves by cement and lack hooks. These are the udonellids (Fig. 16B). Their affinities have long been debated, with arguments in favour of 'turbellarian', monogenean or separate status (see Rohde, 1994). However, recent molecular data place them among the monopisthocotylean monogeneans (Littlewood, Rohde & Clough, 1998). Like microbothriids, udonellids attach themselves to a hard substrate, in this case the carapace or another external site of a parasitic caligid copepod (see Kearn, 1998). From this platform, udonellids probably feed on the epidermis of the host fish, but this has not been unequivocally demonstrated (see references in Kearn, 1998). The haptor of udonellids is well supplied with gland cells (Fig. 16B) and Ivanov (1952, in Kearn, 1998) had no doubt that attachment is achieved by secreting cement rather than by suction.

Reference has been made above to the disintegration and resorption of marginal hooklets during the development of *Anoplodiscus spari* (Fig. 16C). Work on other species of *Anoplodiscus* has shed more light on this unusual development. *A. australis* attaches itself to the caudal fin (more rarely the

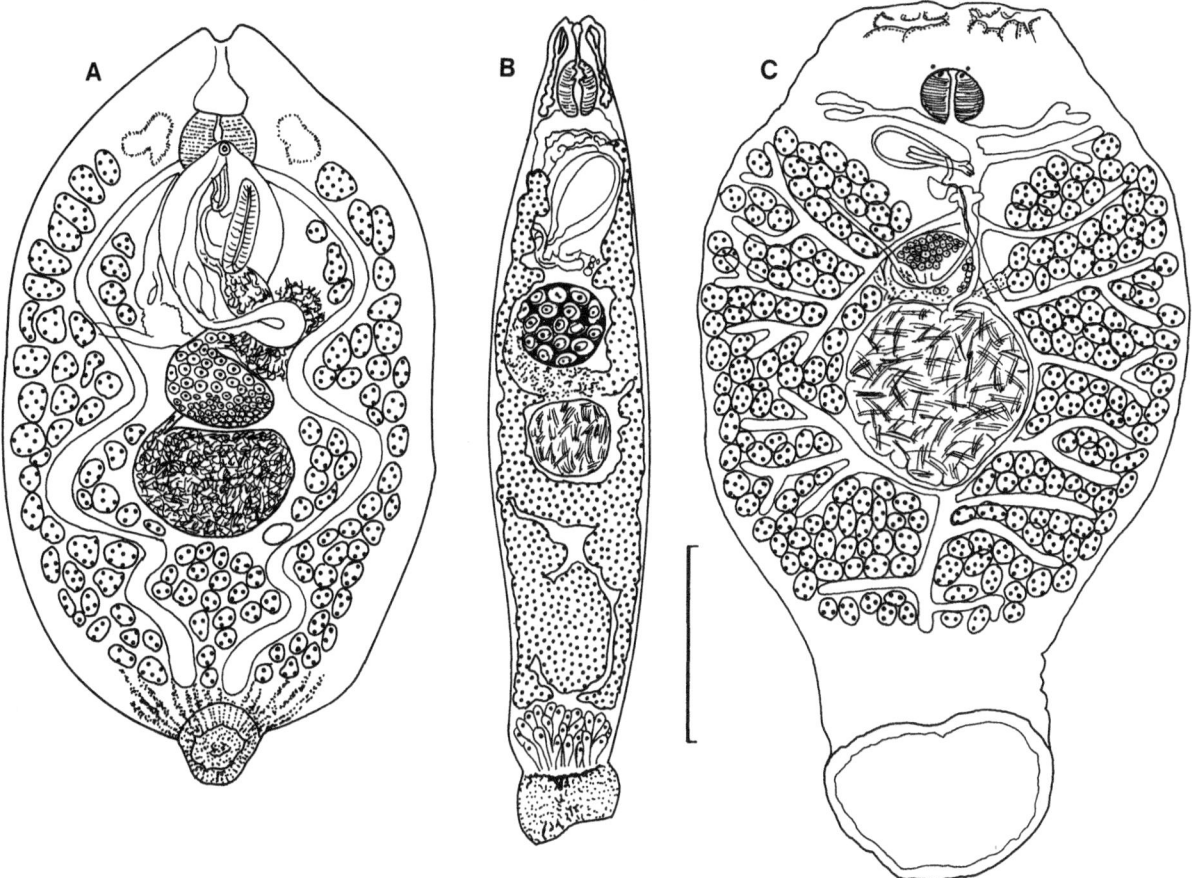

Fig. 16. Skin-parasitic monogeneans lacking haptoral hooks and other sclerites. (A) The microbothriid *Leptocotyle minor*; (B) *Udonella caligorum*; (C) *Anoplodiscus spari*. Drawn to the same scale; scale bar, 500 μm. Redrawn from the following: (A, B) Sproston (1946); (C) Ogawa & Egusa (1981).

pectoral fin) of the yellowfin bream, *Acanthopagrus australis* (see Roubal & Whittington, 1990). *A. cirrusspiralis* infects the caudal and pectoral fins of its host, the snapper *Pagrus auratus*, but has also been reported from the nares of the fish (Roubal, Quartararo & West, 1992). At the haptoral attachment sites of both of these parasites there is no underlying host epidermis. The basement membrane of the host epidermis is exposed and the parasites attach themselves to this basement membrane by an adhesive secretion (Roubal & Whittington, 1990; Roubal *et al.* 1992). It has been argued above that marginal hooklets have evolved in response to the special problems of attachment to fish epidermis, and absorption of marginal hooklets in *Anoplodiscus* may be a reflection of the absence of epidermis beneath the attached adult haptor.

According to Roubal & Whittington (1990) the adhesive secretion of *A. australis* is produced by subtegumentary cytons and is passed to the ventral tegument of the haptor disc. The secretory bodies then pass intact through the apical membrane of the tegument and coalesce to form the cement (a specialized glycocalyx as in *Neocalceostomatoides*? – see above) at the parasite/host interface. How the host epidermis is eroded is unknown, although Roubal & Whittington (1990) have pointed out that other gland cells occur in the haptor and might produce a histolytic secretion. Roubal & Whittington (1990) indicated that *A. australis* adults do not change their attachment site once established, but West & Roubal (1998) observed that *A. cirrusspiralis* adults are able to relocate and do so once or twice a week. Each site of attachment experienced epidermal erosion, but after parasite relocation this wound healed. Specimens of *A. cirrusspiralis* living in the nares erode the host epidermis and attach to the basement membrane by cement in the same way (Roubal *et al.* 1992).

Sensory structures. No cilia-based sensory structures have been reported so far from the haptors of skin-parasitic monogeneans, although they are present on the anterior adhesive areas of skin parasites (see below) and on the haptors of some of the gill-parasitic monogeneans that attach themselves between secondary gill lamellae (see, for example, Kearn & Gowing, 1989). However, the ventral surfaces of the haptors of *Entobdella soleae* and *E. hippoglossi* are almost entirely covered with papillae (Lyons, 1973). *E. soleae* has more than 800 of these, ranging in diameter from 2·5 to 19 μm. The location

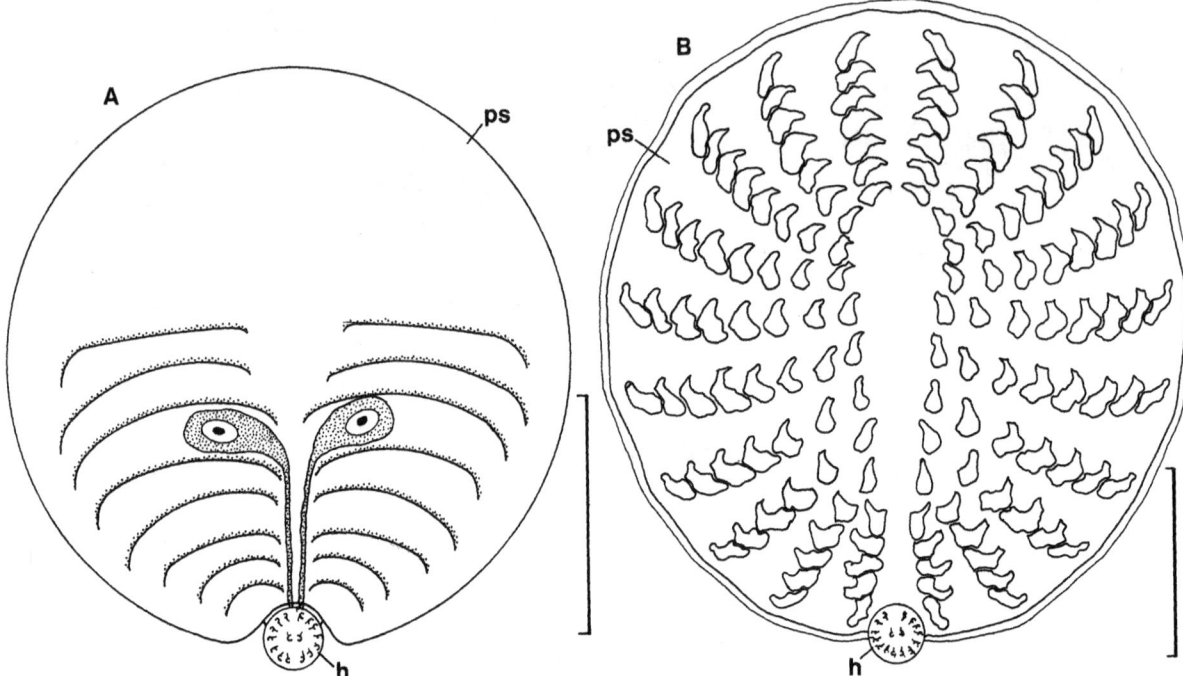

Fig. 17. The pseudohaptors of (A) *Myxinidocotyle californica* and (B) *Acanthocotyle lobianchi*. Scale bars, 250 μm. h, Haptor; ps, pseudohaptor. (A) Redrawn from Malmberg & Fernholm (1989); (B) original.

and abundance of these papillae brings to mind the anti-slip 'tread' of a car tyre, but the haptor is sometimes known to slide relative to the host's skin (see above) and Lyons (1973) claimed that these papillae are sensory. There is no evidence of derivation from cilia, but a nerve that folds back and forth to form a stack of thin lamellae, penetrates each papilla.

Lyons (1973) favoured the idea that the papillae might function as contact receptors. She envisaged that they might record the changing pattern of contact of the haptor surface with the host during locomotion and that they may be involved in the co-ordination of muscle contractions. However, why these papillae should be present in some species of *Entobdella* and absent in others (e.g. in *E. australis*; see Kearn, 1978) is not clear. They are also absent in other capsalids with suctorial haptors and well developed sclerites, such as *Benedenia* spp. Haptor papillae are present in *Trimusculotrema* spp. (see Whittington & Barton, 1990), in which the sclerites are reduced, but it is not yet known whether these papillae are structurally similar to those of *E. soleae*.

The pseudohaptor

There is no evidence that acanthocotylid monogeneans have ever had hamuli. Acanthocotylids are found on the skin of agnathans and cartilaginous fishes (rays), but they supplement their marginal hooklets by transforming the posterior region of the body proper into a false haptor or pseudohaptor (Fig. 17) (Kearn, 1967a; Malmberg & Fernholm, 1989).

The true haptor, carrying 14 peripheral and two central hooklets but no trace of hamuli or any other sclerites, persists right through to the adult stage without further growth, so that in the adult the haptor is dwarfed by the pseudohaptor (Fig. 17). In spite of the small size of the haptor, Malmberg & Fernholm (1991) found evidence that the haptoral hooklets are functional in adults of *Myxinidocotyle californica* (see Fig. 8). However, their observation that the living adult haptor can be moved sideways over the host skin indicates that the hooklets are not always attached. They also found two gland cells (Fig. 17A) in *M. californica* with ducts running into the haptor and opening ventrally, one on each side of the two central hooklets (Malmberg & Fernholm, 1989). Malmberg & Fernholm (1991) suggested that it is only the haptor that provides attachment for the posterior end of acanthocotylids during locomotion, but this has not been confirmed.

The fully developed roughly circular pseudohaptor of an acanthocotylid is attached to the substrate by means of suction (Malmberg & Fernholm, 1991; Kearn, 1998). Its concave ventral surface is not loculate, but is equipped with transverse or radial ridges. In *Myxinidocotyle* spp. these ridges are more or less transverse, restricted to the posterior half or two thirds of the haptor and directed anteriorly or anterolaterally (Fig. 17A). In *Lophocotyle novaezeelandica* and in *Acanthocotyle* spp. the ridges are radially arranged, each ridge being unarmed in *Lophocotyle* but in *Acanthocotyle* possessing a row of curved sclerites, some of which appear to articulate with one another (Fig. 17B). On the flat skin of its hagfish host the suctorial haptor of

M. californica creates a convex circular 'footprint' with deep impressions made by the pseudohaptoral ridges. These ridges would strongly resist any forces produced by water currents travelling anteriorly over the parasite and tending to dislodge the pseudohaptor or slide it forwards. According to Malmberg & Fernholm (1991), the pseudohaptoral sclerites of *Acanthocotyle* create pits but not wounds in the host's skin. The higher level of complexity represented by the pseudohaptoral ridges of *Acanthocotyle* with their armature of linked sclerites suggests that these sclerites may do more than resist shear forces, but what this additional role might be is at present unknown.

TEMPORARY ATTACHMENT

Morphology of anterior adhesive areas

Most, perhaps all, monogenean skin parasites are able to change their location on the host. This is achieved by leech-like locomotory 'steps' requiring temporary attachment of the head region while the haptor is detached and relocated (see above and Fig. 1). Locomotion promotes cross insemination and may permit parasites to escape the consequences of localized immune responses generated by the host's skin (see below), but the brief period during which the haptor is detached and moved to a new site is potentially dangerous. This is the time when the parasite is most likely to be dislodged by water currents and there must have been strong selection pressure for the parasite to shorten this vulnerable period and to ensure that temporary attachment to the host is secure enough to resist shear forces.

In view of the wet, slimy and hence slippery surface of living fish skin, it is surprising that most monogeneans employ adhesive secretions to fulfil the critical task of attaching the head region temporarily. Apparently against all odds, these parasites have produced an adhesive that holds the head region firmly in place on a mucus-covered and water-current-swept surface for as long as the haptor is detached. Even more remarkable is the ability of the parasite to sever the adhesive bond instantly, as soon as the haptor is firmly established at its new site. This contrasts with the length of time needed to slowly strip away the sticky haptor of *L. minor* from the hard surface of a host denticle (see above), and indicates that detachment of the anterior adhesive areas of monogeneans is fundamentally different from haptor detachment in microbothriids.

The gland duct openings are concentrated on the anterolateral borders of the head region, but these openings are arranged differently in different kinds of skin parasites. In gyrodactylids, each anterolateral side of the head forms a projecting lobe, which bears on its ventrolateral terminal aspect the single opening of an adhesive sac into which the gland ducts open (Fig. 18A). In *Gyrodactylus*, groups of gland ducts are said to open on the surfaces of papillae within the sac (Kritsky, 1978; El-Naggar, 1992). In the capsalid *Entobdella soleae*, the gland duct openings are permanently exposed on each anterolateral surface of the head region and cover an extensive area (Fig. 18B). The scanning electron microscope has revealed that, in *E. soleae*, each anterolateral sticky pad consists of three separate zones, which are separated by narrow strips of general body tegument (Kearn & Evans-Gowing, 1998). A tripartite arrangement is also encountered in *Acanthocotyle* (see below), and there is a link, perhaps indicating common ancestry, with gill-parasitic dactylogyroidean monogeneans that have three separate eversible sacs on each side of the head (see Kearn, 1998).

In species of *Entobdella*, such as *E. diadema* from cartilaginous flatfishes, the exposed surface of each elongated pad is subdivided by deep clefts to form a 'diadem' of numerous transverse rays (Fig. 18D). These clefts undoubtedly greatly increase the exposed surface area of each pad when compared with the undivided tripartite pads of *E. soleae*, but Whittington & Cribb (1998) found that the gland duct openings were restricted to the exposed surfaces of the rays and did not occur inside the clefts. Hence the significance of the diadems of these ray parasites remains obscure.

In most of the capsalids that parasitize bony fishes (e.g. *Benedenia*, *Neobenedenia*, *Pseudobenedenia*, *Capsala*) and in some that parasitize cartilaginous fishes (e.g. *Trimusculotrema*), the anterolateral borders of the head region are modified to form ovoid or circular, ventrally directed discs (Fig. 18E). These discs are commonly referred to as 'suckers' (see, for example, Jahn & Kuhn, 1932). Whittington & Barton (1990) observed that the anterior discs of *Trimusculotrema uarnaki* readily attach to glass, slide across the glass surface and leave behind no obvious extruded secretion, indicating that attachment is by suction. However, Kearn *et al.* (1992) found that the glandular anterolateral adhesive pads of the oncomiracidium of *Benedenia seriolae* are retained during post-oncomiracidial development of the discs. The discs are well defined at an early stage of development and they continue to grow. The area occupied by each pad also increases as the disc grows. The pads are still recognizable on the ventral surface of the fully developed discs of juveniles (Fig. 18E) and adults, where their microvillous surface reveals their tripartite nature as in *E. soleae*. Whittington & Cribb (1999) found similar tripartite pads on the anterior attachment discs of *Benedenia rohdei* from the gills and *B. lutjani* from the pelvic fins of *Lutjanus carponotatus*.

The continued growth of these pads indicates that, in the adult, they may still have an adhesive function independent of the suctorial function of the muscular

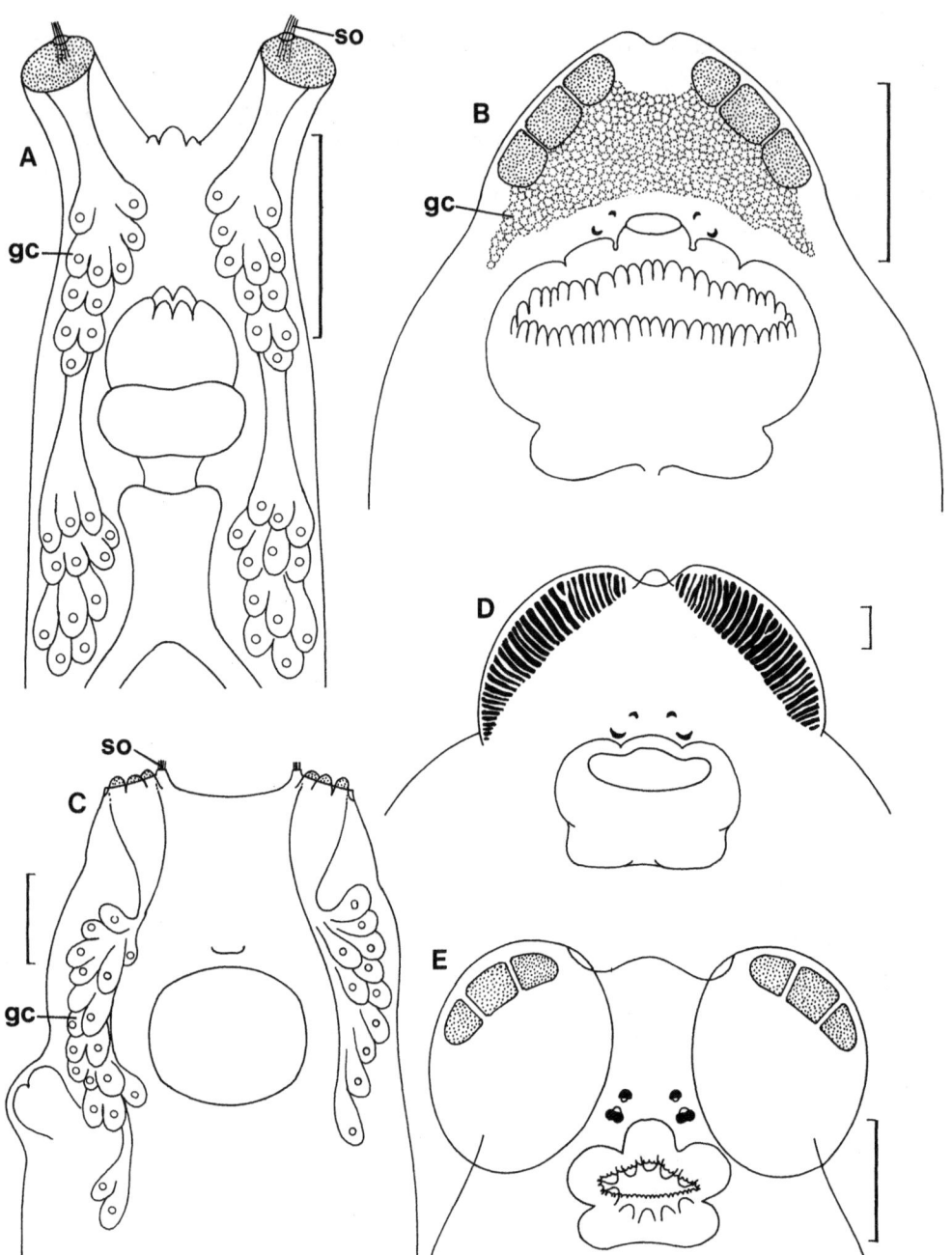

Fig. 18. The anterior adhesive apparatus of some skin-parasitic monogeneans. (A) *Macrogyrodactylus*; (B) *Entobdella soleae*; (C) *Acanthocotyle lobianchi*; (D) *Entobdella diadema*; (E) juvenile *Benedenia seriolae*. Sticky areas are shown stippled in A–C, E and black in D. Scale bars for A, B, D, 250 μm; scale bars for C, E, 100 μm. gc, Gland cells; so, sense organ. Redrawn from the following: (A) El-Naggar & Serag (1987); (B) Kearn (1994); (C) Rees & Kearn (1984); (D) Llewellyn & Euzet (1964); (E) Kearn *et al.* (1992).

disc. This is supported by Whittington & Cribb (1999) who found that the adhesive areas on the attachment discs of *B. rohdei* and *B. lutjani*, like the pads of *E. soleae* (see below), are provided with two secretions, one comprising rod-shaped bodies and the other spherical bodies. Moreover, Whittington & Cribb (1999) observed that the glandular borders of the discs attached independently by cement to a suitable substrate, including a needle, and the bonding was sufficiently robust to resist dislodgement by a jet of water from a pipette. Whether these parasites rely on suction, cement or both may depend on the strength of water currents passing over the body of the host.

How adult monocotylids of the genus *Dendromonocotyle* attach their anterior regions to host skin, if indeed they are able to do so, remains obscure. In *D. taeniurae*, Euzet & Maillard (1967) illustrated what they called a glandular zone, comprising six small and inconspicuous groups of glands on each side of the head region close to its anterior border. However, Olson & Jeffries (1983) did not illustrate these glands in *D. californica* and Chisholm & Whittington (1995) failed to find them in other species of the genus. They found lateral head glands opening on the anterolateral margins of the head region in the oncomiracidium of *D. ardea* (Fig. 9E), but were unable to find anterior gland cells in adults, even in histological sections. They did find circular and radial muscle fibres associated with the mouth and buccal cavity, and it is possible that buccal suction is used for temporary adhesion.

Like the gyrodactylids, *Acanthocotyle lobianchi* also possesses single anterolaterally located adhesive sacs, but the gland ducts opening inside each of them are deployed on the surfaces of three lobes (Fig. 18C). Rees & Kearn (1984) observed the dilatation of the openings of the sacs, followed by eversion of the sacs. The protruding lobes were then applied to the substrate. By manipulating the attached parasite with needles, they observed sticky threads between the adhesive surfaces and the substrate and found no evidence that suction was involved.

Malmberg & Fernholm (1989) described a single pair of adhesive protuberances in the head region of acanthocotylids from hagfishes. In *Myxinidocotyle*, they also described flask-shaped gland cells distributed along the body margins and claimed that these endowed the ventral and the dorsal body margins with adhesive properties. Living specimens were said to attach to host skin, but the function of these adhesive fields remains obscure.

Sensory structures associated with anterior adhesive areas

Conspicuous structures that are probably sensory in function are associated with some of these adhesive areas. Each lateral head lobe of *Gyrodactylus* is equipped with a single terminal retractable 'spike organ' (Fig. 18A) comprising a cluster of individual modified cilia (Lyons, 1969). Multiciliated retractable organs are also associated with the openings of the adhesive sacs of *Acanthocotyle lobianchi* (Fig. 18C; Rees & Kearn, 1984). A circular pit, the rim of which was observed opening and closing, lies close to the anterior border of each attachment disc of *Benedenia seriolae* and may be the site of a similar sensory structure (Kearn *et al.* 1992). The function of these organs is a matter for speculation, but a tangoreceptive role related to locomotion seems likely, although Lyons (1969) did not favour this suggestion in *Gyrodactylus*. She observed that the spike is retracted when the head lobes make contact with the substrate during locomotion, and she pointed out that each spike is surrounded by at least five individual cilia which she believed were more likely to be tangoreceptors. The position of the spikes on the anteriormost tips of the head lobes and the fact that the spikes were seen in extended mode on a stationary parasite, the head region of which was performing 'searching' movements, led her to believe that they were more likely to be chemoreceptors. Gyrodactylids need to be selective when transferring to new hosts (see below) and responsiveness to chemical cues from prospective new hosts may be important.

Function of adhesive pads

The structure and function of the anterior adhesive apparatus of *Entobdella soleae* from the skin of the common sole, *Solea solea*, have been intensively studied (El-Naggar & Kearn, 1983; Kearn & Evans-Gowing, 1998). Two different kinds of secretion are supplied via gland ducts penetrating the special tegument covering the adhesive pads of *E. soleae* (Fig. 19). One of these two secretions is packaged in rods, which usually appear electron-dense in the transmission electron microscope (TEM); the other secretion is packaged in spheroidal bodies and appears finely granular with a low electron density in the TEM. The termination of each of the rod-carrying ducts has many apertures like a pepper pot, each aperture permitting the exit of a single rod, while each of the spheroid-carrying ducts has a single terminal opening, like a salt cellar.

The anterolateral adhesive pads on the head can be attached instantly to host skin during locomotion. According to Kearn & Evans-Gowing (1998), the cement layer is 4 or 5 μm thick and is in intimate contact with both host and parasite surfaces, penetrating between the microridges of the host's epidermal cells (see Fig. 4) and between the microvilli of the parasite's adhesive pads (Fig. 19). Preliminary chemical analysis of the cements of *E. soleae* and other monogeneans indicates that they are novel proteins, displaying some differences between species (Whittington *et al.* 2000). Some component of the secretion extruded by the pads may displace host mucus.

The pads remain firmly cemented to this wet and possibly slimy surface for as long as the haptor is detached, even in the face of strong water currents from a pipette, but, remarkably, the pads can be detached instantly as soon as the haptor is firmly established at its new site.

Kearn & Evans-Gowing (1998) preserved specimens of *E. soleae* at intervals during locomotion and processed these parasites for observation with the TEM and the scanning electron microscope. They found evidence that, immediately prior to attachment of the pads to the skin, both secretions are released and interact to produce the adhesive secretion. The rods provide the bulk of the cement and their persistent boundary membranes may impart added strength to the cement layer.

At the appropriate time, when the haptor is firmly reattached, the anterior adhesive pads are rapidly released. Kearn & Evans-Gowing (1998) showed

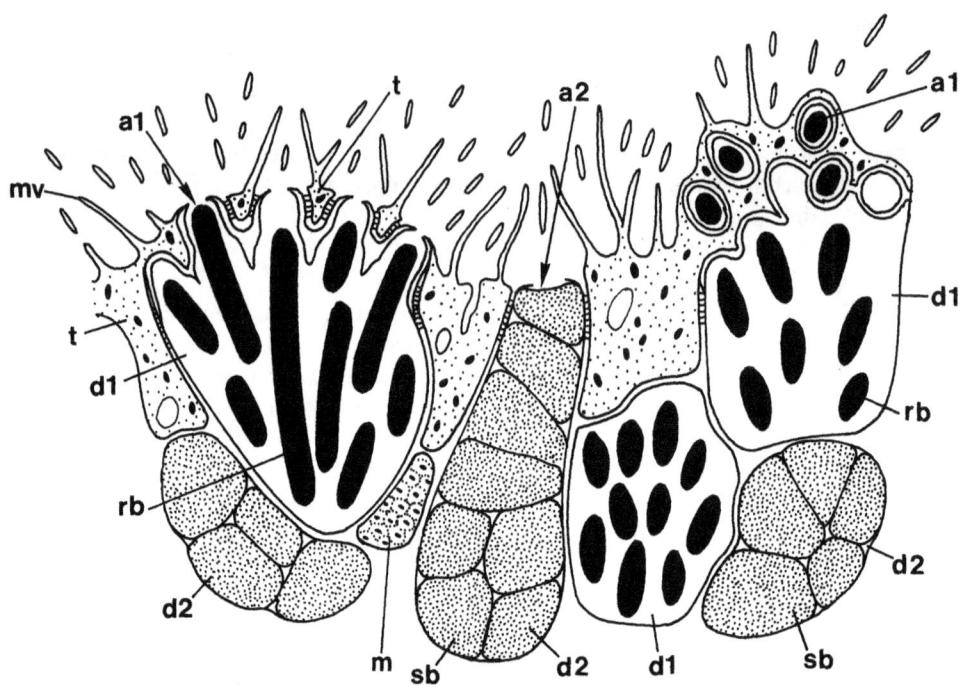

Fig. 19. Diagram drawn from transmission electron microscope sections through the surface of an adhesive pad of *Entobdella soleae*. a1, Aperture of duct ('pepper-pot' duct; d1) carrying rod-shaped secretory bodies (rb); a2, aperture of duct ('salt cellar' duct; d2) carrying spheroidal secretory bodies (sb); m, muscle fibre; mv, microvillus; t, pad tegument. Reproduced from Kearn & Evans-Gowing (1998) with permission from the Australian Society for Parasitology.

that this is unlikely to be achieved by physical force, i.e. by muscle contraction tearing the pads away from the skin. The fact that the parasite leaves a layer of residual cement (a 'pad-print') when it detaches naturally from a glass surface, indicates that release involves the dissolution of a thin layer of cement adjacent to the pad surface, but not dissolution of all of the cement layer. The evidence accumulated by Kearn & Evans-Gowing (1998) has directed attention towards the specialized pad tegument (Fig. 19) as the possible instrument of detachment. This pad tegument is separated from the general tegumentary syncytium covering the body of the parasite by a cell boundary and has a denser covering of branched and unbranched microvilli. The pad tegument covers the whole surface of the pad except for the duct openings and even surrounds each individual opening of the pepper pot ducts (Fig. 19). The microvillous surface of this network of pad tegument is extensively and intimately in contact with the cement. It is therefore in an ideal position to break the bond instantaneously by a physical or chemical change mediated via the tegumentary apical membrane. The pad tegument contains secretory bodies that could by exocytosis be the instruments of this change.

The anterior adhesive areas of *Acanthocotyle lobianchi*, like those of *E. soleae*, are supplied with two different secretions, one of which comprises rod-shaped bodies and the other granules (Rees & Kearn, 1984). On the other hand, three different secretions are reported to be associated with the anterior adhesive sacs of gyrodactylids (Kritsky, 1978; El-Naggar & Serag, 1987), although according to El-Naggar & Serag (1987) two of these secretions in *Macrogyrodactylus clarii* are produced in a single type of gland cell. Furthermore, *Monocotyle spiremae*, a monocotylid parasite from the gill chamber of a ray, *Himantura fai*, is reported to have only one type of anterior secretion (Cribb, Whittington & Chisholm, 1997). How these triple- and single-secretion systems achieve attachment and detachment remains to be determined.

Feeding

Feeding has been studied in relatively few skin parasites. In those in which it has been observed (e.g. *E. soleae*, *A. lobianchi*, see Kearn (1963c) and *L. minor*, see Kearn (1965)), attachment to the host's epidermis is achieved by the everted pharynx, apparently without the participation of the anterior attachment areas. The everted pharynx remains in contact with host epidermis for as little as 20 s in some specimens of *L. minor* (see Kearn (1965)) and for as long as 8 min in some specimens of *E. soleae* (see Kearn (1963c)). It probably maintains its position by suction, but the production of a complementary adhesive or sealant cannot be ruled out. However, the demands on the pharynx as an attachment organ are likely to be less than those imposed on the anterior attachment areas during locomotion, since the haptor is attached throughout a feeding episode.

In *E. soleae*, parasites separated from the host and allowed to reattach themselves by the haptor to a glass surface will sometimes evert the pharynx and attach it to this inert surface (Kearn, 1998). This indicates that no positive chemical stimulus from the host's skin is required to initiate a feeding episode.

Parasite/host contact of the most intimate kind occurs between the haptor and the host's epidermis. It is worth asking whether this is another potential route for the entry of nutrients into the parasite's body. This has not been directly explored experimentally, although Halton (1978), based on an *in vitro* study, claimed that the gill-parasitic monogenean *Diclidophora merlangi* has a tegumental transport system capable of absorbing certain neutral amino acids from seawater. Absorption of nutrients from host epidermis might be significant in monogeneans such as *E. soleae* that have extensive haptoral contact with the host and lack haptoral gut branches. If nutrient uptake via the haptor tegument does not take place then nutrients must diffuse over considerable distances through the dense parenchyma: from the gut terminations in the body through the narrow peduncle and out to the marginal valve, which is a living structure.

SUSCEPTIBILITY AND DEFENSIBILITY OF FISH SKIN

The eel *Anguilla anguilla* is notoriously free of monogenean skin parasites (Buchmann, 1988 in Bakke *et al.* 1992), although several species infect the gills, and this raises the question of whether the eel has an innate resistance to the establishment of monogenean skin parasites. However, according to Bakke, Jansen & Hansen (1991), *Gyrodactylus salaris*, a skin parasite of the salmon, *Salmo salar*, is capable of attaching itself to eel skin, although it remains attached for no more than 8 days, a period which is too short to permit reproduction. The reason why the duration of attachment of *G. salaris* is limited is unknown, but it has been suggested by Bakke *et al.* (1991) that the parasite is unable to feed on eel skin. In the context of innate resistance it is also worth recalling that agnathan fishes were regarded as free of monogenean skin parasites until the relatively recent discovery by Malmberg & Fernholm (1989) of three acanthocotylid monogeneans on the skin of hagfishes.

Most monogeneans display some degree of host specificity, which means that their range of hosts is narrow, often restricted to one or two closely related fishes, even though apparently suitable but unrelated hosts may be available in the natural environment. The reasons for this host restriction are largely unknown, although a few clues have been obtained from studies of *Entobdella soleae*.

E. soleae is a common skin parasite of the Dover sole, *Solea solea*. It has also been reported from the closely related soles *S. lascaris* and *S. senegalensis* (Sproston, 1946; Carvalho-Varela & Cunha-Ferreira, 1987, respectively). Kearn (1967b) examined several hundred soleid, pleuronectid, bothid and cartilaginous flatfishes at Plymouth (UK) over a period of 6 years, and collected many hundreds of adult specimens of *E. soleae*. With the exception of two specimens each found on the skin of a ray (*Raja montagui*), all of these parasites came from *Solea solea*. Since these rays had been brought up from the sea bottom in a tightly packed trawl together with infected soles, it is possible that the parasites transferred themselves from the soles to the rays.

Experimentally-detached adult specimens of *E. soleae* readily reattached when placed on another sole (Kearn, 1967b) and fish-to-fish transfers occur in aquaria (and probably also in the sea) when soles make contact (Kearn, 1988b). Experimentally-detached adult parasites also reattached to rays and remained attached for 2–8 days, demonstrating that the adult haptor of *E. soleae* functions well on ray skin, at least for a time. However, when rays were kept for about 2 months in the same tank with infected soles, where they would be exposed to free-swimming oncomiracidia as well as to adults, they acquired only one adult parasite and no larvae (Kearn, 1988b). Thus, in spite of the fact that the adult haptor functions well, there seems little incentive for adult parasites or oncomiracidia to establish themselves on ray skin. Adult parasites attached to the bottom of a glass or plastic container remain attached by the haptor for approximately the same time as adult parasites on ray skin. This suggests that the parasite is unable to feed on ray skin with the consequence that, ultimately, the haptor is weakened by starvation and becomes detached.

The haptor of *E. soleae* functioned initially when adult parasites were experimentally transferred from *S. solea* to another soleid fish, *Buglossidium luteum* and to a pleuronectid flatfish, *Pleuronectes platessa*, but these parasites became detached within 24 h (Kearn, 1967b). Perhaps there is a chemical or physical reason why the haptor of *E. soleae* fails so quickly on the skin of these alien hosts. Curiously, *B. luteum* acquires some immature parasites when confined for long periods in the same tank as infected soles, suggesting that the fish is not unattractive to oncomiracidia and that young parasites do not have the same attachment difficulty as adults. It is interesting that these larvae do not appear to reach sexual maturity on *B. luteum* (see Kearn, 1967b).

Thus, there appears to be no correlation between the ability of adult parasites to survive on alien hosts and the response of oncomiracidia to these fishes. On the one hand, *B. luteum* has some attraction for oncomiracidia, but the fish is not tolerated by, or does not tolerate, adult parasites. On the other hand, oncomiracidia show no apparent interest in rays while the adults attach to them readily and remain attached until weakened by their inability to feed. Clearly the haptor of *E. soleae* is not so highly modified that it will only operate on the skin of *Solea* spp. and it will function, at least for a time, on the skin of pleuronectid fishes, rays and even on glass.

So, host specificity seems to be created by a complex interplay of factors which include as yet unknown properties of fish skin that influence the response of oncomiracidia and whether the haptor and/or feeding apparatus are able to function.

Host specificity does not always correspond with host phylogenetic relationships. *Gyrodactylus salaris*, for example, establishes itself on salmon (*Salmo salar*), on species of the genus *Salvelinus* and on *Salmo trutta*, but reproduces effectively only on salmon and on *Salvelinus* (see Bakke, Jansen & Harris, 1996 and references therein). Such inconsistencies may be more dramatic. For example, the squid *Alloteuthis subulata* is parasitized by gyrodactylid monogeneans of the genus *Isancistrum* (see Llewellyn, 1984). Other gyrodactylids are parasites of fishes with a few on amphibians, so their presence on an invertebrate undoubtedly represents a host-switching event. This takes place when a parasite finds itself on the skin of an unrelated host that by chance provides all the right chemical and/or physical signals to permit establishment of the parasite.

The narrow host specificity of *E. soleae*, encompassing only species of the genus *Solea*, contrasts sharply with the remarkably broad specificity displayed by another morphologically similar and closely related capsalid, *Neobenedenia melleni* (originally called *Epibdella melleni*). Jahn & Kuhn (1932) found that 47 species of bony fish belonging to 17 families were susceptible to infection with this parasite in the New York Aquarium. Sixty years later the number of fish species from which *N. melleni* has been recorded, including records from freshly caught as well as from aquarium hosts, had risen to over 100 (Whittington & Horton, 1996), these fishes belonging to 30 families of teleosts from five different orders. The factors that permit *N. melleni* to parasitize such a wide range of hosts are no better understood than those that are responsible for the restricted host range of *E. soleae*. However, even this broad specificity of *N. melleni* appears to have its limits. Cartilaginous fishes (various sharks and rays) in the New York Aquarium did not become infected in spite of continued exposure to infection (Jahn & Kuhn, 1932), and all the susceptible fishes listed by Whittington & Horton (1996) are acanthopterygian (percomorphan) teleosts, with the exception of one record from a siluriform fish.

The many teleost fishes that are susceptible to *N. melleni* are not entirely at the mercy of this versatile parasite. Jahn & Kuhn (1932) and Nigrelli & Breder (1934) were the first to recognize that the skin of some fishes maintained in the New York Aquarium is able to mount a defence against monogenean parasites. Nigrelli & Breder (1934) found that the degree of resistance that develops varies greatly between members of the same family of fishes. Some fishes acquired partial resistance, some acquired a total resistance lasting for long periods and some fishes were always susceptible.

Neobenedenia girellae also has many teleost hosts (Ogawa et al. 1995), and the relationship between this parasite and the marine Japanese flounder (*Paralichthys olivaceus*) has been studied experimentally by Bondad-Reantaso et al. (1995). A 10-day-old primary infection was removed from the host by immersion in fresh water, which is lethal to the parasites. These primed fishes were then exposed to oncomiracidia. Parasites collected after 10 days from these secondarily infected hosts were significantly fewer and smaller than those from control fishes. Bondad-Reantaso et al. (1995) showed that there was no significant difference in serum antibody levels between primed and control fishes. They also collected parasites from two groups of fishes, both of which had been challenged with oncomiracidia after intra-peritoneal injections, but one group had received antigen prepared from sonicated *N. girellae* and the other phosphate-buffered saline. Antibody levels in the sera of the antigen-injected fishes were significantly higher than in the sera from buffer-injected fishes, but no significant difference was detected between numbers of parasites of the challenge infection collected from fishes of these two groups. Thus, limited acquired protection against *N. girellae* seems to develop in the Japanese flounder, but serum antibody apparently plays no part in this immunity. This supports the notion that there is some kind of defence mechanism localized in the skin (see above and also below).

Studies of gyrodactylid monogeneans reveal more about the responses of fish skin to metazoan skin parasites. This is because of the unique reproductive biology of gyrodactylids which often leads to rapid generation of an anti-parasite response (see Kearn, 1998). Most gyrodactylids are viviparous, retaining an embryo inside the body until it is fully developed (Fig. 10A). After birth this embryo establishes itself on the same host individual as the parent (autoinfection). Because the parent is unable to nurture another embryo until the first is born, this greatly reduces fecundity, but some degree of compensation is achieved by permitting a second embryo to develop inside the first and often a third embryo inside the second. Thus, when the first embryo is born it already contains partly developed smaller embryos, one inside the other. Since the parasite no longer faces the hazards associated with host finding by free-swimming oncomiracidia, the infrapopulation increases rapidly and many hosts respond to this challenge by generating a response that leads to loss of parasites. Parasites spread to new hosts when hosts make contact with each other or by using the bottom as a staging post.

Typically, primary infections with gyrodactylids show an initial increase followed by a decline to extinction (see, for example, Scott, 1985), although in some circumstances a low-intensity infection may persist (see, for example, Richards & Chubb, 1998). In juvenile guppies infected with *Gyrodactylus turnbulli*, there may be considerable resistance to a challenge infection given after the decline of the primary and several weeks may pass before full susceptibility to reinfection is restored (Scott, 1985; Scott misidentified *G. turnbulli* as *G. bullatarudis* according to Harris, 1986). This pattern is consistent with the expression of a response generated by the primary infection.

The way in which gyrodactylids are eliminated from fish skin remains unclear. Lester (1972) claimed that *G. alexanderi* are carried away from their stickleback host on rafts of solidified mucus or 'cuticle' (glycocalyx) which are secreted and shed by the host's skin at one- or two-day intervals. However, Scott & Anderson (1984) did not observe continual shedding of cuticle by guppies infected with *G. bullatarudis* (incorrectly identified as *G. turnbulli*; see Harris, 1986). They questioned the role of this phenomenon in the control of parasite numbers but were unable to offer an alternative explanation for parasite losses. Richards & Chubb (1996) observed no cuticular shedding in guppies infected with *G. bullatarudis* or *G. turnbulli*.

As long ago as 1937, Nigrelli found that *Neobenedenia* (= *Epibdella*) *melleni* survived for different lengths of time in skin mucus from host species with different susceptibilities to the parasite. If, as this observation indicates, skin mucus has a central role in host defences against skin parasites, then susceptibility or resistance of hosts might be reflected in parameters such as densities of epidermal mucous cells (goblet cells) and thickness of the epidermis. There is also the possibility that secretions from other sources in the skin (epithelial cells, sacciform cells, club cells; see above) may be concerned with defence against parasites.

Evidence is accumulating to indicate that goblet mucous cell density falls in susceptible fishes (Cusack & Cone, 1986; Wells & Cone, 1990; Sterud, Harris & Bakke, 1998). The study of Sterud *et al.* (1998) is particularly revealing since they focused on the effect of *Gyrodactylus salaris* on epidermal structure and thickness in brook trout (*Salvelinus fontinalis*) and in Norwegian Atlantic salmon (*Salmo salar*). Typically, brook trout respond to and eliminate the parasite while Norwegian Atlantic salmon are unable to control the increase in numbers of *G. salaris* and eventually die. Skin samples from brook trout were taken at a stage when the hosts had responded to and almost eliminated the parasites and this revealed a slight increase in epidermal thickness but no change in mucous cell density. Sterud *et al.* (1998) explained these changes by suggesting that an increase in host epidermal cell proliferation occurred while the parasite population was high. They suggested that this high proliferation rate was maintained after decline of the parasite population and the corresponding reduction in grazing pressure on epidermal cells, leading to an increase in epidermal thickness. In contrast, susceptible salmon showed a general reduction in mucous cell density and a decrease in epidermal thickness during the increasing phase of the infection. These changes may simply reflect intense grazing pressure by the parasites as their numbers increase, leading to a reduction in overall numbers of maturing epidermal and mucous cells. On the other hand, they may be related in some way to the inability of salmon to respond to the parasite. One possibility is that the parasite itself may inhibit the production or differentiation of new mucous cells.

If these arguments are on the right lines, then one would expect high mucous cell densities to be associated with resistance of fishes to gyrodactylids. Buchmann & Uldal (1997) found a positive correlation between epidermal mucous cell density in various salmonids and resistance to *G. derjavini* and, according to Buchmann & Bresciani (1998), during the later phases of infection this parasite avoids areas of the host with high densities of mucous cells. These observations support the notion that goblet mucous cells play an essential part in the host response to gyrodactylids.

Sterud *et al.* (1998) made the interesting point that resistance against skin parasites might be akin to that of the intestinal mucosa of rats against the nematode *Nippostrongylus brasiliensis*. In the rat, changes in binding of lectins to goblet mucous cells reflect changes in the specific carbohydrate moieties of the mucus, which may be related to nematode expulsion (Ishikawa, Horii & Nawa, 1993). According to Fujino *et al.* (1998), rejection by the mouse of the gut-parasitic digenean *Echinostoma trivolvis* corresponds with a peak in intestinal goblet cell hyperplasia and they regarded goblet cells as the instruments of rejection, not the immune system.

Moore, Kaattari & Olson (1994) found evidence to suggest that both serum and skin mucus from the flatfish *Pleuronectes vetulus* contained factors involved in resistance to *Gyrodactylus stellatus*. Survival of parasites was generally shorter in serum and mucus samples from flatfishes at the later recovery stages of infection. They also found that serum from fishes which had experienced intraperitoneal injection with a preparation of whole, killed *G. stellatus* had a significant effect on parasite survival, while mucus from these fishes had no such effect.

Several observations point to the lack of specificity of the anti-parasite response. Richards & Chubb (1996) reported that the primary response of guppies provided some protection against challenge from *G. bullatarudis* as well as from *G. turnbulli*, irrespective of which of these two parasites was responsible for the initial infection. Harris, Soleng & Bakke (1998) found that *G. decorus*, a parasite of roach (*Rutilus rutilus*) that is not closely related to *G. salaris*, is also susceptible to salmon serum and that serum and mucus from trout (*Salmo trutta*) is as effective as salmon serum and mucus in killing *G. salaris*.

There are no indications at present that specific immunoglobulins play any part in the anti-parasite response (Buchmann, 1999). According to Buchmann (1998) rainbow trout immunoglobulins from infected fishes did not bind to the tegument of *G. derjavini*. However, Harris, Soleng & Bakke (2000) demonstrated a higher susceptibility to *G. salaris* in salmonids treated with the immunosuppressant hydrocortisone acetate, indicating that the immune system has some part to play in the control of this parasite.

On some hosts, *N. melleni* appears to have a preference for attachment to the eyes (Jahn & Kuhn, 1932; Nigrelli & Breder, 1934). This led Llewellyn (1957) to suggest that this site preference might be related to the lack of vascularization of the cornea, thereby creating a refuge for parasites from blood-borne antibodies (see also below). However, the apparent lack of anti-parasite activity by the host cornea may have more to do with lack of mucous cells than lack of blood vessels. As Buchmann (1999) has pointed out, host blood is much closer to the surface in fish gill lamellae and antibodies may be more effective against gill parasites.

In the search for a mechanism for control of fish skin parasites attention has now turned to non-specific humoral factors. Several such factors have been detected in fish skin, such as lectins, lysozyme, proteases, C-reactive protein, haemolysins, peptides and complement (see Yano, 1996; Buchmann, 1999). Peptides deserve a special mention since monogeneans possess peptidergic innervation and peptides may influence parasite behaviour (see Buchmann, 1999). However, there is no evidence as yet that any of these substances, with the exception of complement, has an anti-parasite role (Buchmann, 1999). In fact, Harris *et al.* (1998) found no residual activity against *G. salaris* in serum from *Salmo salar* that had been heat- or EDTA-treated. Since these treatments are known to inactivate complement it suggests that other non-complement mechanisms such as those listed above, are not involved.

Two independent studies, one by Buchmann (1998), working with *G. derjavini* on rainbow trout (*Oncorhynchus mykiss*), and another by Harris *et al.* (1998), working with *G. salaris* on Norwegian Atlantic salmon (*Salmo salar*), have established the anti-parasite effectiveness of complement *in vitro*. Their conclusions are similar: that these gyrodactylids are highly susceptible to killing by host serum, that serum from uninfected fishes is as effective as serum from infected fishes and that heat-treated serum is no longer lethal to the parasites. Other evidence collected in both studies indicated that the effect on gyrodactylids is antibody-independent and is compatible with activation of the alternative complement pathway rather than the classical pathway. Harris *et al.* (1998) found that skin mucus from salmon is also effective against *G. salaris*, but less so than host serum, the former having about 1/20th of the potency of the latter.

Complement is said to be present in fish mucus (Sakai, 1992) and Harris *et al.* (1998) tentatively suggested that host attack might be focused on the haptor. They speculated that the great mobility of gyrodactylids might be useful in this situation, since the parasites might limit exposure to complement and curtail further damage by changing hosts or taking refuge temporarily on the substrate. If the host attack is localized, as some of the evidence cited above suggests, parasites might also escape by changing location on the same host (see also below). It is particularly interesting in this context that Buchmann (1998), using an immunocytochemical assay, found that complement factor C3 binds directly to the hamulus sheath of *G. derjavini*, as well as to the openings and glands of the anterior adhesive system and to the body surface of the parasite. He suggested that some carbohydrate epitopes on the parasite are involved in C3 complement activation in this *in vitro* system.

Buchmann & Bresciani (1999) have established that macrophages from rainbow trout (*Oncorhynchus mykiss*) attach themselves to the surface of living *Gyrodactylus derjavini*, and especially to the openings of the anterior adhesive system. They claimed that the adverse effect of the activated macrophages towards the parasites was associated with a heat-labile component, which may be complement factor C3 released by the macrophages. They suggested that macrophages in fish skin may contribute to elevated levels of complement in trout skin during the host response and that interleukin (Il-1), also released by the macrophages, might play an important part in the host response by promoting hyperplasia of host epidermis and enhanced mucus secretion. However, as Buchmann & Bresciani (1999) pointed out, this is *in vitro* work and there is, as yet, no evidence that macrophages bind to the parasites *in vivo*.

It also remains to be seen whether complement-mediated killing is significant *in vivo*. In spite of the effectiveness of complement-mediated killing of *G.*

salaris by salmon skin mucus *in vitro*, some strains of Norwegian Atlantic salmon are unable to limit population growth of *G. salaris* and are subsequently killed by the infection, often in conjunction with a fungal infection (*Saprolegnia*). There is no adequate explanation for this paradox at the present time.

Recent immunological studies suggest that the skin of fishes has a secretory defence system independent of the systemic system and that the route of administration of the antigen may determine whether or not the skin defence system responds. Confirmation of this has come from the work of Moore et al. (1994) and Bondad-Reantaso et al. (1995) on *Gyrodactylus stellaris* and on *Neobenedenia girellae*, respectively, and strongly indicates that multicellular parasites like monogeneans, as well as micro-organisms, are vulnerable to this secretory system. There are also indications that, as far as monogeneans are concerned, the secretory defence response of the skin may be localized. Nigrelli & Breder (1934) noted that when a moonfish (*Vomer setapinnis*) infected with *N. melleni* was chemically deparasitized and then reinfected, the new parasites occupied a different site on the host and did not return to the site of the original infection. Richards & Chubb (1996) found evidence to suggest that the host response to gyrodactylid monogeneans by guppies is largely limited to areas of heavy infection. The possible influence of this localized response on the distribution of monogeneans on the host will be considered below.

SITE SPECIFICITY AND MIGRATION

Few monogenean skin parasites appear to be randomly distributed on their hosts, although it is only recently that the extent of site specificity has been recognized (see Whittington, 1996). Adult specimens of the capsalid *Entobdella soleae* are resident on the lower surface of the common sole *Solea solea*, where they have a scattered distribution (Kearn, 1998). However, adult parasites are capable of attachment to the upper surface of the sole and probably do so after transfer from another sole. Most newly invaded oncomiracidia also alight on the upper surface. None of these invaders, larvae or adults, remain for more than a few days on the upper surface, migrating forwards on the host and eventually reaching the lower surface. Nevertheless, the area of the host available to *E. soleae* for permanent residence is large compared with some of its capsalid relatives. For example, Whittington & Kearn (1993) found that *Benedenia lutjani* prefers the pelvic fins of its round-bodied host *Lutjanus carponotatus* and Horton & Whittington (1994) found *Metabenedeniella parva* only on the dorsal fin of *Diagramma pictum* (now *D. labiosum*).

Site preferences also occur in gyrodactylids and these may change during the course of an infection. Newly acquired infections of the guppy, *Poecilia reticulata*, with *Gyrodactylus turnbulli* occur predominantly on the fins, which are likely to be the main points of contact between hosts and hence the most likely route of invasion (Harris, 1988). As the infrapopulation increases there is evidence of a migration from the fins to the caudal peduncle and then, in the declining phase of the infection, a return migration to the fins, perhaps as a prelude to transfer to new hosts. *G. derjavini* has a similar pattern of migration, initially infecting the pectoral, pelvic and anal fins of rainbow trout, *Oncorhynchus mykiss*, and later concentrating on the cornea and on the caudal fin (Buchmann & Bresciani, 1998).

Site preferences may have several possible advantages for monogenean skin parasites. Kearn (1998) suggested that *E. soleae* may be less vulnerable to predators on the lower surface (see also below) and would be more likely to find a mating partner with the breeding population restricted to the lower surface rather than scattered over the whole fish. Moreover, a pheromonal system for finding a mating partner, if such a system exists, is more likely to be effective on the sheltered underside of the fish than on its exposed upper surface, and eggs can be attached more readily to sand ballast by adults on the lower surface. According to Kearn (1998), adult parasites readily feed on the upper skin, so there is no obvious nutritional reason for the restriction to the lower surface. It is not known whether there is a difference between the upper and lower skin of the common sole in the anti-parasite defences that these surfaces can mobilize, but the work of Buchmann & Bresciani (1998) on *G. derjavini* focuses attention on this phenomenon as a possible reason for site-specificity. They observed that the cornea and caudal fins of rainbow trout have relatively few or no mucous cells and Buchmann (1997) regarded these sites as a safer haven from the marked host response.

How gyrodactylids find their way to these specific sites is unknown, but there is evidence that *E. soleae* uses its tactile sense to identify the orientation of the host's scales, which project backwards above the general skin surface in the common sole (Kearn, 1998). The parasite is thought to be able to respond to these 'signposts' and find its way forwards on the host.

PREDATION

The preference of adult specimens of *Entobdella soleae* for the lower surface of their bottom-dwelling flatfish host (*Solea solea*) (see above) may be a means of avoiding predation. When the host is resting or buried, parasites on the lower surface would be inaccessible to small, free-swimming predators such as fishes or crustaceans, while white parasites on the exposed upper surface of the host would be visible and vulnerable against the host's dark skin. However, there is no evidence to support this explanation for site preference in *E. soleae* and no potential 'cleaner' organism has been identified in association with soles. Nevertheless, cleaner organisms are well known in coral reef communities and may be present, but unrecorded, in colder temperate habitats less attractive to divers (see Kearn, 1998, p. 101).

Round-bodied fishes do not provide their skin parasites with the opportunity for concealment that is provided by the blind sides of flatfishes. However, Whittington (1996) made the interesting point that some sites on the body of a parasitized fish (e.g. some fins, some regions of the gills) may receive less attention from cleaner organisms than others. Thus, foraging intensity by cleaner organisms may be one of the factors leading to site-specificity in monogenean skin parasites.

Some parasites of round-bodied hosts have features that may reduce their conspicuousness on the fish's exposed flanks. Whittington (1996) has reviewed these features and their possible role in parasite camouflage. They include pigmentation, which may be coloured as well as black (e.g. in the capsalids *Capsala martinieri* and various species of *Benedenia*, the monocotylid *Dendromonocotyle kuhlii*, the microbothriid *Pseudoleptobothrium aptychotremae* and *Anoplodiscus* spp.) and body transparency (e.g. in *Benedenia seriolae* and *Anoplodiscus* spp.). There is evidence that the kind of selection pressure exists that might generate such adaptations. Sclerites from the haptors of monogenean skin parasites have been found in the guts of aquarium-held cleaner fishes and inside cleaners taken directly from the field (Deveney, Whittington & Grutter, personal communication). Moreover, it has been demonstrated that in a coral reef environment pigmented benedeniine species are less likely to be eaten by cleaner fish than unpigmented species (Deveney, Whittington & Grutter, personal communication).

SUMMARY AND CONCLUSIONS

Llewellyn (1982) believed that the major groups of monogeneans were already established as parasites of ancient fishes in Palaeozoic times, before the modern lineages of fish-like vertebrates had emerged. If this is so then monogeneans have been 'fine-tuned' by an enormously long period of natural selection and it is not surprising that they are probably the most diverse and highly specialized modern group of fish skin parasites. Of special interest is their exploitation of proteinaceous hooks embedded in a posterior haptor for attachment to fish skin. They have adopted a two-tier system of hook attachment, taking advantage of the keratinous tonofilaments in the superficial cells of the fish epidermis and the collagenous matrix of the deeper dermis for anchorage. For attachment to the former, skin-parasitic monogeneans typically utilize 14 or 16 tiny marginal hooklets with blades that pierce the apical cell membrane and are embedded in the terminal web of tonofilaments beneath it. Much larger hooks or hamuli (one or two pairs) may penetrate through the epidermis and its collagenous basement membrane into the fibrous dermis beneath. Marginal hooklets are used essentially for attachment of larvae, although some small adults also rely on them. Larger post-larvae and adult monogeneans employ hamuli.

The haptors of many large monogeneans generate suction. Some, like *Entobdella soleae*, achieve this in two ways, by relative movements of haptoral sclerites (hamuli and persistently growing marginal hooklets – accessory sclerites) or by means of intrinsic haptor musculature. Dermal scales contribute to the efficiency of the former in *E. soleae*. There is still much to learn about the relative contributions of these suctorial systems, particularly in relation to changes in the strength of water currents generated by the host's activities.

Some skin parasites have developed the intrinsic suctorial capability of the haptor at the expense of the sclerites, which may be reduced or entirely lost. Others compensate for reduction of sclerites by exploiting adhesive secretions for attachment, which may be an energetically cheaper option than generating suction.

Neocalceostomatoides adheres to living skin within the buccal and gill cavities and the adhesive secretion seems to be the modified glycocalyx of the ventral haptor tegument. *Anoplodiscus* produces a similar tegumentary adhesive secretion but is exceptional in cementing itself to the epidermal basement membrane, which it exposes by eroding the epidermis in some unknown way. The microbothriids and the udonellids use cement for attachment to hard surfaces (the denticles of cartilaginous fishes and copepod carapaces, respectively).

Hamuli are absent in acanthocotylids. It is not clear whether they have never been present or whether they have been lost. Marginal hooklets serve for attachment of the oncomiracidium and the haptor carrying these hooklets is retained without further growth in the adult, where it is dwarfed by a new attachment organ or pseudohaptor derived from the posterior region of the body. Suction seems to make a major contribution to the attachment of the pseudohaptor.

Adhesive secretions are also used by many skin-parasitic monogeneans for temporary attachment of the anterior end during leech-like locomotion. This is a remarkable system permitting the parasite to attach the head region to wet slimy fish skin with sufficient strength to resist being detached by water currents during the brief period during which the haptor is separated from the host. Moreover, detachment of the head is instantaneous when the haptor is reattached at the end of the locomotory step. Little is known about the chemical nature of these remarkable adhesives, but the mechanism of attachment and detachment has been studied in one skin parasite, *E. soleae*. Apparently, the mixing together of two separate secretions, reminiscent of the two interactive components of commercial epoxy resins, generates stickiness. How these secretions interact with host skin mucus, if indeed they do, is unknown, but there has been speculation that such an interaction may contribute to host recognition and host specificity (Whittington *et al.* 2000). The evidence suggests that the tegument of the adhesive pad brings about instant detachment of *E. soleae*, but precisely how this is achieved is not known. This provides a clue as to how unsticking might occur in parasites such as *Monocotyle ijimae* that have only one adhesive secretion.

The anterior adhesive areas of some capsalid skin parasites have adopted a disc-shape with suctorial potential, but in those that have been studied, this seems to be in addition to, rather than at the expense of, their cement-secreting ability.

Unknown properties of fish skin determine its attractiveness for oncomiracidia and its suitability as an attachment platform and as a source of food for skin-parasitic monogeneans. The properties offered by a potential host may not suit all stages of the life cycle of a particular parasite, being sometimes compatible with larvae and not adults and vice versa. It is these properties that contribute to the width of host specificity, which may range from one or two species of the same genus (as in *Entobdella soleae*) to representatives of several fish families (as in *Neobenedenia melleni*). On the basis that related hosts might have a related suite of properties, host specificity usually follows host phylogeny. However, these properties may by chance be shared with a host that is distantly related, on rare occasions even by a member of another phylum, and this may permit host switching and establishment of the parasite on an unrelated host.

Skin-parasitic monogeneans, especially gyrodactylids, have proved to be excellent models for the study of the defensive responses of fish skin against multicellular skin parasites. That the skin is capable of mounting such a response has been clearly demonstrated. Moreover, there is evidence of enhancement of this response following an initial infection of a naïve host, leading to speedier rejection of a challenge infection. This pattern is reminiscent of the familiar blood-mediated response to foreign antigens of the vertebrate immune system and increased susceptibility of immunosuppressed salmonids to *Gyrodactylus salaris* implicates the host immune system in anti-parasite responses. However, so far, there is no strong evidence to implicate immunoglobulins in the rejection of monogeneans. In fact, the anti-parasite response of fish skin may be non-specific and possibly localized to areas of the body where parasites occur.

Fish skin secretions are the obvious vehicles for chemical agents capable of dislodging multicellular skin parasites, and mucous (goblet) cells appear to have a central role in this process. There appears to be a parallel here with the rejection of some platyhelminth parasites by the intestinal mucosa. The mechanism of the enhanced response of fish skin to challenge infection by monogeneans has not been explained, but changes in the abundance of mucous cells and possibly in the carbohydrate moieties of the mucus itself may be involved. It is interesting that microbothriids and udonellids escape these problems by attaching themselves semi-permanently to a hard, non-living surface, namely host denticles and the carapace of a copepod, respectively.

It has been demonstrated *in vitro* that serum complement, activated via the alternative pathway, is an effective tool for killing skin-parasitic monogeneans. Since complement is present in fish skin mucus, contacts between this non-specific agent and parasites living on the skin surface are likely to occur. However, complement in skin mucus is less effective as a killing agent *in vitro* than serum complement, and there is no evidence that complement is effective *in vivo*. In fact, paradoxically, although *Gyrodactylus salaris* is vulnerable to attack by activated complement in salmon serum and mucus *in vitro*, *in vivo* the parasite multiplies unchecked and often overwhelms the host.

Regional variation in the density of mucous cells and corresponding regional variation in the strength of the defensive response provide one possible explanation for the preferences of monogeneans for specific sites on the host. These sites may be extensive in area (the left – lower – surface of the sole infected with *E. soleae*) or highly restricted (the dorsal fin infected with *Metabenedeniella parva*). Apart from evidence that *E. soleae* uses the host's projecting scales as tactile signposts, nothing is known about how the parasites find their way to these preferred sites. Regional changes in mucous cell density during the course of an infection of a single host may also provide the driving force for temporal changes in preferred site.

The influence on skin parasites of external predation by 'cleaner' organisms may have been underestimated. There is evidence that some monogeneans are eaten by cleaners and some skin parasites appear to have responded to such pressures by acquiring camouflage (pigmentation) and/or body translucency and possibly by seeking out sites on the host that are less frequently searched by cleaners.

In conclusion, survival of semi-permanent metazoan parasites like monogeneans on the exposed body surface of fishes is greatly influenced by the microenvironment created by the host skin and especially by the epidermis. However, in spite of recent research on gyrodactylids and capsalids we still have no clear view of how fishes defend themselves against skin-parasitic monogeneans. Until we improve this understanding a thorough appreciation of the intriguing phenomenon of host specificity is likely to remain elusive. There may also be influences from the external environment, particularly the possibility of selection pressure exerted by 'cleaner' organisms. We are beginning to appreciate that selection pressure from predation is a significant evolutionary force in coral reef communities, but more research is needed to determine how widespread its influence is. A deeper understanding of how skin-parasitic monogeneans survive on the exposed and often hostile body surface of fishes may throw light on the factors that have lead to the colonization by monogeneans of internal sites such as gills, nasal fossae and cloacae.

REFERENCES

BAKKE, T. A., HARRIS, P. D., JANSEN, P. A. & HANSEN, L. P. (1992). Host specificity and dispersal strategy in gyrodactylid monogeneans, with particular reference to *Gyrodactylus salaris* (Platyhelminthes, Monogenea). *Diseases of Aquatic Organisms* **13**, 63–74.

BAKKE, T. A., JANSEN, P. A. & HANSEN, L. P. (1991). Experimental transmission of *Gyrodactylus salaris* Malmberg, 1957 (Platyhelminthes, Monogenea) from the Atlantic salmon (*Salmo salar*) to the European eel (*Anguilla anguilla*). *Canadian Journal of Zoology* **69**, 733–737.

BAKKE, T. A., JANSEN, P. A. & HARRIS, P. D. (1996). Differences in susceptibility of anadromous and resident stocks of Arctic charr to infections of *Gyrodactylus salaris* under experimental conditions. *Journal of Fish Biology* **49**, 341–351.

BEREITER-HAHN, J., RICHARDS, K. S., ELSNER, L. & VOTH, M. (1980). Composition and formation of flame cell caps: a substratum for the attachment of micro-organisms to sea horse epidermis. *Proceedings of the Royal Society of Edinburgh* **79B**, 105–111.

BONDAD-REANTASO, M. G., OGAWA, K., YOSHINAGA, T. & WAKABAYASHI, H. (1995). Acquired protection against *Neobenedenia girellae* in Japanese flounder. *Fish Pathology* **30**, 233–238.

BUCHMANN, K. (1997). Population increase of *Gyrodactylus derjavini* on rainbow trout induced by testosterone treatment of the host. *Diseases of Aquatic Organisms* **30**, 145–150.

BUCHMANN, K. (1998). Binding and lethal effect of complement from *Oncorhynchus mykiss* on *Gyrodactylus derjavini* (Platyhelminthes: Monogenea). *Diseases of Aquatic Organisms* **32**, 195–200.

BUCHMANN, K. (1999). Immune mechanisms in fish skin against monogeneans – a model. *Folia Parasitologica* **46**, 1–9.

BUCHMANN, K. & BRESCIANI, J. (1998). Microenvironment of *Gyrodactylus derjavini* on rainbow trout *Oncorhynchus mykiss*: association between mucous cell density in skin and site selection. *Parasitology Research* **84**, 17–24.

BUCHMANN, K. & BRESCIANI, J. (1999). Rainbow trout leucocyte activity: influence on the ectoparasitic monogenean *Gyrodactylus derjavini*. *Diseases of Aquatic Organisms* **35**, 13–22.

BUCHMANN, K. & ULDAL, A. (1997). *Gyrodactylus derjavini* infections in four salmonids: comparative host susceptibility and site selection of parasites. *Diseases of Aquatic Organisms* **28**, 201–209.

BULLOCK, A. M., MARKS, R. & ROBERTS, R. J. (1978a). The cell kinetics of teleost fish epidermis: mitotic activity of the normal epidermis at varying temperatures in plaice (*Pleuronectes platessa*). *Journal of Zoology* **184**, 423–428.

BULLOCK, A. M., MARKS, R. & ROBERTS, R. J. (1978b). The cell kinetics of teleost fish epidermis: epidermal mitotic activity in relation to wound healing at varying temperatures in plaice (*Pleuronectes platessa*). *Journal of Zoology* **185**, 197–204.

BYCHOWSKY, B. E. (1957). *Monogenetic Trematodes – Their Systematics and Phylogeny*. English translation by Hargis, W. J. (1961). Washington, American Institute of Biological Sciences.

CARVALHO-VARELA, M. & CUNHA-FERREIRA, V. (1987). Helminth parasites of the common sole, *Solea solea*, and the Senegalese sole, *Solea senegalensis*, on the Portuguese continental coast. *Aquaculture* **67**, 135–138.

CHISHOLM, L. A. & WHITTINGTON, I. D. (1995). A revision of *Dendromonocotyle* Hargis, 1955 (Monogenea: Monocotylidae) with a description of a new species from *Pastinachus sephen* Forsskål (Myliobatiformes: Dasyatididae) from the Great Barrier Reef, Australia. *Journal of Natural History* **29**, 1093–1119.

CONE, D. & BURT, M. D. B. (1981). The invasion route of the gill parasite *Urocleidus adspectus* Mueller, 1936 (Monogenea: Ancyrocephalinae). *Canadian Journal of Zoology* **59**, 2166–2171.

CONE, D. K. & ODENSE, P. H. (1984). Pathology of five species of *Gyrodactylus* Nordmann, 1832 (Monogenea). *Canadian Journal of Zoology* **62**, 1084–1088.

CONE, D. K. & WILES, M. (1989). Ultrastructural study of attachment of *Gyrodactylus colemanensis* (Monogenea) to fins of fry of *Salmo gairdneri*. *Proceedings of the Helminthological Society of Washington* **56**, 29–32.

CRIBB, B. W., WHITTINGTON, I. D. & CHISHOLM, L. A. (1997). Observations on ultrastructure of the anterior adhesive areas and other anterior glands in the monogenean, *Monocotyle spiremae* (Monocotylidae), from the gills of *Himantura fai* (Dasyatididae). *International Journal for Parasitology* **27**, 907–917.

CUSACK, R. & CONE, D. K. (1986). *Gyrodactylus salmonis* (Yin and Sproston, 1948) parasitizing fry of *Salvelinus fontinalis* (Mitchill). *Journal of Wildlife Diseases* **22**, 209–213.

DOLLFUS, R. P. & EUZET, L. (1964). Complément à la description de *Pseudobenedenia nototheniae* T. H. Johnston, 1931 (Trematoda Monogenea) parasite d'un téléostéen du genre *Notothenia* Richardson des Kerguelen (Mission Jean-Claude Hureau, 1963–1964). *Bulletin du Muséum National d'Histoire Naturelle* (2nd Series) **36**, 849–857.

EL-NAGGAR, M. M. (1992). Scanning electron microscope studies on the head lobes and haptor of the monogenean *Gyrodactylus groschafti* Ergens, 1973. *Journal of the Egyptian German Society of Zoology* **8 (B)**, *Anatomy & Embryology*, 435–445.

EL-NAGGAR, M. M. & KEARN, G. C. (1983). Glands associated with the anterior adhesive areas and body margins in the skin-parasitic monogenean *Entobdella soleae*. *International Journal for Parasitology* **13**, 67–81.

EL-NAGGAR, M. M. & SERAG, H. M. (1987). Redescription of *Macrogyrodactylus clarii* Gusev 1961, a monogenean gill parasite of *Clarias lazera* in Egypt. *Arab Gulf Journal of Scientific Research. Agricultural and Biological Sciences* **5**, 257–271.

EUZET, L. & MAILLARD, C. (1967). Parasites de poissons de mer ouest-africains, récoltés par J. Cadenat. VI. Monogènes de sélaciens. *Bulletin de l'Institut Fondamental d'Afrique Noire* **29**, 1435–1493.

FLETCHER, T. C. (1978). Defence mechanisms of fish. In *Biochemical and Biophysical Perspectives in Marine Biology*, Vol. 4 (ed. Malins, D. C. & Sargent, J. R.), pp. 189–222. London, Academic Press.

FLETCHER, T. C. & GRANT, P. T. (1969). Immunoglobulins in the serum and mucus of the plaice (*Pleuronectes platessa*). *Biochemical Journal* **115**, 65P.

FLETCHER, T. C., JONES, R. & REID, L. (1976). Identification of glycoproteins in goblet cells of epidermis and gill of plaice (*Pleuronectes platessa* L.), flounder (*Platichthys flesus* (L.)) and rainbow trout (*Salmo gairdneri* Richardson). *Histochemical Journal* **8**, 597–608.

FLETCHER, T. C. & WHITE, A. (1973). Antibody production in the plaice (*Pleuronectes platessa* L.) after oral and parenteral immunization with *Vibrio anguillarum* antigens. *Aquaculture* **1**, 417–428.

FUCHS, E. & HANUKOGLU, I. (1986). Epidermal α-keratins: structural diversity and changes during tissue differentiation. In *Biology of the Integument. 2. Vertebrates* (ed. Bereiter-Hahn, J., Matoltsy, G. M. & Richards, K. S.), pp. 644–665. Berlin, Springer-Verlag.

FUJINO, T., ICHIKAWA, H., FUKUDA, K. & FRIED, B. (1998). The expulsion of *Echinostoma trivolvis* caused by goblet cell hyperplasia in severe combined immunodeficient (SCID) mice. *Parasite* **5**, 219–222.

HALTON, D. W. (1978). Trans-tegumental absorption of L-alanine and L-leucine by a monogenean, *Diclidophora merlangi*. *Parasitology* **76**, 29–37.

HARRIS, P. D. (1982). Variations in the mechanism of attachment in the Gyrodactyloidea (Monogenea). *Parasitology* **85**, lviii.

HARRIS, P. D. (1986). Species of *Gyrodactylus* von Nordmann, 1832 (Monogenea Gyrodactylidae) from poeciliid fishes, with a description of *G. turnbulli* sp. nov. from the guppy, *Poecilia reticulata* Peters. *Journal of Natural History* **20**, 183–191.

HARRIS, P. D. (1988). Changes in the site specificity of *Gyrodactylus turnbulli* Harris, 1986 (Monogenea) during infections of individual guppies (*Poecilia reticulata* Peters, 1859). *Canadian Journal of Zoology* **66**, 2854–2857.

HARRIS, P. D., SOLENG, A. & BAKKE, T. A. (1998). Killing of *Gyrodactylus salaris* (Platyhelminthes, Monogenea) mediated by host complement. *Parasitology* **117**, 137–143.

HARRIS, P. D., SOLENG, A. & BAKKE, T. A. (2000). Increased susceptibility of salmonids to the monogenean *Gyrodactylus salaris* following administration of hydrocortisone acetate. *Parasitology* **120**, 57–64.

HIDALGO ESCALANTE, E. (1958). Hallazgo de una nueva especie de Capsala, *Capsala pricei* n. sp. (Trematoda, Monogenea) en un pez marino del puerto de Mazatlan Sinaloa, Mexico. *Anales del Instituto de Biologia, México* **29**, 209–217.

HOBSON, E. S. (1974). Feeding relationships of teleostean fishes on coral reefs in Kona, Hawaii. *Fishery Bulletin, U.S. Fish and Wildlife Service* **72**, 915–1031.

HORTON, M. A. & WHITTINGTON, I. D. (1994). A new species of *Metabenedeniella* (Monogenea: Capsalidae) from the dorsal fin of *Diagramma pictum* (Perciformes: Haemulidae) from the Great Barrier Reef, Australia with a revision of the genus. *Journal of Parasitology* **80**, 998–1007.

IGER, Y., JENNER, H. A & WENDELAAR BONGA, S. E. (1994). Cellular responses in the skin of rainbow trout (*Oncorhynchus mykiss*) exposed to Rhine water. *Journal of Fish Biology* **45**, 1119–1132.

ISHIKAWA, N., HORII, Y. & NAWA, Y. (1993). Immune-mediated alteration of the terminal sugars of goblet cell mucins in the small intestine of *Nippostrongylus brasiliensis*-infected rats. *Immunology* **78**, 303–307.

JAHN, T. L. & KUHN, L. R. (1932). The life history of *Epibdella melleni* MacCallum 1927, a monogenetic trematode parasitic on marine fishes. *Biological Bulletin* **62**, 89–111.

JARA, C., AN, L. & CONE, D. (1991). *Accessorius peruensis* gen. et sp. n. (Monogenea: Gyrodactylidea) from *Lebiasina bimaculata* (Characidae) in Peru. *Journal of the Helminthological Society of Washington* **58**, 164–166.

KAYTON, R. J. (1983). Histochemical and X-ray elemental analysis of the sclerites of *Gyrodactylus* spp. (Platyhelminthes: Monogenoidea) from the Utah chub, *Gila atraria* (Girard). *Journal of Parasitology* **69**, 862–865.

KEARN, G. C. (1963a). The oncomiracidium of *Capsala martinieri*, a monogenean parasite of the sun fish (*Mola mola*). *Parasitology* **53**, 449–453.

KEARN, G. C. (1963b). The egg, oncomiracidium and larval development of *Entobdella soleae*, a monogenean skin parasite of the common sole. *Parasitology* **53**, 435–447.

KEARN, G. C. (1963c). Feeding in some monogenean skin parasites: *Entobdella soleae* on *Solea solea* and *Acanthocotyle* sp. on *Raia clavata*. *Journal of the Marine Biological Association of the United Kingdom* **43**, 749–766.

KEARN, G. C. (1964). The attachment of the monogenean *Entobdella soleae* to the skin of the common sole. *Parasitology* **54**, 327–335.

KEARN, G. C. (1965). The biology of *Leptocotyle minor*, a skin parasite of the dogfish, *Scyliorhinus canicula*. *Parasitology* **55**, 473–480.

KEARN, G. C. (1967a). The life-cycles and larval development of some acanthocotylids (Monogenea) from Plymouth rays. *Parasitology* **57**, 157–167.

KEARN, G. C. (1967b). Experiments on host-finding and host-specificity in the monogenean skin parasite *Entobdella soleae*. *Parasitology* **57**, 585–605.

KEARN, G. C. (1968). The development of the adhesive

organs of some diplectanid, tetraonchid and dactylogyrid gill parasites (Monogenea). *Parasitology* **58**, 149–163.

KEARN, G. C. (1974). A comparative study of the glandular and excretory systems of the oncomiracidia of the monogenean skin parasites *Entobdella hippoglossi*, *E. diadema* and *E. soleae*. *Parasitology* **69**, 257–269.

KEARN, G. C. (1978). *Entobdella australis*, sp. nov., a skin-parasitic monogenean from the Queensland stingrays *Taeniura lymma* and *Amphotistius kuhlii*. *Australian Journal of Zoology* **26**, 207–214.

KEARN, G. C. (1988*a*). Orientation and locomotion in the monogenean parasite *Entobdella soleae* on the skin of its host (*Solea solea*). *International Journal for Parasitology* **18**, 753–759.

KEARN, G. C. (1988*b*). The monogenean skin parasite *Entobdella soleae*: movement of adults and juveniles from host to host (*Solea solea*). *International Journal for Parasitology* **18**, 313–319.

KEARN, G. C. (1990). The rate of development and longevity of the monogenean skin parasite *Entobdella soleae*. *Journal of Helminthology* **64**, 340–342.

KEARN, G. C. (1993). A new species of the genus *Enoplocotyle* (Platyhelminthes: Monogenea) parasitic on the skin of the moray eel *Gymnothorax kidako* in Japan, with observations on hatching and the oncomiracidium. *Journal of Zoology* **229**, 533–544.

KEARN, G. C. (1994). Evolutionary expansion of the Monogenea. *International Journal for Parasitology* **24**, 1227–1271.

KEARN, G. C. (1998). *Parasitism and the Platyhelminths*. London, Chapman & Hall.

KEARN, G. C. & EVANS-GOWING, R. (1998). Attachment and detachment of the anterior adhesive pads of the monogenean (platyhelminth) parasite *Entobdella soleae* from the skin of the common sole (*Solea solea*). *International Journal for Parasitology* **28**, 1583–1593.

KEARN, G. C. & GOWING, R. (1989). Glands and sensilla associated with the haptor of the gill-parasitic monogenean *Tetraonchus monenteron*. *International Journal for Parasitology* **19**, 673–679.

KEARN, G. C. & GOWING, R. (1990). Vestigial marginal hooklets in the oncomiracidium of the microbothriid monogenean *Leptocotyle minor*. *Parasitology Research* **76**, 406–408.

KEARN, G. C., OGAWA, K. & MAENO, Y. (1992). Egg production, the oncomiracidium and larval development of *Benedenia seriolae*, a skin parasite of the yellowtail, *Seriola quinqueradiata*, in Japan. *Publications of the Seto Marine Biological Laboratory* **35**, 351–362.

KEARN, G. C., WHITTINGTON, I. D. & EVANS-GOWING, R. (1995). Use of cement for attachment in *Neocalceostomoides brisbanensis*, a calceostomatine monogenean from the gill chamber of the blue catfish, *Arius graeffei*. *International Journal for Parasitology* **25**, 299–306.

KRIEG, T. & TIMPL, R. (1986). Protein components of the epidermal basement membrane. In *Biology of the Integument. 2. Vertebrates* (ed. Bereiter-Hahn, J., Matoltsy, A. G. & Richards, K. S.), pp. 788–799. Berlin, Springer-Verlag.

KRITSKY, D. C. (1978). The cephalic glands and associated structures in *Gyrodactylus eucaliae* Ikezaki and Hoffman, 1957 (Monogenea: Gyrodactylidae). *Proceedings of the Helminthological Society of Washington* **45**, 37–49.

KRITSKY, D. C. & FENNESSY, C. J. (1999). *Calicobenedenia polyprioni* n. gen., n. sp. (Monogenoidea: Capsalidae) from the external surfaces of wreckfish, *Polyprion americanus* (Teleostei: Polyprionidae), in the north Atlantic. *Journal of Parasitology* **85**, 192–195.

KRITSKY, D. C. & FRITTS, T. H. (1970). Monogenetic trematodes from Costa Rica with the proposal of *Anacanthocotyle* gen. n. (Gyrodactylidae: Isancistrinae). *Proceedings of the Helminthological Society of Washington* **37**, 63–68.

LAMOTHE-ARGUMEDO, R. (1997). Nuevo arreglo taxonómico de la subfamilia Capsalinae (Monogenea: Capsalinae), clave para los géneros y dos combinaciones nuevas. *Anales del Instituto de Biologia, México* **68**, 207–223.

LESTER, R. J. G. (1972). Attachment of *Gyrodactylus* to *Gasterosteus* and host response. *Journal of Parasitology* **58**, 717–722.

LITTLEWOOD, D. T. J., ROHDE, K. & CLOUGH, K. A. (1998). The phylogenetic position of *Udonella* (Platyhelminthes). *International Journal for Parasitology* **28**, 1241–1250.

LLEWELLYN, J. (1957). Host-specificity in monogenetic trematodes. In *First Symposium on Host-specificity Among Parasites of Vertebrates*, pp. 199–212. Neuchâtel, Paul Attinger.

LLEWELLYN, J. (1982). Host-specificity and corresponding evolution in monogenean flatworms and vertebrates. *Mémoires du Muséum national d'Histoire naturelle, Série A, Zoologie* **123**, 289–301.

LLEWELLYN, J. (1984). The biology of *Isancistrum subulatae* n.sp., a monogenean parasitic on the squid, *Alloteuthis subulata*, at Plymouth. *Journal of the Marine Biological Association of the United Kingdom* **64**, 285–302.

LLEWELLYN, J. & EUZET, L. (1964). Spermatophores in the monogenean *Entobdella diadema* Monticelli from the skin of sting-rays, with a note on the taxonomy of the parasite. *Parasitology* **54**, 337–344.

LOBB, C. J. & CLEM, L. W. (1981). The metabolic relationships of the immunoglobulins in fish serum, cutaneous mucus, and bile. *Journal of Immunology* **127**, 1525–1529.

LYONS, K. M. (1966). The chemical nature and evolutionary significance of monogenean attachment sclerites. *Parasitology* **56**, 63–100.

LYONS, K. M. (1969). Compound sensilla in monogenean skin parasites. *Parasitology* **59**, 625–636.

LYONS, K. M. (1973). Scanning and transmission electron microscope studies on the sensory sucker papillae of the fish parasite *Entobdella soleae* (Monogenea). *Zeitschrift für Zellforschung* **137**, 471–480.

MALMBERG, G. & FERNHOLM, B. (1989). *Myxinidocotyle* gen. n. and *Lophocotyle* Braun (Platyhelminthes, Monogenea, Acanthocotylidae) with descriptions of three new species from hagfishes (Chordata, Myxinidae). *Zoologica Scripta* **18**, 187–204.

MALMBERG, G. & FERNHOLM, B. (1991). Locomotion and attachment to the host of *Myxinidocotyle* and

Acanthocotyle (Monogenea, Acanthocotylidae). *Parasitology Research* **177**, 415–420.

MATOLTSY, A. G. & BEREITER-HAHN, J. (1986). Introduction. In *Biology of the Integument. 2. Vertebrates* (ed. Bereiter-Hahn, J., Matoltsy, A. G. & Richards, K. S.), pp. 1–7. Berlin, Springer-Verlag.

MOORE, M. M., KAATTARI, S. L. & OLSON, R. E. (1994). Biologically active factors against the monogenetic trematode *Gyrodactylus stellatus* in the serum and mucus of infected juvenile English soles. *Journal of Aquatic Animal Health* **6**, 93–100.

NIGRELLI, R. F. (1937). Further studies on the susceptibility and acquired immunity of marine fishes to *Epibdella melleni*, a monogenetic trematode. *Zoologica* **22**, 185–192.

NIGRELLI, R. F. & BREDER, C. M. (1934). The susceptibility and immunity of certain marine fishes to *Epibdella melleni*, a monogenetic trematode. *Journal of Parasitology* **20**, 259–269.

OGAWA, K., BONDAD-REANTASO, M. G., FUKUDOME, M. & WAKABAYASHI, H. (1995). *Neobenedenia girellae* (Hargis, 1955) Yamaguti, 1963 (Monogenea: Capsalidae) from cultured marine fishes of Japan. *Journal of Parasitology* **81**, 223–227.

OGAWA, K. & EGUSA, S. (1981). The systematic position of the genus *Anoplodiscus* (Monogenea: Anoplodiscidae). *Systematic Parasitology* **2**, 253–260.

OLSON, A. C., JR. & JEFFRIES, M. (1983). *Dendromonocotyle californica* sp. n. (Monogenea: Monocotylidae) from the bat ray, *Myliobatis californica*, with a key to species. *Journal of Parasitology* **69**, 602–605.

PICKERING, A. D. & RICHARDS, R. H. (1980). Factors influencing the structure, function and biota of the salmonid epidermis. *Proceedings of the Royal Society of Edinburgh* **79B**, 93–104.

REES, J. & KEARN, G. C. (1984). The anterior adhesive apparatus and an associated compound sense organ in the skin-parasitic monogenean *Acanthocotyle lobianchi*. *Zeitschrift für Parasitenkunde* **70**, 609–625.

RICHARDS, G. R. & CHUBB, J. C. (1996). Host response to initial and challenge infections, following treatment, of *Gyrodactylus bullatarudis* and *G. turnbulli* (Monogenea) on the guppy (*Poecilia reticulata*). *Parasitology Research* **82**, 242–247.

RICHARDS, G. R. & CHUBB, J. C. (1998). Longer-term population dynamics of *Gyrodactylus bullatarudis* and *G. turnbulli* (Monogenea) on adult guppies (*Poecilia reticulata*) in 50-l experimental arenas. *Parasitology Research* **84**, 753–756.

ROBERTS, R. J. & BULLOCK, A. M. (1980). The skin surface ecosystem of teleost fishes. *Proceedings of the Royal Society of Edinburgh* **79B**, 87–91.

ROGERS, W. A. (1967). *Polyclithrum mugilini* gen. et sp. n. (Gyrodactylidae: Polyclithrinae subfam.n.) from *Mugil cephalus* L. *Journal of Parasitology* **53**, 274–276.

ROGERS, W. A. (1968). *Swingleus polyclithroides* gen. et sp. n. (Monogenea: Gyrodactylidae) from *Fundulus grandis* Baird and Girard. *Tulane Studies in Zoology and Botany* **16**, 22–25.

ROHDE, K. (1994). The minor groups of parasitic Platyhelminthes. *Advances in Parasitology* **33**, 145–234.

ROMBOUT, J. H. W. M., TAVERNE, N., van de KAMP, M. & TAVERNE-THIELE, A. J. (1993). Differences in mucus and serum immunoglobulin of carp (*Cyprinus carpio* L). *Developmental and Comparative Immunology* **17**, 309–317.

ROUBAL, F. R., QUARTARARO, N. & WEST, A. (1992). Infection of captive *Pagrus auratus* (Bloch & Schneider) by the monogenean, *Anoplodiscus cirrusspiralis* Roubal, Armitage & Rohde (Anoplodiscidae) in Australia. *Journal of Fish Diseases* **15**, 409–415.

ROUBAL, F. R. & WHITTINGTON, I. D. (1990). Observations on the attachment by the monogenean, *Anoplodiscus australis*, to the caudal fin of *Acanthopagrus australis*. *International Journal for Parasitology* **20**, 307–314.

SAKAI, D. K. (1992). Repertoire of complement in immunological defense mechanisms of fish. *Annual Review of Fish Diseases* **1992**, 223–247.

SCHLIWA, M. (1975). Cytoarchitecture of surface layer cells of the teleost epidermis. *Journal of Ultrastructure Research* **52**, 377–386.

SCOTT, M. E. (1985). Dynamics of challenge infections of *Gyrodactylus bullatarudis* Turnbull (Monogenea) on guppies, *Poecilia reticulata* (Peters). *Journal of Fish Diseases* **8**, 495–503.

SCOTT, M. E. & ANDERSON, R. M. (1984). The population dynamics of *Gyrodactylus bullatarudis* (Monogenea) within laboratory populations of the fish host *Poecilia reticulata*. *Parasitology* **89**, 159–194.

SENGEL, P. (1986). Epidermal–dermal interactions. In *Biology of the Integument. 2. Vertebrates* (ed. Bereiter-Hahn, J., Matoltsy, A. G. & Richards, K. S.), pp. 374–408. Berlin, Springer-Verlag.

SHINN, A. P., GIBSON, D. I. & SOMMERVILLE, C. (1995). A study of the composition of the sclerites of *Gyrodactylus* Nordmann, 1832 (Monogenea) using X-ray elemental analysis. *International Journal for Parasitology* **25**, 797–805.

SKERROW, D. (1986). Epidermal α-keratin: structure and chemical composition. In *Biology of the Integument. 2. Vertebrates* (ed. Bereiter-Hahn, J., Matoltsy, A. G. & Richards, K. S.), pp. 621–643. Berlin, Springer-Verlag.

SPEARMAN, R. I. C. (1973). *The Integument: a Textbook of Skin Biology*. London, Cambridge University Press.

SPROSTON, N. G. (1946). A synopsis of the monogenetic trematodes. *Transactions of the Zoological Society of London* **25**, 185–600.

STERUD, E., HARRIS, P. D. & BAKKE, T. A. (1998). The influence of *Gyrodactylus salaris* Malmberg, 1957 (Monogenea) on the epidermis of Atlantic salmon, *Salmo salar* L., and brook trout, *Salvelinus fontinalis* (Mitchill): experimental studies. *Journal of Fish Diseases* **21**, 257–263.

ST. LOUIS-CORMIER, E. A., OSTERLAND, C. K. & ANDERSON, P. D. (1984). Evidence for a cutaneous secretory immune system in rainbow trout (*Salmo gairdneri*). *Developmental and Comparative Immunology* **8**, 71–80.

UITTO, J. (1986). Interstitial collagens. In *Biology of the Integument. 2. Vertebrates* (ed. Bereiter-Hahn, J., Matoltsy, A. G. & Richards, K. S.), pp. 800–809. Berlin, Springer-Verlag.

WELLS, P. R. & CONE, D. K. (1990). Experimental studies on the effect of *Gyrodactylus colemanensis* and

G. salmonis (Monogenea) on density of mucous cells in the epidermis of fry of *Oncorhynchus mykiss*. *Journal of Fish Biology* **37**, 599–603.

WEST, A. J. & ROUBAL, F. R. (1998). Experiments on the longevity, fecundity and migration of *Anoplodiscus cirrusspiralis* (Monogenea) on the marine fish *Pagrus auratus* (Bloch & Schneider) (Sparidae). *Journal of Fish Diseases* **21**, 299–303.

WHITEAR, M. (1970). The skin surface of bony fishes. *Journal of Zoology* **160**, 437–454.

WHITEAR, M. (1986a). Epidermis. In *Biology of the Integument. 2. Vertebrates* (ed. Bereiter-Hahn, J., Matoltsy, A. G. & Richards, K. S.), pp. 8–38. Berlin, Springer-Verlag.

WHITEAR, M. (1986b). Dermis. In *Biology of the Integument. 2. Vertebrates* (ed. Bereiter-Hahn, J., Matoltsy, A. G. & Richards, K. S.), pp. 39–64. Berlin, Springer-Verlag.

WHITEAR, M., MITTAL, A. K. & LANE, E. B. (1980). Endothelial layers in fish skin. *Journal of Fish Biology* **17**, 43–65.

WHITEAR, M. & MOATE, R. (1998). Cellular diversity in the epidermis of *Raja clavata* (Chondrichthyes). *Journal of Zoology* **246**, 275–285.

WHITTINGTON, I. D. (1996). Benedeniine capsalid monogeneans from Australian fishes: pathogenic species, site-specificity and camouflage. *Journal of Helminthology* **70**, 177–184.

WHITTINGTON, I. D. & BARTON, D. P. (1990). A new genus of monogenean parasites (Capsalidae: Benedeniinae) from stingrays (Rajiformes: Dasyatidae) with a description of a new species from the long-tailed stingray *Himantura uarnak* Forsskål from Queensland, Australia. *Journal of Natural History* **24**, 327–340.

WHITTINGTON, I. D. & CRIBB, B. W. (1998). Glands associated with the anterior adhesive areas of the monogeneans, *Entobdella* sp. and *Entobdella australis* (Capsalidae) from the skin of *Himantura fai* and *Taeniura lymma* (Dasyatididae). *International Journal for Parasitology* **28**, 653–665.

WHITTINGTON, I. D. & CRIBB, B. W. (1999). Morphology and ultrastructure of the anterior adhesive areas of the capsalid monogenean parasites *Benedenia rohdei* from the gills and *B. lutjani* from the pelvic fins of *Lutjanus carponotatus* (Pisces: Lutjanidae). *Parasitology Research* **85**, 399–408.

WHITTINGTON, I. D., CRIBB, B. W., HAMWOOD, T. E. & HALLIDAY, J. A. (2000). Host-specificity of monogenean (platyhelminth) parasites: a role for anterior adhesive areas? *International Journal for Parasitology* **30**, 305–320.

WHITTINGTON, I. D. & HORTON, M. A. (1996). A revision of *Neobenedenia* Yamaguti, 1963 (Monogenea: Capsalidae) including a redescription of *N. melleni* (MacCallum, 1927) Yamaguti, 1963. *Journal of Natural History* **30**, 1113–1156.

WHITTINGTON, I. D. & KEARN, G. C. (1993). A new species of skin-parasitic benedeniine monogenean with a preference for the pelvic fins of its host, *Lutjanus carponotatus* (Perciformes: Lutjanidae) from the Great Barrier Reef. *Journal of Natural History* **27**, 1–14.

WHITTINGTON, I. D. & KEARN, G. C. (1995). A new calceostomatine monogenean from the gills and buccal cavity of the catfish *Arius graeffei* from Moreton Bay, Queensland, Australia. *Journal of Zoology* **236**, 211–222.

YANO, T. (1996). The nonspecific immune system: humoral defense. In *The Fish Immune System: Organism, Pathogen, and Environment* (ed. Iwama, G. & Nakanishi, T.), pp. 105–157. San Diego, Academic Press.

Pentastomids and the tetrapod lung

J. RILEY[1] and R. J. HENDERSON[2]

[1] *Department of Biological Sciences, The University of Dundee, DD1 4HN*
[2] *Institute of Aquaculture, The University of Stirling, Stirling FK9 4LA*

SUMMARY

Pentastomids comprise a highly specialized taxon of arthropod-like parasites that probably became adapted to the lungs of amphibians and reptiles early in their long evolutionary history. Few other macroparasites exploit this particular niche. Pentastomids are often large, long-lived and yet they cause little observable pathology in lungs, despite being haematophagous. The lungs of all tetrapods are lined with pulmonary surfactant, a remarkable biological material consisting of a complex mixture of phospholipids, neutral lipids and proteins that has the unique ability to disperse over the air-liquid lining of the lung. In the lower tetrapods it acts as an anti-glue preventing adhesion of respiratory surfaces when lungs collapse during swallowing prey or upon expiration. In mammals, pulmonary surfactant also plays a critical role regulating the activity of alveolar macrophages, the predominant phagocytes of the lower airways and alveoli. This review outlines the evidence suggesting that lung-dwelling pentastomids, and also nymphs encysted in the tissues of mammalian intermediate hosts, evade immune surveillance and reduce inflammation by coating the chitinous cuticle with a their own stage-specific surfactant. The lipid composition of surfactant derived from lung instars of the pentastomid *Porocephalus crotali* cultured *in vitro* is very similar to that recovered from the lung of its snake host. Pentastomid surfactant, visualised as lamellate droplets within sub-parietal cells, is delivered to the cuticle via chitin-lined efferent ducts that erupt at a surface density of < 400 mm^{-2}. The fidelity of the system, which ensures that every part of the cuticle surface is membrane-coated, testifies to its strategic importance. Two other extensive glands discharge membrane-associated (hydrophobic?) proteins onto the hooks and head; some have been purified and partly characterized but their role in minimising inflammatory responses is, as yet, undetermined.

Key words: Pentastomids, pulmonary surfactant, lipids, reptile lungs, immune evasion.

INTRODUCTION

The phylum Pentastomida comprises a highly specialized taxon of about 110 arthropod-like parasites whose adults infect the respiratory tracts of reptiles and other vertebrates (for general reviews see Riley, 1986, 1994: Table 1). The only pentastomids infecting non-reptile hosts include three species of *Raillietiella* in the lungs of amphibians (Krishnasamy *et al.* 1995), two species of *Reighardia* inhabiting the air sacs of birds (Dyck, 1975), and six species, all belonging to the genus *Linguatula*, from the nasal sinuses of mammals (Table 1). Clearly, the reptile respiratory tract is the favoured site of parasitism by pentastomids.

Reptiles have a simple respiratory system comprising a short uncomplicated nasal cavity, devoid of functional nasal sinuses, a long trachea or paired bronchi and a multicompartmented lung (Duncker, 1978: Fig. 1). Thus, in contrast to the bronchoalveolar lungs of mammals and the parabronchial (lung-airsac) system of birds, reptile lungs can be broadly classified as unicameral, paucicameral, or multicameral based on the degree of infolding of the lung wall (Perry, 1983). In most snakes and lizards, the anteriorly located respiratory surface of the simple bag-like unicameral lung (Fig. 1 A, C, D, E) is followed by an extensive non-respiratory air sac (Daniels, Barr & Nicholas, 1989; Daniels, Smits & Orgeig, 1995). At the other extreme, the multicameral lungs of crocodilians possess numerous large chambers, each of which is subdivided, and these lungs have a spongy appearance due to pockets of trapped air (Fig. 1B).

An examination of the sites occupied by pentastomids reveals that a total of 95 species are recorded as growing to maturity in the lumen of the lungs of reptiles, living in intimate association with the respiratory surface. Exceptions to this general pattern are two genera, *Leiperia* and *Subtriquetra*, which inhabit the trachea and nasal sinuses respectively of their crocodilian hosts (Riley & Huchzermeyer, 1996), and at least one member of the genus *Elenia* which infests the throat of its varanid host (Bosch & Frank, 1986). In four other lung-dwelling genera, *Waddycephalus*, *Parasambonia*, *Kiricephalus* and *Cubirea*, the swollen and club-shaped head of females is permanently anchored deep in the lung wall, enclosed within a fibrous capsule of host origin, although in all cases males are freely mobile (Riley, 1992*a*).

Baer (1952) considered the distribution of extant pentastomids amongst their hosts and speculated about their evolutionary history. He concluded that reptiles were the ancestral hosts and that the association was probably very ancient, possibly originating in the Mesozoic. Recently discovered fossil evidence, however, indicates the association

Table 1. An outline classification of the phylum Pentastomida (from Riley, 1992a)

Order	Family	Genus	No. of species	Definitive host	Intermediate host
Cephalobaenida	Cephalobaenidae	*Cephalobaena*	1	Snakes	?
		Raillietiella	> 35	Snakes, lizards, amphisbaenians, amphibians	Direct (?), insects, amphibians, lizards
Porocephalida	Reighardiidae	*Reighardia*	2	Marine birds	Direct (1 sp.)
	Sebekidae	*Sebekia*	12	Crocodilians (Chelonians)	Fish (snakes, lizards?)
		Alofia	5	Crocodilians	Fish
		Leiperia	2	Crocodilians	Fish
		Diesingia	1	Chelonians	?
		Selfia	1	Crocodilians	Fish
		Agema	1	Crocodilians	Fish
	Subtriquetridae	*Subtriquetra*	3 (?)	Crocodilians	Fish
	Sambonidae	*Sambonia*	4	Monitor lizards	Direct (1 sp.)
		Elenia	2	Monitor lizards	Amphibians, mammals
		Waddycephalus	10	Snakes	Amphibians, reptiles, mammals
		Parasambonia	2	Snakes	Amphibians, reptiles, mammals
	Porocephalidae	*Porocephalus*	8	Snakes	Snakes and/or mammals
		Kiricephalus	5	Snakes	Amphibians or lizards or mammals, and snakes
	Armilliferidae	*Armillifer*	7	Snakes	Mammals
		Cubirea	2	Snakes	?
		Gigliolella	1	Snakes	Mammals
	Linguatulidae	*Linguatula*	6	Mammals	Direct (1 sp.), mammals

may be considerably older even than that (Walossek & Müller, 1994; Walossek, Repetski & Müller, 1994). Baer (1952) concluded "It is therefore clear that conditions of so non-specialized a nature as occur in the lungs of reptiles have saved pentastomids from extinction and, consequently, this archaic group owes its present day existence to having adapted itself to parasitism". While the last part of Baer's statement is undoubtedly true, the intricacies of lung physiology were largely unknown in 1952 and Baer was unaware of the unique suite of adaptations that enable pentastomids to chronically infect the lung lumen (Riley, 1992b). Consequently, it is almost the converse of his statement that actually obtains; probably the main reason for pentastomid survival lies in the complex adaptations they show to the highly specialized conditions within the reptile lung. Throughout life pentastomids produce membrane-dominated excretory/secretory (E/S) products which coat the cuticle. We will argue that these probably evolved very early in the evolutionary history of pentastomids, as a direct response to the selective pressures of life in the membrane-delimited environment of the reptile lung. Subsequently, these mechanisms were further selected for life in the tissues of intermediate hosts, constituting one of the more unusual examples of immune evasion by a parasite.

Lung-dwelling pentastomids are typically large, some reaching 5–10 cm or more in length and < 1 cm in diameter (Fig. 1C, E). They are long-lived (many months to many years) and all imbibe blood pumped from capillaries in the lung epithelium (Thomas & Böckeler, 1992a, b). They crawl using the body wall in a manner analogous to that of dipteran maggots (von Haffner, 1972), through a microhabitat – the air/water interface of the lung lining – which is one of the most carefully regulated environments in the vertebrate body. Self & Kuntz (1967), Self (1972) and Self & Cosgrove (1972) commented upon the remarkable degree of compatibility between tissue-dwelling pentastomids and their mammalian and reptilian hosts, and concluded that the adaptation of the host to the parasite is so complete as to result in little observable pathological effect. The deployment of surfactant by pentastomids, a key adaptation resulting from a long association with the lungs of lower tetrapods, lies at the heart of this apparent stability. The elaborate structural and biochemical adaptations that underpin the prolonged existence of pentastomids in this extreme and delicate environment are the subjects of this paper.

Macroparasites and lungs

The majority of the vertebrate helminth endoparasites are enteric and, although the respiratory tract has been successfully invaded by a number of quite disparate taxa (Rose, 1976), relatively few species actually reside in or on the respiratory epithelium. The exceedingly small size of alveoli, of the order of tens of micrometers, is clearly an

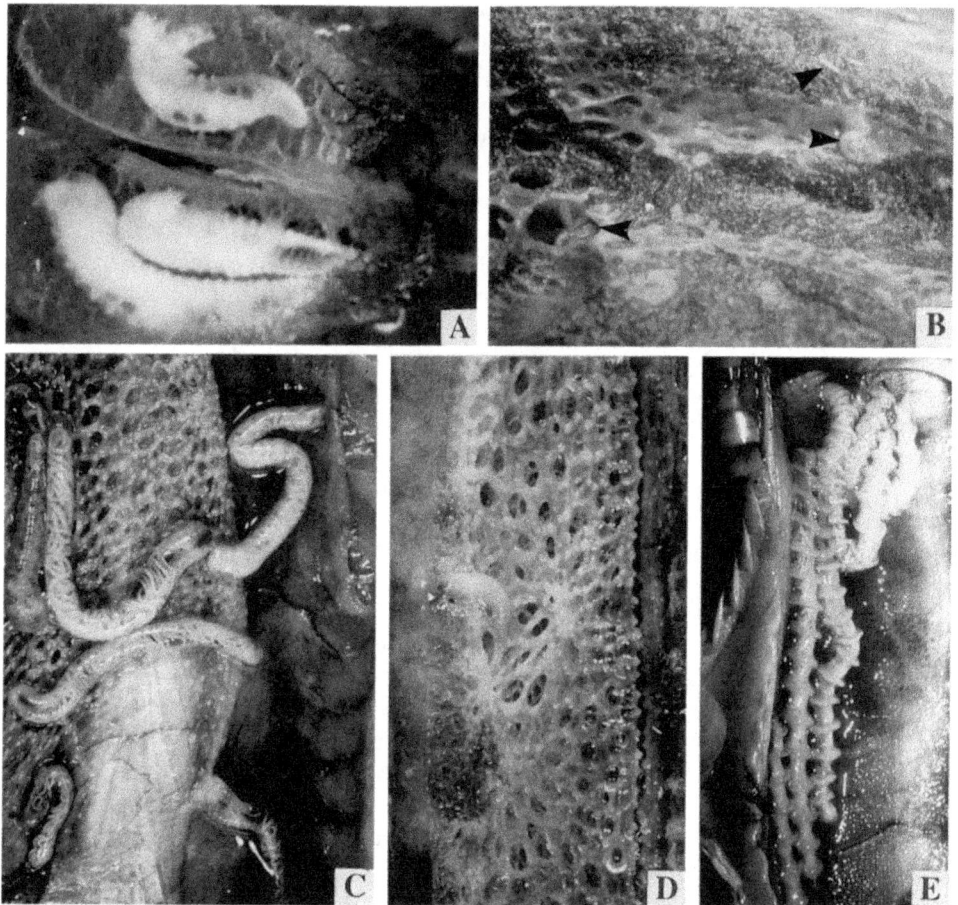

Fig. 1. Living pentastomids in the lungs of freshly dissected reptilian hosts. (A) The lungs of a gecko, *Hemidactylus brookii*, containing three large females and one small male *Raillietiella* sp. The heads of the females are orientated towards the fundus of the lung (left) and their tails extend towards the trachea (right). The individual gas exchange units (faveoli) carry substantial blood vessels and are clustered mainly around the top of the lung (right). ×56. (B) Part of the lung of an African dwarf crocodile (*Osteolaemus tetraspis*), which has been sliced open to reveal pentastomids (*Sebekia* and *Alofia*, arrows) in the bronchioles and air sacs of these paucicameral lungs. ×2·1. (C & D) *Porocephalus crotali* within the lung of a rattlesnake (*Crotalus atrox*); on the left are several females showing coils of the uterus through the thin chitinous cuticle and these mostly obscure the black, haematin-filled intestine. The right panel shows a single male that has penetrated into a single faveolus. Many faveoli are plainly visible, as are the smooth muscle bands (trabeculae) which encircle the neck of each of these respiratory units. ×0·9 and ×1·2 respectively. (E) *Armiller armillatus* within the lung of a gaboon viper (*Bitis gabonica*); the elevated annuli of the four females are plainly visible and two much smaller males are attached to the faveolar lining of the lung towards the top of the photograph. ×0·5.

important limiting factor in the case of mammals. Thus, the slender trichurid, metastrongylid, dictyocaulid and protostrongylid nematode lungworms of mammals, which measure 2–10 cm in length (Mehlhorn, 1988), actually lodge within the fine branches of the bronchi, as do certain syngamid nematodes in birds and mammals (Smyth, 1976). The ontogenetic migrations of many species of larval nematodes, ultimately destined for adulthood in the gut, may involve penetration of the lungs but passage to the bronchioles is usually transient (Rose, 1976).

Adult flukes of the genus *Paragonimus*, found deep within the lungs of a number of mammals species, are invariably located within a conspicuous thick-walled cyst composed of fibrous granulation tissue of host origin (Miyazaki, 1991). Granulation tissue also invests lung-infecting hydatid cysts of the dog tapeworm *Echinococcus granulosus*, and these are further protected by a thick striated cuticular layer enclosing the germinal epithelium (Miyazaki, 1991). Several digeneans, including *Haplometra*, *Haematoloechus*, *Pneumonocoeces* and *Pneumobites* infect the lungs of frogs and are truly intraluminal (reviewed by Tinsley & Earle, 1983), as are three digenean species infesting the lungs of turtles (Crandall, 1960) and snakes (Krull, 1930, 1931). However, these few helminth examples appear to represent only isolated successes in the colonisation of the lung environment.

Two monogeneans, species of *Pseudodiplorchis* and *Neodiplorchis*, infecting the spadefoot toad *Scaphiopus* also have an obligatory lung stage associated with ontogenetic migration. However, this is strictly short term (3–4 weeks) and occurs when developing worms migrate from the nares and mouth of the toad to the definitive site for maturation, the urinary bladder (Tinsley & Earle, 1983).

Several groups of arthropods (the taxon most closely related to the pentastomids (Walossek & Müller, 1994)) also inhabit the vertebrate respiratory tract. These include the larvae of dipteran head bots (Oestridae and Cuterebridae) in the nasal cavities of mammals, and various acarines (Rhinonyssidae and Halarachnidae) in the nasal passages of birds and bronchi and bronchioles of mammals (Chandler & Read, 1981). However, only one family of mites, the Entonyssidae, are specialist parasites of the snake lung (Turk, 1947).

In order to determine the exact *in situ* location of pentastomids in lungs, hosts must be examined very soon after death because worms undertake lengthy *post-mortem* migrations (Riley, 1986). Unsurprisingly, such information is limited. Female raillietiellids in gecko lungs are nearly always orientated with their heads towards the fundus of the lung and their caudal extremities towards the trachea (Fig. 1 A), which often contains long slicks of haematin-contaminated faecal material. One group of raillietiellids which infect small lizards possess large blunt-tipped posterior hooks (Ali, Riley & Self, 1985), and the head and projecting hooks are precisely contoured to fit snugly inside individual gas exchange units termed faveoli (Duncker, 1978: Fig. 1 A). This arrangement holds the worm securely in place and the blunt tips of the hooks presumably minimise damage to the delicate respiratory epithelium. *Porocephalus crotali* and *Armillifer armillatus* mostly inhabit the faveolar region of the lung of their snake hosts (Fig. 1 C, E) and males are small enough to occupy partly individual faveoli (Fig. 1 D). *Sebekia*, *Agema* and *Alofia* species in dwarf crocodile lungs can be seen occasionally through the pleura, but deep incisions into the lung reveal that most worms live within the air spaces and fine bronchioles often very close to the pleura (Riley & Huchzermeyer, 1995: Fig. 1 B).

The above inventory, although not exhaustive, emphasises that the phylum Pentastomida constitutes the highest-ranking taxon that has become irrevocably associated with the tetrapod lung. An important corollary is that for most pentastomids the parasite tegument is likely to interface intimately with the host respiratory surface.

The lung and pulmonary surfactant in endothermic tetrapods

The tetrapod respiratory system is equipped with protective mechanisms essential for the maintenance of lung homeostasis. A better understanding of the problems faced by reptilian systems can be obtained by comparing them with the protective mechanisms of mammals (Fowler, 1980). In the first instance, inhaled air must be warmed or cooled and cleansed. In mammals this begins as air is inhaled when particles may be trapped either by nasal hairs or on the sticky surface of the nasal mucosa and turbinates. Any particles penetrating the trachea, bronchi and bronchioles are caught by mucus and swept out of the tract by cilia at a rate of about $1\ cm\ min^{-1}$, a process which may be greatly assisted by coughing. Particles less than $10\ \mu m$ in diameter can enter alveoli where they are removed by phagocytosis. The honeycomb arrangement of faveoli in reptile lungs (Fig. 1 A, C, D, E) inhibits fluid removal and reptiles are also unable to sneeze or cough. These attributes alone may be sufficient to explain why bulky lung-dwelling pentastomids are mostly restricted to reptiles and amphibians (Table 1).

Virtually all of our knowledge of the interrelationships between the chemical constituents of pulmonary surfactant, the lung-lining material of tetrapods, is derived from mammalian studies (Farrell, 1982). Therefore, mammalian examples will be used to introduce the biology of the delicate interface to which pentastomids have become adapted. Note, however, that surfactant composition (Table 2) and function is broadly similar across all tetrapods (Daniels *et al.* 1998 a, b).

The respiratory surface of tetrapod lungs is richly supplied by alveolar type-II cells, which synthesise and secrete pulmonary surfactant, a remarkable biological material that has physical properties critical for normal lung function. The only other alveolar epithelial cell, the type-I cell, covers 95 % of the alveolar surface area but has no known role in surfactant metabolism (Hawgood & Poulain, 1995). Pulmonary surfactant consists of a complex mixture of phospholipids, neutral lipids and proteins and has the unique ability to disperse as a monolayer over the liquid inner lining of the lung (Harwood, 1987). In mammals and birds its primary function is to lower the surface tension of the thin fluid layer between the tissue and air, thereby reducing the work associated with inspiration and preventing alveolar collapse at low lung volumes (Farrell, 1982).

Thus, surfactant is present in two major pools; in the extracellular compartment it spreads as a film over the liquid lining of the lung whereas intracellularly it is sequestered within distinctive lamellar bodies that characterise alveolar type II cells (Smith, Smith & Ryan, 1972). Surfactant isolated from either source contains about 85–90 % lipid by weight, of which 90–95 % is composed of phospholipids, with cholesterol as the main neutral lipid (Harwood, 1987). Phosphatidylcholine (PC) accounts for 70–80 % of total lipid, and about 60 % of that is disaturated, both of the constituent fatty acids being saturated. The resulting molecule, dipalmitoylphosphatidylcholine (DPPC) is the key component that is

Table 2. The phospholipid composition of isolated lung surfactants (% wt/wt). For the phospholipid notations see text and Table 3

Class	Species	PC	PE	PG	PI+PS	Sph	Others	Source
Mammal	Rat	76	6	6	4	3	4	Harwood (1987)
	Rabbit	83	4	4	5	1	3	
	Man	69	4	7	5	5	7	
Bird	Turkey	86	6	—	7	tr.	1	
Reptile								
Snakes	Rattlesnake	68.5	—	—	11.5	11.5	17.7 (LPC)*	Daniels, Smits &
	Garter snake	64	—	4.0	16.9	1.0	14.2 (LPC)*	Orgeig (1995)
Lizard	*Ctenophorus nuchalis*	72.5	—	—	24.4	3.1	—	Daniels, Barr & Nicholas (1989)
Amphibian								
Salamander	*Amphiuma tridactylum*	77.5	1.0	—	17.5	4.0	—	Daniels, Orgeig, Wilson & Nicholas (1994)
Toad	*Bufo marinus*	62.5	9.0	13.0	8.5	6.5	—	

* LPC, lysophosphatidylcholine. tr. = trace.

responsible for the low surface tension and stability of the monolayer (Kuroki & Voelker, 1994). The phosphocholine in DPPC is polar and hydrophilic, while the two palmitic acid residues are non-polar and hydrophobic. The choline part of the molecule associates with the liquid phase in the alveoli and the palmitic acid orientates towards the air. DPPC molecules become closely packed together when the surface area is reduced, which confers great stability on the surface film (Wright & Clements, 1987). Phosphoglycerides comprised 90% of the total phospholipids and in all tetrapods the PC fraction dominates, but there is some variation between species in the composition of the various lipid classes (Harwood, 1987: Table 2). In ranking order of importance, phosphatidylglycerol (PG), phosphatidylethanolamine (PE), phosphatidylinositol (PI) and phosphatidylserine (PS) constitute the bulk of the minor phospholipid component in mammals, and sphingomyelin (Sph) is the only significant sphingolipid (Table 2). Disaturated phospholipids (DSP) account for most of the phospholipid fraction but significant amounts of polyunsaturated fatty acids may be present in some mammals and these probably function as liquifiers (Bangham, 1987)

Although lipids constitute < 90% of surfactant mass in mammals, surfactant proteins (SP) make up the remainder and these are critical to every aspect of surfactant function. SP-A and SP-D are members of the collectin family of proteins that both regulate surfactant phospholipid secretion and uptake, as well as playing an important role in host defence within the lung. In particular, SP-A binds with broad specificity to a variety of microorganisms stimulating phagocytosis and chemotaxis in alveolar macrophages (Kuroki & Voelker, 1994; Wright, 1997). Two small surfactant proteins, SP-B and SP-C, are also synthesised and secreted by alveolar type II pneumocytes. Both are strongly hydrophobic and remain lipid-associated, and both are essential in modulating the processes of phospholipid adsorption and spreading (Perez-Gil, Casals & Marsh, 1995).

Surfactant has a complicated life cycle, first through synthesis, then to a regulated secretion to the alveolus, followed by monolayer formation and subsequent clearance by catabolic and recycling pathways (Wright & Clements, 1987). The change in the alveolar form of surfactant appears to depend on surface exposure of the surfactant and the presence of a serine protease (Gross, 1995), although the exact mechanisms remain poorly understood. A number of lines of evidence suggest that surfactant PC is retrieved from the alveolar pool and recycled by alveolar type II cells (Ikegami & Jobe, 1993). It is even possible for labelled PC, administered via the trachea (of rabbits), to be recycled via these cells (Hallman, Epstein & Gluck, 1981). The continuous recycling of surfactant, which in rabbits has an efficiency of between 25% and 90% (Rider, Ikegami & Jobe, 1990), has important implications for the E/S products of lung-dwelling pentastomids.

Lung surfactant in ectothermic tetrapods

Snakes and most lizards commonly possess simple bag-like lungs with air-sacs that are up to a 1000-fold larger and up to a 100-fold more compliant than those of comparably sized mammals. They are divided structurally and functionally into two regions; a proximal faveolar (respiratory) area, and a much larger non-respiratory saccular region (Daniels et al. 1989, 1995: Fig. 1). Upon inspiration, air passes over a honeycomb of bellows-shaped faveoli, where actual gas exchange occurs, into a saccate region lined with a simple squamous epithelium which stores air that can be made available for gas exchange during apnoea. Both regions contain large amounts of surfactant (Daniels et al. 1995).

The two main functions of surfactant in mammals are to provide alveolar stability and to increase compliance of the relatively stiff bronchoalveolar lung. Since the respiratory units of most non-mammalian vertebrates are > 1000 times larger, surfactant is not required for these functions (Daniels et al. 1998a). In non-mammalian vertebrates, surfactant appears to act as an anti-glue preventing adhesion of respiratory surfaces that may occur when lungs collapse during swallowing prey or upon expiration. Surfactant also controls lung fluid balance and this may represent the primitive function of surface active material in vertebrate lungs (Orgeig et al. 1997; Daniels et al. 1998a).

Lung surfactant has been investigated in a number of ectotherms (McGregor, Daniels & Nicholas, 1993; Daniels, Eskandari-Marandi & Nicholas, 1993; Daniels et al. 1994: Table 2). In toads and salamanders, DSP lower surface tension at the air-water interface in the lung, while cholesterol and unsaturated phospholipids assist spreading. Lizards possess significantly more surfactant per unit area of respiratory surface than do similar sized mammals, and increases in the cholesterol content of surfactant at low temperatures may promote fluidity and adsorption (Daniels et al. 1989). Snake faveolar surfactant contains very little cholesterol (3–8%), but that of rattlesnakes (*Crotalus atrox*) differs from most other vertebrates in that it contains unusually high levels of lysophosphatidylcholine (LPC) and large amounts of Sph, whereas PG and PI are virtually absent (Table 2). However, LPC has detergent properties and can be toxic and consequently it does not normally accumulate in biological systems. Hence, the elevated levels recorded by Daniels et al. (1989) probably reflect some form of degradation of the sample prior to analysis.

Generally the saccular lungs and other gas-holding structures in non-mammals have 7–70% more surfactant/cm^{-2} than occurs in the lungs of mammals. The lipid composition of surfactant is highly conserved within the vertebrates, except that in lungfish it is dominated by cholesterol whereas in higher tetrapods DSP is the dominant fraction (Daniels et al. 1998b). Finally, homology within a surfactant protein (SP-A) across a whole range of the vertebrate gas-holding structures including the swim-bladder of goldfish, points to a single evolutionary origin for the surfactant system (Sullivan et al. 1998).

The lung and regulation of immunity

Apart from the various biophysical properties of pulmonary surfactant outlined above, it is also crucially involved in the protection of the lungs from injuries and infections caused by inhaled particles and microorganisms (Griese, 1999): again our knowledge of these non-physiological functions stems largely from studies of mammalian systems. Alveolar macrophages are the predominant phagocytes of the lower airways and alveoli and it is now well established that pulmonary surfactant regulates their activity, thereby protecting alveoli from potentially damaging immune reactions (Wilsher, Hughes & Haslam, 1988; Geertsma et al. 1993). Although surfactant lipids can suppress a variety of immune cell functions, most notably lymphocyte proliferation, conversely SP-A and SP-D enhance phagocytosis by alveolar macrophages (Mariencheck et al. 1999; Restrepo et al. 1999). Thus surfactant lipids and proteins may be counter-regulatory, and changes in the lipid-to-protein ratio may be important in regulating the immune status of the lung. That these ratios change in certain disease states is clear, but it is not yet established whether this constitutes a cause or an effect (Wright, 1997).

THE ADAPTATIONS OF PENTASTOMIDS FOR LIFE IN THE TETRAPOD LUNG

The pentastomid cuticle and its relation to glands which secrete E/S products

In common with the larvae of many holometabolous insects, pentastomids crawl using their tegument (von Haffner, 1972; Riley, 1986), which remains soft and flexible throughout life. The cuticle comprises three main layers, the endo-, exo- and epicuticles (Mehlhorn, 1988). A number of lipid analogues intercalate with the outermost leaflet of the epicuticle, identified as cuticulin by analogy with the insect cuticle (Riley & Banaja, 1975). The technique of fluorescence recovery after photobleaching reveals that the lateral fluidity of this cuticulin layer is very much like that of a conventional plasma membrane (Ambrose & Riley, 1988c).

A long recognised characteristic of pentastomids is the abundance of large glandular cells suspended in the haemocoel (Doucet, 1965). Three of these glands, the sub-parietal cell (SPC) system, frontal glands (FG) and hook glands (HG), discharge E/S products onto the cuticle via minute chitin-lined ducts (Riley, 1986, 1992b: Fig. 2). FG and HG, restricted to the cephalic region in primitive cephalobaenid pentastomids (Table 1), are diffuse and masses of individual secretory lobules and associated efferent ducts empty secretion onto the anterior margin of the head, the buccal cavity and the hook pits (Riley, 1973; Riley, James & Banaja, 1979). The equivalent gland cells in the advanced Porocephalida are arranged into compact organs within which individual chitin-lined efferent ducts, each serving a secretory lobule, converge into common collecting ducts. Again these breach the cuticle at the hook pits or the anterior margin of the head at the frontal papillae (Riley, 1986; Ambrose & Riley, 1988b: Fig. 2).

By contrast, SPC are distributed similarly in both

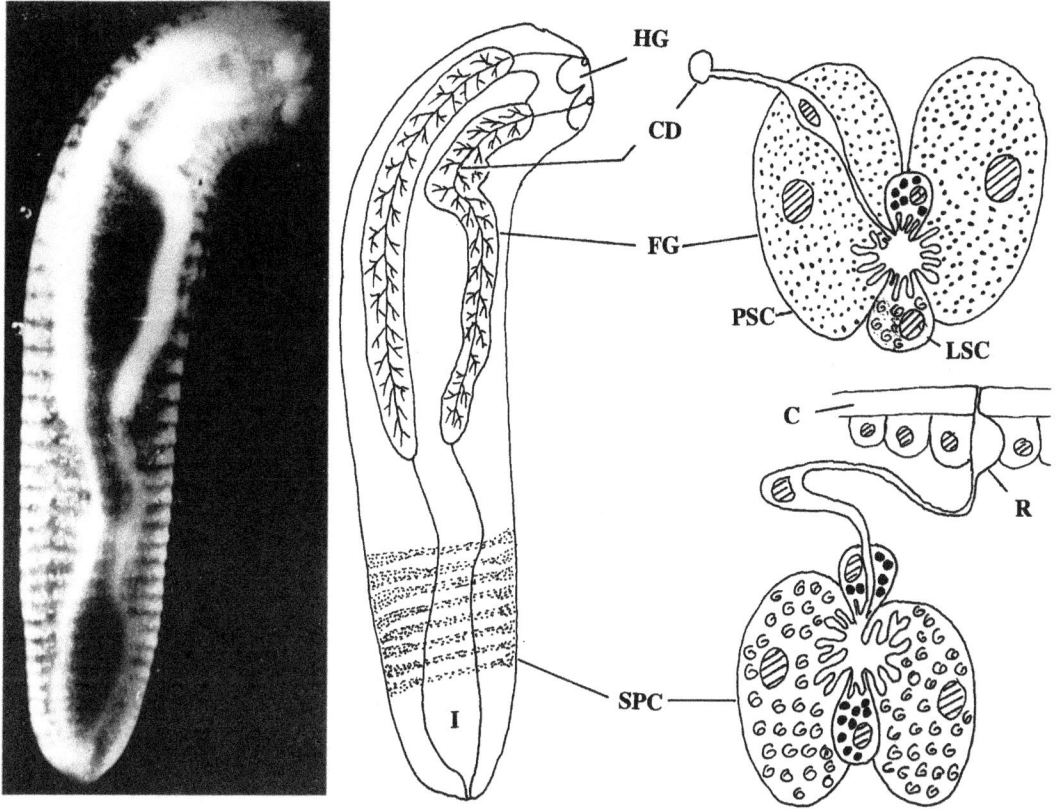

Fig. 2. An 11 mm long, living infective nymph of *Porocephalus crotali* (left), mounted on a slide in saline and flattened slightly by coverslip pressure, to show the disposition of the three major gland systems which produce surface-active material. The glands are drawn on a tracing of the nymph in the middle panel and the arrangement of efferent ducts, which eventually converge onto the collecting duct (CD) before discharging at the anterior margin of the head, have been added to frontal glands (FG). The right-hand panel shows an individual secretory lobule within FG, comprised of two large protein-secreting cells (PSC), a lipid secreting cell (LSC) and another small cell type, apparently also synthesising protein, but of a different type. The efferent duct is shown connecting to the collecting duct. Sub-parietal cells (SPC) are composed of four cells, all linked to the cuticle (C) by a sinuous efferent duct which expands to form a reservoir just below the level of the cuticle. The two large cells in each lobule secrete lamellate droplets and their ultrastructure closely resembles that of LSC in FG. The other two cells are filled with protein (?) droplets. Hook glands (HG) are drawn in outline only. (I = intestine). Diagrammatic and not to scale.

orders and appear as diffuse annular clusters of subtegumental secretory lobules, each connected to a single efferent duct (Fig. 2). The latter erupt over the entire surface of the cuticle, but are more densely concentrated in broad bands over the raised portion of each abdominal annulus (Doucet, 1965: Riley *et al.* 1979; Ambrose & Riley, 1988*b*). The potential of the SPC system to coat simultaneously the entire cuticle with secretion is striking. For example, microscopical examination of the cuticles of living pentastomids under short wavelength blue light generates bright autofluorescence, particularly in the chitin-lined efferent ducts of gland cells. Those serving SPC, though tiny (< 1 μm in diameter), can be easily counted and their surface density has been estimated at about 400 mm^{-2} (Ambrose & Riley, 1988*b*). Most emerge onto the cuticle near to the periphery of chloride cell pore caps (Doucet, 1965; Banaja, James & Riley, 1977), although it should be emphasised that every part of the cuticle receives SPC ducts. The subtlety of the system is quite remarkable and testifies to the strategic importance of the SPC system (Ambrose & Riley, 1989).

Precursor gland cells filled with secretory droplets and corresponding to HG, FG and SPC have been detected by light microscopy in primary larvae of *P. crotali* within the egg (Esslinger, 1962*b*). By the time nymphs have developed in rodent intermediate hosts to become infective to rattlesnake definitive hosts (Esslinger, 1962*a*), FG, for example, are macroscopic measuring 0.3 × 5 mm (Ambrose & Riley, 1988*a*: Fig. 2). Adult instars in snake lung (Fig. 1C, D) possess enormous FG, attaining lengths of 2 and 4 cm (in ♂♂ and ♀♀ respectively: Jones *et al.* 1991). The other two systems also increase in size with each moult and the available evidence suggests that, in both pentastomid orders, all instars are coated with E/S material throughout life. Inevitably, E/S products from each of the three glands must interact with the host's immune system; in the case of pre-adult and adult instars, this occurs necessarily within the environment of a lung.

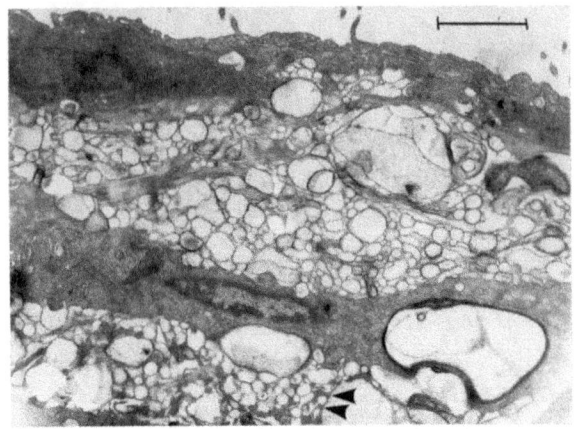

Fig. 3. An electron micrograph of host cells and SPC-derived vesicles on the ensheathing VIth instar cuticle surrounding an encysted VIIth instar of *Porocephalus crotali* (120 d post-infection) in the abdominal fat body of a rat. A fragment of cuticle is visible (bottom left) and the mat of epicuticular hairs (arrowed) holding a foamy layer of vesicles over the cuticle surface can be seen above it. Parts of two macrophages can be visualised and, although both are immersed in vesicles of parasite origin, they merely deform around them and remain totally unreactive. Collagen fibres, which form a very significant component of the final granuloma, are beginning to be deposited (middle left). Scale bar is 1 μm.

Pentastomid SPC-derived surfactant and the host immune response

Sub-parietal cell (SPC) secretion is predominantly membranous, which has the enormous advantage that evidence of SPC activity at the host/parasite interface can be visualised by conventional transmission electron microscopy (Ambrose and Riley, 1988b, c; 1989; Riley, 1992b). In *P. crotali* infections in rodent and rattlesnake hosts, individual SPC secretory lobules consist of four secretory cells, two large and two very small, which share a common efferent duct (Ambrose & Riley, 1988b: Fig. 2). The larger cells synthesise and secrete elaborate lamellate droplets. In terms of their ultrastructure and biosynthesis via Golgi bodies, these droplets are indistinguishable from pulmonary surfactant within alveolar type II cells, even to the extent of forming tubular myelin (Smith *et al.* 1972; Williams, 1977; Beckman & Dierichs, 1984; and see Fig. 17 in Ambrose & Riley, 1988b). Efferent ducts are equipped with bulbous reservoirs measuring approximately 3×4.5 μm which lie entirely within ductule-secreting cells deep in the tegumental epithelium (Fig. 2). Reservoirs store membranous droplets prior to their discharge onto the cuticle surface through exceedingly narrow pores, 60–80 nm in diameter, enclosed within chitinous collars. In rodent hosts the end product of SPC secretory activity blankets the surface of the cuticle of growing nymphs of *P. crotali* in a foam of polymorphic vesicles and stacked membrane systems up to 12 μm thick (Ambrose & Riley, 1988b, c; 1989: Fig. 3). The epicuticle of instars II–VI is elevated to form a dense mat of hair-like processes that effectively anchors this foam layer to the parasite surface. Macrophages, eosinophils and other inflammatory cells freely penetrate the foam, and many become completely submerged within it, but significantly none reacts to it in any obvious way (Fig. 3). Equally, smooth-surfaced infective nymphs, recovered from the body cavity of rattlesnakes *en route* to the lung, are also membrane-clad, but now in sheets up to twenty deep. Membranous profiles can also be recovered by the centrifugation of culture medium in which either infective nymphs or subsequent lung-stages have been grown, and thus it seems that all instars are permanently enveloped in membranes (Riley, 1992b; Buckle, 1998).

Actively moulting larvae of *P. crotali* encysted in tissues of rodent intermediate hosts become the focus of intense granulomatous responses, characterized in the early stages by a pronounced eosinophilia (Ambrose & Riley, 1988a), and later by the recruitment of mast cells (McHardy, Riley & Huntley, 1993). Cytotoxic responses, as evidenced by degranulation and giant cell formation, are very significant particularly following ecdysis but this activity is localised and directed largely, if not exclusively, against the inner surface of cast cuticles (Fig. 4). The continuously exposed hairy membrane-clad outside surface of the cuticle remains immunologically inert, irrespective of whether it is 'living' or 'dead' (i.e. an integral part of a living nymph or cast/ensheathing) (Ambrose & Riley, 1989). Host cells have direct access to the parasite surface of the first six instars – these nymphs actually feed on eosinophils (Fig. 4) – but the smooth-surfaced seventh stage is quite different. Uniquely, it retains the hairy cast cuticle of the previous instar as a protective sheath which affords an unbroken barrier, totally blocking the access of host cells to the living worm within (Ambrose & Riley, 1988a, b). Clearly, membranes attached to the ensheathing cuticle, and indeed those associated with every cast cuticle, cannot be replenished and yet continue to down-regulate effector cells (Fig. 4). An important corollary is that, at least in intermediate hosts, membranes do not need to be continuously secreted in order to exert their protective function.

The VIIth instar is infective to rattlesnake definitive hosts (Esslinger, 1962a) and can be readily dissected free of its ensheathing cuticle (Buckle, Riley & Hill, 1997). When such 'naked' larvae are transplanted parenterally into naive or previously infected rodents, the cuticle bears the brunt of a sustained attack by macrophages and eosinophils and worms are eventually killed (Fig. 5). Marked differences in the onset and intensity of the inflammatory responses in the two categories of hosts point

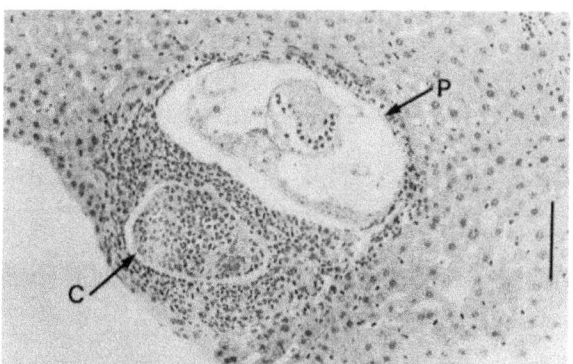

Fig. 4. A light micrograph of a sagittal section through a IIIrd instar of *P. crotali* in mouse liver, showing the onset of granuloma formation. A discrete layer of effector cells, mostly composed of macrophages and eosinophils, surrounds the parasite (P), and large populations of these cells invest a recently cast cuticle (C). The latter, filled with degranulating cells and transformed giant cells, will eventually constitute one of the many independent 'cuticle 'granulomas' within the main lesion. By contrast, cells at the parasite surface appear inactive when observed by electron microscopy. The small round gut of the developing nymph is filled with eosinophils, to the virtual exclusion of other cell types. Reproduced with permission from Ambrose & Riley (1988a). Scale bar = 100 μm.

to some form of specific recognition (Ambrose & Riley, 1989). Since both the ensheathing hairy VIth stage cuticle and the underlying smooth cuticle of the VIIth instar are membrane-clad (Ambrose & Riley, 1988c), stage-specific differences in the surfactants of instars I–VI and VII are thought to underpin these responses.

Ambrose & Riley (1989) postulated that the cuticles of instars I–VI were protected against the rodent phagocytes by so-called 'rodent-compatible' SPC-derived surfactant. Effectively so are VIIth instars in natural rodent infections but only because of protection afforded by the ensheathing VIth stage cuticle, complete with its covering layer of 'rodent-compatible' surfactant. Ambrose & Riley (1989) argued that the encysted VIIth instar in rodents was actually pre-adapted for life in a rattlesnake lung. This hypothesis is based on the observation that larvae remain viable in rodent tissues for several years and, following ingestion by a rattlesnake, penetrate through the stomach wall and into the body cavity within 24 h. Consequently, if VIIth instars are divested of their protective sheath by dissection, and immediately transplanted back into a rodent, only 'snake-compatible' surfactant is available for surface deployment. This is immediately recognised as 'incompatible' and endocytosed by macrophages (Fig. 5A). Subsequently, the worm surface attracts large populations of granulocytes, in various states of activation, which target the now unprotected and vulnerable cuticle (Fig. 5B) (Ambrose & Riley, 1989). By contrast, migrating

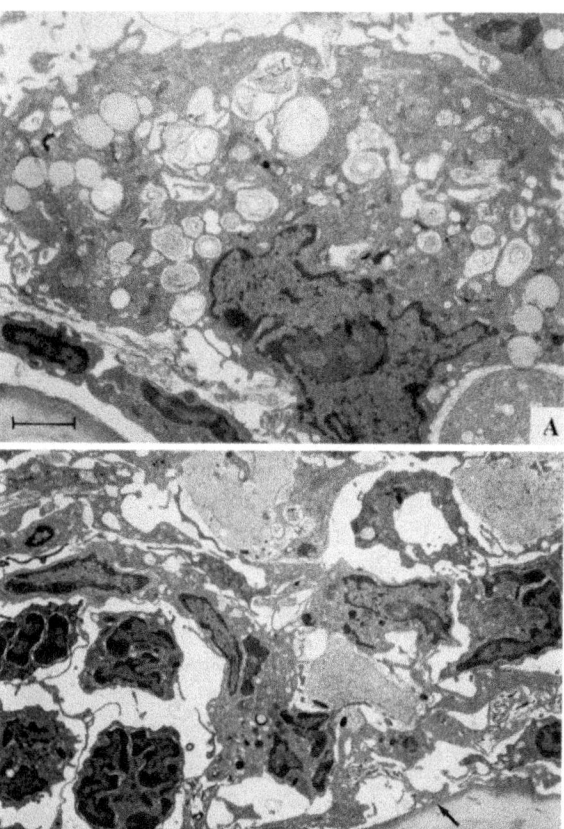

Fig. 5. Electron micrographs of cytotoxic activity around implanted 'naked' nymphs of *Porocephalus crotali* after 3 days in the body cavity of an infected mouse. A. Detail of a macrophage at the surface of the cuticle (bottom left) including numerous endocytotic vesicles, many containing concentric membranes that are highly reminiscent of SPC membranous E/S products. Normally, this material would be destined for deployment at the surface following successful transmission to a snake. However, under these circumstances of experimental manipulation, it appears to be incompatible with the rodent immune system with the result that it is endocytosed. B. The pseudopods of activated macrophages (in the centre of the micrograph) direct lysosomal contents (arrows) against the parasite cuticle (bottom right) amid a welter of granule discharge from eosinophils. The spent remains of the latter are visible as the lighter-staining, flocculent masses of cytoplasm. Reprinted, with permission, from Ambrose & Riley (1989). Scale bars are 2 μm.

infective nymphs, recovered from the body cavity after several days in snakes, are invariably 'clean' (totally free of investing cells (Ambrose & Riley, 1988b)). Following penetration of the lung after about 16 days post infection (Riley, 1981), it is possible that surfactant composition might change once again (Tables 3 and 4; Fig. 7)

Indirect fluorescent antibody tests on sectioned V–VIIth instar nymphs of *P. crotali* probed with infected rat serum (50–150 d.p.i.), or rabbit antiserum raised against purified SPC lamellate droplets

Table 3. The lipid class composition of E/S lipids secreted into MEM by cohorts of cultured *P. crotali* (PC, phosphatidylcholine; PS, phosphatidylserine; PI, phosphatidylinositol; PG, phosphatidylglycerol; PE, phosphatidylethanolamine; CHOL, cholesterol; FFA, free fatty acids; SE, steryl esters)

Instar	Culture age	Worm no.	Extracted lipid μg worm^{-1} d^{-1}	PC	PS	PI	PG	PE	CHOL	FFA	SE
VII	15 h	186	37	33·4	tr	tr	8·4	8	14	20·9	13·5
VII	2 d	125	10·9	34·8	5·3	6·6	0·3	5·1	36·2	5·7	5·9
VIII	21 d	125	97·9	5·6**	6·6	—	—	1·6	22·4	17	46·7
IX	36 d	95	163·2	11·3**	0·7	1·6	2·3	2·1	25	11·4	45·7
X	114 d	90	16·9	29·1	2·6	4·5	—	7·6	36·6	16·2	3·4

** These samples contain relatively low PC and high SE compared to the other instars. tr = trace.

Table 4. Fatty acid composition (wt%) of lipids extracted from MEM in which 122 *P. crotali* infective nymphs had been cultured at the times indicated (blood medium fatty acids are included as a control). Chloroform–methanol extracts were dried down and methylated prior to analysis by gas chromatography. The identifications were confirmed by GC-mass spectrometry

Age of culture	Blood medium	0–22 h	6 d	10 d	20 d	37 d 0–80 min $\longrightarrow$	22 h
Amount of lipid (mg)	0·988	0·404	0·036	0·227	0·048	0·063	0·046
Fatty acid							
16:0	19·6	13·7	21·4	19·0	15·7	21·4	18·0
16:1 n-7	0·6	1·3		1·4			1·3
18:0	19·3	6·8	20·4	18·4	16·0	21·0	10·6
18:1 n-9	14·8	4·3	11·5	6·5	11·6	11·2	10·4
18:2 n-6	22·2	2·3	6·3	3·3	7·7	10·4	8·6
18:3 n-3	3·9					2·0	1·6
20:0	0·3	7·1	7·2	4·7	9·5	2·9	8·5
20:1 n-9	0·2	10·4	13·0	12·0	8·4	1·3	11·5
20:3 n-6	2·1	0·3				1·5	
20:4 n-6	7·2	0·9		1·2	4·4	6·1	5·5
22:0	0·9	6·8	2·4	4·0	5·1	4·2	3·0
22:1 n-9	0·6	11·2	3·2	15·8	6·2	2·2	4·8
22:5 n-6	2·3					1·7	1·1
22:6 n-3	1·9					3·2	3·3
24:0	1·9	13·6	4·9	6·7	5·5	5·9	5·0
24:1 n-9	2·3	21·3	9·8	7·0	10·1	5·0	6·9

from VIIth instar nymphs, shed further light on the hypothesis of the stage-specificity of SPC surfactant (Jones, Henderson & Riley, 1992). Significantly, rabbit antiserum uniformly labelled SPC lamellate droplets and also morphologically identical cells in FG and HG but only in VIIth instars. Pooled serum from infected rats consistently failed to identify similar cells in any instars, although protein-secreting cells in FG and gut enzyme-secretory cells and gut contents labelled consistently (Jones & Riley, 1991; Jones et al. 1991). Ambrose & Riley (1988b, c, 1989) and Riley (1992b) concluded that although the parasite cuticle was potentially vulnerable to immune attack, this rarely occurred because of protective host compatible surfactant.

Jones et al. (1992) partially characterized the lipid and protein composition of SPC-derived lamellate droplets recovered by density gradient centrifugation of pooled homogenised cuticles dissected from the infective nymphs of *P. crotali*. Very small amounts of lipid were recovered and their purity was checked only by the electron microscopical examination of centrifuged pellets. Unsurprisingly there was some inconsistency between the three samples. Nonetheless, neutral lipids comprising cholesterol, cholesterol esters (= steryl esters) and triacylglycerol were dominant (65% of the total lipid), with PC being the most important phospholipid, together with significant amounts of Sph (5%) and PS/PI (5·6%). Palmitic acid (16:0) was the commonest fatty acid but overall there was nothing unusual about the lipid class composition. Anti-lamellate droplet antiserum raised in rabbits labelled both SPC and lipid-secreting cells in FG with equal

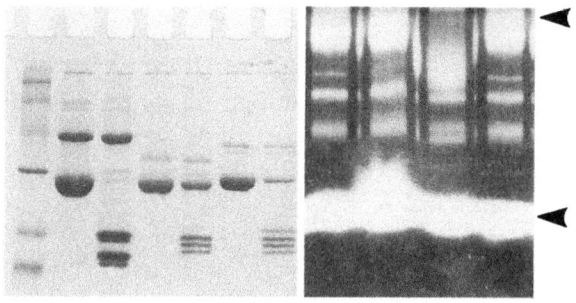

Fig. 6. The left panel is an SDS-PAGE analysis of *P. crotali* (lanes 2 and 3) and *Porocephalus clavatus* (lanes 4–7) frontal gland proteins recovered from pooled, dissected frontal glands. The low molecular weight marker lane proteins (lane 1) ascend through molecular masses of 21, 31, 45, 66, 97, 116 and 200 kDa. Note that in the unboiled lanes (3, 5 and 7), three of four bands of polypeptides are present in the 14–22 kDa range which aggregate to reinforce a band at approximately 30 kDa when boiled (lanes 2, 4 and 6). The metalloproteinase enzyme is at 48 kDa (in *P. crotali*) and 39 kDa (in *P. clavatus*). Lanes 2–5 non-reduced, Lanes 6 and 7 reduced. The Coomassie-stained gelatin substrate gel (right hand panel) shows enzyme activity in *P. crotali* associated with the metalloproteinase at 48 kDa (lower arrow) and a variety of other less active proteins of up to 200 kDa molecular mass (upper arrow).

intensity (Jones *et al.* 1992). Triton-X phase separation of the proteins associated with lamellate droplets sorted them into an aqueous phase (with a major band at 60 kDa), a detergent phase (with two hydrophobic proteins at 16 and 24 kDa) and an insoluble pellet.

Non-lipid E/S products of HG and FG

Morphologically identical cells to SPC are located in every secretory lobule within HG and FG (Fig. 2), and therefore membranes are likely to be implicated also in the function of the non-lipid E/S products discharged from these glands (Ambrose & Riley, 1988a). Nothing is known about the activity of the other secretory cell types present in HG, but since efferent ducts serving this gland discharge over the hooks – effectively the point of most intimate contact between host and parasite – this gland is likely to have important, but as yet undetermined, functions.

SDS-PAGE analysis of whole nymphal homogenate of *P. crotali* followed by Western-blotting and probing with immune rat serum revealed that only two bands labelled consistently. Both are metallo-proteinases, with molecular masses of 48 and 150 kDa, but only the former labelled strongly (Jones & Riley, 1991: Fig. 6). Subsequent double antibody labelling of sectioned nymphs demonstrated that the 48 kDa molecule originated as 2 μm diameter secretory droplets in the dominant cell type in FG (Jones *et al.* 1991: Fig. 2). 70% of cytoplasmic volume is occupied by this secretion (Ambrose & Riley, 1988b). Antibody responses to this enzyme, detected in the serum of infected rats from as early as 12 d.p.i., peaked at about 100–120 d.p.i., coincident with the termination of moulting. Characterization of proteins purified from pooled FG dissected from VIIth instars shows the 48 kDa species to be an elastase-like metallo-proteinase with an alkaline pH dependency. In addition to this proteinase, four low molecular weight peptides (16–22 kDa) were resolved as two doublets which, following boiling in SDS, with or without dithiothreitol, aggregated to form a single band at 31 kDa (Fig. 6). Aggregation may be explained in terms of a conformational change or, more likely, some form of heat-induced cross-linking. In either event, it is a very unusual phenomenon. Antibodies to these small peptides could not be demonstrated in the serum of *P. crotali*-infected rats. However, all of these proteins, together with the 150 kDa species, can be harvested from saline in which living infective nymphs, freed of their ensheathing cuticles by dissection, have been temporarily cultured (Jones *et al.* 1991; Buckle, 1998). The presence of proteases (Fig. 6) is consistent with a tissue-penetrating post-mortem role in transmission. Snakes swallow their prey whole (Riley, 1992b) and < 24 h at 28 °C are required for nymphs to escape from a cyst, migrate through tissues and then penetrate out through the skin of a rodent. However, it is difficult to reconcile this alleged *post-mortem* role with evidence of sustained secretion in rodent tissues and, because glands continue to grow and remain active in lung-stages (see below), with ongoing secretory activity in lungs.

Interestingly, Mariencheck *et al.* (1999) reported that SP-A opsonises a mucoid strain of *Pseudomonas aeruginosa*, thereby enhancing phagocytosis by alveolar macrophage. Moreover, it seems that the bacterium has the capacity to counter this activity by secreting a metalloproteinase that degrades SP-A (Mariencheck & Wright, 1999).

IN VITRO LIPID PRODUCTION BY *P. CROTALI* AND A COMPARISON WITH RATTLESNAKE LUNG SURFACTANT

A number of experiments have been conducted to study the formation and release of surfactant by *P. crotali* and all take advantage of the fact that these parasites can be grown *in vitro* from the VIIth instar to the adult stage (Buckle *et al.* 1997). The basic protocol was modified so that E/S lipids from putative 'lung worms' could be recovered from the culture medium. Two quite different methods were used to determine the nature of the lipids released *in vitro*, and each had its problems (Buckle, 1998). In the first case, $[1-^{14}C]$ palmitic acid was added to

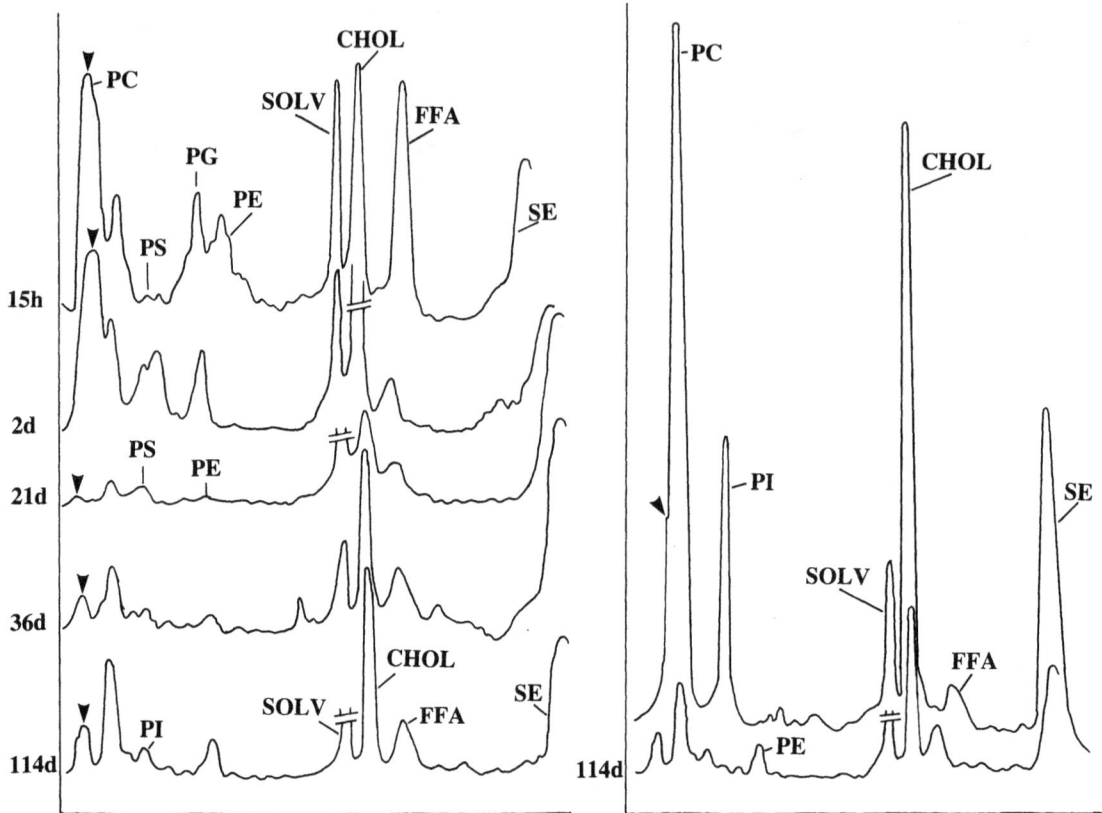

Fig. 7. The left hand panel comprises HPTLC scanning densitometer profiles showing the lipid class composition of E/S lipids, secreted into MEM by cohorts of cultured *P. crotali*. Each was traced from the original scan and then aligned in vertical register (these data are quantified in Table 3). The PC peaks are left, and a gradual shift in relative abundance from the unsaturated form (small arrow) to the saturated species immediately to its right, is apparent with increasing age (from 15 h to 114 d in culture). The abbreviations are explained in Table 3. (SOLV = solvent front). The right hand panel is the identical 114 d profile (left), but this time aligned below a densitometer trace of rattlesnake lung lavage; the latter was far more abundant and hence the big difference in the height of the peaks. The unsaturated PC peak (small arrow) appears as a tiny shoulder on the side of the saturated PC peak in the rattlesnake.

20 ml of blood culture medium – this was known to be the main fatty acid in both lung – and *P. crotali* surfactant (Buckle, 1998). Two batches of worms (80 newly-excysted VIIth instars and 100, 78 d old 'lung worms') were allowed to imbibe blood and thereby acquire label over 48 h. Worms were then washed thoroughly and transferred to normal blood medium. At two day intervals thereafter, 5–10 worms were removed and dissected in order to determine the tissue-localisation of label. All remaining worms were cultured normally, the medium changed every two days and lipids bulk-extracted. The radio-labelled lipid fraction was analysed by a combination of high performance thin-layer chromatography (HPTLC) and scintillation counting. After two days in cold medium, 74–80% of label was located in the head and cuticle of both groups, 13% resided in the gut, whereas the remainder was distributed though FG, body fluids and gut contents (Buckle, 1998). Significant loss of label occurred over the experimental period in both groups, but this was mostly limited to the head/cuticle. Approximately 35% of label was lost from this site after 14 d and 66% after 36 d. It was assumed that, at first, SPC in the head/cuticle acquired palmitate preferentially but that this was subsequently metabolised and exported across the cuticle to the bathing medium. Elsewhere in the worm, label was mostly locked into somatic tissues as part of the normal lipid pool

Extracting ^{14}C-labelled lipid metabolites from whole blood culture media proved difficult. Quenching affected subsequent scintillation counting and this precluded quantification of lipid fluxes. Adequate amounts of labelled lipid metabolites were recovered for analysis, but only from the cohort of 80 infective instars. Among polar lipids, 64% of label initially resided in PC, 18% in PI, 8% in PS, 6·5% in PE and 2·5% in PG. However, after 12 d, PC was the only secreted phospholipid to retain significant label – down to 9% of the original level (Buckle, 1998). At first, neutral lipid profiles were dominated by the release of free fatty acids, almost certainly composed of unabsorbed palmitic acid voided from the worm intestine. The remaining label was located in metabolites such as triacylglycerols (50%), sterols/partial acylglycerols (30%) and cholesterol (= steryl) esters (19%), in all of which label was lost rapidly over 2–6 d.

The other study of *in vitro* lipid release followed cohorts of 90–186 cultured worms continuously from the infective to the Xth instar over a period of 114 d (Table 3). At various intervals worms sieved from blood were washed thoroughly in saline and transferred for periods of 5–18 h into Minimum Essential Medium (MEM), from which lipids were extracted and analysed (Buckle, 1998). The major problem with the interpretation of MEM-derived lipids was that transcuticular E/S surfactant tended to be overwhelmed by faecal blood contaminants. Blood faecal material is indicated by the high proportion of sterol/cholesterol esters (the 21 and 36 d samples in Table 3). Cholesterol and cholesterol esters are known to be major components of the lipids of erythrocytes and plasma lipoproteins in vertebrates and the absorption of dietary cholesterol is slower and less complete than that of other lipids (Gurr & Harwood, 1991). This source of contamination is probably insignificant in the 15 h sample because infective instars do not have access to rodent cells and consequently their intestines would contain few digestible residues (Table 3). However, despite this problem, when the HPTLC scanning densitometer profiles of these various lipids are aligned in vertical register, a significant switch in relative abundance of two forms of PC is strongly associated with increasing worm age (Fig. 7, Table 3). The PC profile is composed of two peaks, the left being the predominantly unsaturated form and the right the disaturated species (Fig. 7). The former dominated the PC profile in the 15 h and 2 d worms, but declined relative to the saturated species thereafter. From 21 d, around the time worms enter the rattlesnake lung during natural infection (Riley, 1981), DSP is the dominant form and by 114 d it is 2·5 times more abundant than the other fraction. This apparent shift in the relative abundance of the two forms with increasing worm age is unlikely to have been influenced by faecal contamination (see below).

Finally, we lavaged a rattlesnake lung (following the protocol of Daniels *et al.* 1995), and analysed it by HPTLC and scanning densitometry. When its densitometer profile is aligned with that of the 114 d lung worms (mentioned above), a close correspondence in lipid peaks is evident (Buckle, 1998: Fig. 7). This observation lends credence to the hypothesis that, in lungs, pentastomids may reduce the antigenicity of the cuticle by coating it with a surfactant which is immunologically compatible with pulmonary surfactant (Riley, 1992*b*).

FUNCTIONAL CONSIDERATIONS

Rattlesnake faveolar lung contains more visible pulmonary surfactant than almost any other vertebrate although its exact function remains unclear (Daniels *et al.* 1995). Moreover, in common with surfactant in other tetrapods it will contain surfactant proteins that enable it to adsorb as a monolayer over the air/water interface of the lung, and much of it will be retrieved and recycled, as happens in mammals (Wright & Clements, 1987; Sullivan *et al.* 1998). Pentastomid surfactant will, presumably, contribute indirectly to the lung surfactant pool but there is no good reason to suppose that it is has to be presented to the host as a monolayer. Lipids in mammalian lung surfactants undergo remote assembly into an "active" form following secretion from type II pneumocytes, a process dependent upon a combination of surfactant proteins SP-A and SP-B (Hawgood & Poulain, 1995). *Porocephalus crotali* surface membranes apparently contain proteins (Jones *et al.* 1992; Buckle, 1998) but it remains to be discovered whether any of these are structurally or functionally analogous with the ubiquitous SP-A present in the pulmonary surfactants of tetrapods (Sullivan *et al.* 1998).

Certainly in the case of *P. crotali* in rodents, pentastomid surfactant is semi-permanent – it remains active even on cast cuticles, physically bound to the tegument by a combination of epicuticular hairs, host cells and connective tissue (Ambrose & Riley, 1988*a*: Fig. 3). Electron microscopy reveals that this surfactant exists as fluid-filled vesicles delimited by one or more bilayers that are somehow extruded onto the cuticle through exceedingly narrow pores (Ambrose & Riley, 1988*c*). Therefore, once pentastomid 'rodent-compatible' surfactant has been secreted, it does not behave like lung surfactant which passes through an unusual intermediate geometry comprising interwoven membranes known as tubular myelin in the liquid layer of the alveolar epithelium (Williams, 1977). We do not know whether surfactant produced by lung-dwelling pentastomids behaves similarly within the lung. Nonetheless, membranes enveloping these instars (Ambrose & Riley, 1988*c*) may require constant renewal as evidenced by the recovery of E/S lipids and membranes from *in vitro* culture medium up to 114 d post-excystation (Table 3).

We have repeatedly analysed fatty acids assumed to be associated with E/S lipids produced *in vitro* by cultured *P. crotali*, but no clear pattern has emerged (Jones *et al.* 1992; Buckle, 1998). This may have been the result of a combination of small sample size and contamination by faecal material. In a recent attempt to resolve this problem, a cohort of 122 *P. crotali* was cultured, but worms were very carefully sieved and gently washed in warm saline to minimise any trauma engendered during transfer to MEM. The 0–22 h and 6 d samples contained very little faecal contamination (easily visualised as black granular haematin) and although copious faecal material was liberated by the later stages, many worms appeared to retain their intestinal contents for extended periods (< 18 h). The fatty acid composition of E/S lipids and that of fresh blood medium which was lipid-extracted as a control were compared. The blood extract was typical of mammalian somatic cell membranes in general in that it contained a high proportion of polyunsaturated fatty acids, especially the essential fatty acids for animal species (18:2 n-6, 18:3 n-3) and the products derived from them (20:4 n-6, 22:5 n-6, 22:6 n-3) (Table 4). In contrast, the 0–22 h VIIth instars had a high proportion of saturated and monoenoic fatty acids, with up to 24 carbons, which presumably must have been synthesised by the worms (Table 4). It is highly unusual to find these in abundance in any type of tissue – certain lipids in nervous tissues are the exception. Possibly, the comparatively long acyl tails on these fatty acids form bridges between opposed hydrophobic domains within the lipid bilayers constituting pentastomid surfactant (Fig. 3), conferring stability and durability.

The prime function of pentastomid surfactant may be simply to lower the antigenic profile of the cuticle but this is unlikely to be the sole function. For example, rodent phagocytes and granulocytes freely penetrate the vesicle layer around encysted *P. crotali* (Fig. 3) and yet, despite their being completely enveloped in material of parasite origin, they remain totally unreactive to it (Ambrose & Riley, 1989). The contrast between the behaviour of eosinophils and macrophages on either side of a cast cuticle is remarkable; on the vesicle-coated side essentially normal cells are found whereas on the underside a veritable orgy of activation and degranulation is evident. During moulting the hairy epicuticle appears to hold inflammatory cells at bay until a new protective vesicle layer is secreted (Ambrose & Riley, 1989).

These observations suggest that it is physical contact between surfactant and granulocytes that renders the latter inactive and thereby minimises inflammation. The PC and PG components of mammalian pulmonary surfactant phospholipids are immunosuppressive (Ansfield & Benson, 1980) and have a key role in down-regulating the activity of alveolar macrophages (Wilsher *et al.* 1988; Geertsma *et al.* 1993). It is tempting to speculate that pentastomid SPC system evolved to exploit this property but the evidence is, at best, circumstantial. There remains a great deal to be learnt about pentastomid surfactant. In concert with the membrane-associated (?) proteins emanating from HG and FG, it has a key role in immune-evasion and reducing inflammation, both of which are critical to the long-term survival of these bizarre and ancient inhabitants of the delicate environment that is the tetrapod lung.

Accepting the risk of appearing anthropomorphic, we leave you with this thought. The next time that you inadvertently bypass the first protective barrier of your respiratory tract, by inhaling something solid that should have been swallowed; when the paroxysms of coughing subside, consider the following (from Self & Kuntz, 1967). "We have one collection of over 100 adult *Porocephalus crotali* taken from the lung of a healthy diamond-back rattlesnake ... This supports the contention of host-parasite compatibility as a result of a long evolutionary association between them" (Fig. 1C).

REFERENCES

ALI, J. H., RILEY, J. & SELF, J. T. (1985). A review of the taxonomy and systematics of the pentastomid genus *Raillietiella* Sambon, 1910 with a description of a new species. *Systematic Parasitology* **7**, 111–123.

AMBROSE, N. C. & RILEY, J. (1988a). Light microscope observations of granulomatous reactions against developing *Porocephalus crotali* (Pentastomida: Porocephalida) in mouse and rat. *Parasitology* **97**, 1–16.

AMBROSE, N. C. & RILEY, J. (1988b). Fine structural aspects of secretory processes in a pentastomid arthropod parasite in its mouse and rattlesnake hosts. *Tissue and Cell* **20**, 381–404.

AMBROSE, N. C. & RILEY, J. (1988c). Studies on the host/parasite interface during the development of *Porocephalus crotali* (Pentastomida: Porocephalida) in rodent intermediate hosts with observations of protective surface membranes. *Tissue and Cell* **20**, 721–744.

AMBROSE, N. C. & RILEY, J. (1989). Further evidence for the protective role of sub-parietal cell membranous secretory product on the cuticle of a pentastomid arthropod parasite developing in its rodent intermediate host. *Tissue and Cell* **21**, 699–722.

ANSFIELD, M. J. & BENSON, B. J. (1980). Identification of the immunosuppressive components of canine pulmonary surface active material. *Journal of Immunology* **125**, 1093–1098.

BAER, J. G. (1952). *Ecology of Animal Parasites*. Urbana, The University of Illinois Press.

BANAJA, A. A., JAMES, J. L. & RILEY, J. (1977). Observations on the osmoregulatory system of pentastomids: the tegumental chloride cells. *International Journal for Parasitology* **7**, 27–40.

BANGHAM, A. D. (1987). Lung surfactant: how it does and does not work. *Lung* **165**, 17–25.

BECKMAN, H. J. & DIERICHS, R. (1984). Extramembranous particles and structural variations of tubular myelin figures in rat lung surfactant. *Journal of Ultrastructural Research* **86**, 57–66.

BOSCH, H. & FRANK, W. (1986). The Australian pentastomid genus *Elenia*, Heymons, 1932 – development in the intermediate and definitive host. *Proceedings of the Sixth International Congress of Parasitology*, p. 97.

BUCKLE, A. C. (1998). A study of the *in vitro*-released products (IVRP) of the pentastomid *Porocephalus crotali*. PhD Thesis, University of Dundee

BUCKLE, A. C., RILEY, J. & HILL, G. F. (1997). The in vitro development of the pentastomid *Porocephalus crotali* from the infective instar to the adult stage. *Parasitology* **115**, 503–512.

CHANDLER, A. C. & READ, C. P. (1981). *Introduction to Parasitology*. 10th Edition. New York and London, John Wiley & Sons.

CRANDALL, R. B. (1960). The life history and affinities of the turtle lung fluke, *Heronimus chelydrae* McCallum, 1902. *Journal of Parasitology* **46**, 63–64.

DANIELS, C. B., BARR, H. A. & NICHOLAS, T. E. (1989). A comparison of the surfactant associated lipids derived from reptilian and mammalian lungs. *Respiratory Physiology* **75**, 335–348.

DANIELS, C. B., ESKANDRI-MARANDI, B. D. & NICHOLAS, T. E. (1993). The role of surfactant in the static lung mechanics of the lizard *Ctenophorus nuchalis*. *Respiration Physiology* **94**, 11–23.

DANIELS, C. B., LOPATKO, O. V. & ORGEIG, S. (1998a). Evolution of surface activity related functions of vertebrate pulmonary surfactant. *Clinical and Experimental Pharmacology and Physiology* **25**, 716–721.

DANIELS, C. B., ORGEIG, S., WILSON, J. & NICHOLAS, T. E. (1994). Pulmonary-type surfactants in the lungs of terrestrial and aquatic amphibians. *Respiration Physiology* **95**, 249–258.

DANIELS, C. B., ORGEIG, S., WOOD, P. G., SULLIVAN, L. C., LOPATKO, O. V. & SMITS, A. W. (1998b). The changing state of surfactant lipids: new insights from ancient animals. *American Zoologist* **38**, 305–320.

DANIELS, C. B., SMITS, A. W. & ORGEIG, S. (1995). Pulmonary surfactant lipids in the faveolar and saccular lung regions of snakes. *Physiological Zoology* **68**, 812–830.

DOUCET, J. (1965). Contribution à l'étude anatomique, histologique et histochimique des pentastomes (Pentastomida). *Mémoires de l'Office de la Recherche Scientifique et Technique Outre-Mer, Paris* **14**, 1–50.

DUNCKER, H. R. (1978). Funktionsmorphologie des Atemapparates und Coelomgliederung bei Reptilien, Vögeln und Säugern. *Verhandlungen der Deutschen Zoologischen Gesellschaft*, 99–132.

DYCK, J. (1975). *Reighardia sternae* sp. nov., a new pentastomid from the guillemot. *Norwegian Journal of Zoology* **23**, 97–109.

ESSLINGER, J. H. (1962a). Development of *Porocephalus crotali* (Humboldt, 1808) (Pentastomida) in experimental intermediate hosts. *Journal of Parasitology* **48**, 452–456.

ESSLINGER, J. H. (1962b). Morphology of the egg and larva of *Porocephalus crotali* (Pentastomida). *Journal of Parasitology* **48**, 457–462.

FARRELL, P. M. (1982). *Lung Development: Biological and Clinical Perspectives*. London, Academic Press.

FOWLER, M. E. (1980). Differential diagnosis of pneumonia in reptiles. In *Reproductive Biology and Diseases of Captive Reptiles* (ed. Murphy, J. B. & Collins, J. T.), pp. 227–223. Society for the Study of Amphibians and Reptiles.

GEERTSMA, M. F., BROOS, H. R., VAN DEN BARSELAAR, P. H. & VAN FURTH, R. (1993). Lung surfactant suppresses oxygen-dependent bactericidal functions of human blood monocytes by inhibiting the assembly of the NADPH oxidase. *The Journal of Immunology* **150**, 2391–2400.

GRIESE, M. (1999). Pulmonary surfactant in health and human lung disease: state of the art. *European Respiratory Journal* **13**, 1455–1476.

GROSS, N. J. (1995). Extracellular metabolism of pulmonary surfactant: the role of a new serine protease. *Annual Review of Physiology* **57**, 135–150.

GURR, M. I. & HARWOOD, J. L. (1991). *Lipid Biochemistry: An Introduction*. 4th Edition (ed. Gurr, M. I. & Harwood, J. L.). London, Chapman and Hall.

HALLMAN, M., EPSTEIN, B. L. & GLUCK, L. (1981). Analysis of labelling and clearing of lung surfactant phospholipids in rabbit. Evidence of bidirectional surfactant flux between lamellar bodies and alveolar lavage. *Journal of Clinical Investigation* **68**, 742–751.

HARWOOD, J. L. (1987). Lung surfactant. *Progress in Lipid Research* **26**, 211–256.

HAWGOOD, S. & POULAIN, F. R. (1995). Functions of the surfactant proteins: a perspective. *Pediatric Pulmonology* **19**, 99–104.

IKEGAMI, M. & JOBE, A. H. (1993). Surfactant metabolism. *Seminars in Perinatology* **17**, 233–240.

JONES, D. A. C., HENDERSON, J. R. & RILEY, J. (1992). Preliminary characterization of the lipid and protein components of the protective surface membranes of a pentastomid *Porocephalus crotali*. *Parasitology* **104**, 469–478.

JONES, D. A. C. & RILEY, J. (1991). An ELISA for the detection of pentastomid infections in the rat. *Parasitology* **103**, 331–337.

JONES, D. A. C., RILEY, J., KERBY, N. W. & KNOX, D. P. (1991). Isolation and preliminary characterization of a 48-kilodalton metalloproteinase of the excretory/secretory components of the frontal glands of *Porocephalus* pentastomids. *Molecular and Biochemical Parasitology* **46**, 61–72.

KRISHNASAMY, M., JEFFERY, J., INDER SINGH, K. & OOTHUMAN, P. (1995). *Raillietiella rileyi*, a new species of pentastomid from the lung of toad, *Bufo melanostictus* from Malaysia. *Tropical Biomedicine* **12**, 31–38.

KRULL, W. H. (1930). The life history of two North American frog lung flukes. *Journal of Parasitology* **16**, 207–212.

KRULL, W. H. (1931). Life history studies on two frog lung flukes, *Pneumonoeces medioplexus* and *Pneumobites parviplexus*. *Transactions of the American Microscopical Society* **50**, 215–277.

KUROKI, Y. & VOELKER, D. R. (1994). Pulmonary surfactant proteins. *Journal of Biological Chemistry* **269**, 25943–25946.

MARIENCHECK, W. I., SAVOR, J., DONG, Q., TINO, M. J. & WRIGHT, J. R. (1999). Surfactant protein A enhances alveolar macrophage phagocytosis of a live, mucoid strain of *Pseudomonas aeruginosa*. *American Journal of Physiology* **277**, 777–786.

MARIENCHECK, W. I. & WRIGHT, J. R. (1999). A metalloproteinase secreted by *Pseudomonas aeruginosa* degrades pulmonary surfactant protein A (SP-A). *American Journal of Respiratory and Clinical Care Medicine* **159**, No. 355, p. A506.

MCGREGOR, L. K., DANIELS, C. B. & NICHOLAS, T. E. (1993).

Lung ultrastructure and the surfactant-like system of the central netted dragon, *Ctenophorus nuchalis*. *Copeia* **2**, 326–333.

MCHARDY, P., RILEY, J. & HUNTLY, J. F. (1993). The recruitment of mast cells, exclusively of the mucosal phenotype, into granulomatous lesions caused by the pentastomid parasite *Porocephalus crotali*: recruitment is irrespective of site. *Parasitology* **106**, 47–54.

MEHLHORN, H. (1988). *Parasitology in Focus*. Berlin, Heidelberg, New York, London, Paris, Tokyo, Springer Verlag.

MIYAZAKI, I. (1991). *Helminthic Zoonoses*. South Eastern Medical Information Center, Publication No. 62. Tokyo, International Medical Foundation of Japan.

ORGEIG, S., SMITS, A. W., DANIELS, C. B. & HERMAN, J. K. (1997). Surfactant regulates pulmonary fluid balance in reptiles. *American Journal of Physiology – Regulatory Integrative and Comparative Physiology* **42**, 2103–2021.

PEREZ-GIL, J., CASALS, C. & MARSH, D. (1995). Interactions of hydrophobic lung surfactant proteins SP-B and SP-C with dipalmitoylphosphatidylcholine and dipalmitoylphosphatidylglycerol bilayers studied by electron spin resonance spectroscopy. *American Chemical Society* **34**, 3964–3971.

PERRY, S. F. (1983). Reptilian lungs: functional anatomy and evolution. *Journal of Advances in Anatomy, Embryology and Cell Biology* **79**, 1–80.

RESTREPO, C. I., DONG, Q., SAVOV, J., MARIENCHECK, W. I. & WRIGHT, J. R. (1999). Surfactant protein D stimulates phagocytosis of *Pseudomonas aeruginosa* by alveolar macrophages. *American Journal of Respiratory Cell and Molecular Biology* **21**, 576–585.

RIDER, E. D., IKEGAMI, M. & JOBE, A. H. (1990). Intrapulmonary catabolism of surfactant-saturated phosphatidylcholine in rabbits. *Journal of Applied Physiology* **69**, 1856–1862.

RILEY, J. (1973). A redescription of *Reighardia sternae* Diesing 1864 (Pentastomida: Cephalobaenida) with some observations on the glandular systems of pentastomids. *Zeitschrift für Morphologie der Tiere* **76**, 243–259.

RILEY, J. (1981). Some observations on the development of *Porocephalus crotali* (Pentastomida: Porocephalida) the Western Diamondback Rattlesnake (*Crotalus atrox*). *International Journal for Parasitology* **11**, 127–131.

RILEY, J. (1986). The biology of pentastomids. In *Advances in Parasitology* **25**, 46–128 (ed. Baker, J. R. & Muller, R.). London, Academic Press.

RILEY, J. (1992a). Sexual differentiation and behaviour of pentastomids. In *Reproductive Biology of Invertebrates* Vol. V. (ed. Adiyodi, K. G. and R. G.), pp. 401–411. Oxford and IBH Publishing Co.

RILEY, J. (1992b). Pentastomids and the immune response. *Parasitology Today* **8**, 133–137.

RILEY, J. (1994). Reproductive strategies of pentastomids. In *Reproductive Biology of Invertebrates* Vol. VI. (ed. Adiyodi, K. G. and R. G.), pp. 293–307. Oxford and IBH Publishing Co.

RILEY, J. & BANAJA, A. A. (1975). Some ultrastructural observations on the integument of a pentastomid. *Tissue and Cell* **7**, 33–50.

RILEY, J. & HUCHZERMEYER, F. W. (1995). Pentastomid parasites of the family Sebekidae Fain, 1961 in West African dwarf crocodiles *Osteolaemus tetraspis* Cope, 1851 from the Congo, with a description of *Alofia parva* n. sp. *Onderstepoort Journal of Veterinary Research* **62**, 157–162.

RILEY, J. & HUCHZERMEYER, F. W. (1996). A reassessment of the pentastomid genus *Leiperia* Sambon, 1922, with a description of a new species from both the Indopacific crocodile *Crocodylus porosus* and Johnston's crocodile *C. johnsoni* in Australia. *Systematic Parasitology* **34**, 53–66.

RILEY, J., JAMES, J. L. & BANAJA, A. A. (1979). The possible role of the frontal and sub-parietal gland systems of the pentastomid *Reighardia sternae* (Diesing, 1864) in the evasion of the host immune response. *Parasitology* **78**, 53–66.

ROSE, J. H. (1976). Lungs. In *Ecological Aspects of Parasitology* (ed. Kennedy, C. R.), pp. 227–242. Amsterdam, North-Holland Publishing Company.

SELF, J. T. (1972). Pentastomiasis: host responses to larval and nymphal infections. *Transactions of the American Microscopical Society* **91**, 2–8.

SELF, J. T. & COSGROVE, G. E. (1972). Pentastomida. In *Pathology of Simian Primates* (ed. Fiennes, R. N. T-W.), pp. 194–204. Basel, Karger.

SELF, J. T. & KUNTZ, R. E. (1967). Host-parasite relationships in some Pentastomida. *Journal of Parasitology* **53**, 202–206.

SMITH, D. S., SMITH, U. & RYAN, J. W. (1972). Freeze-fractured lamella body membranes of the rat lung great alveolar cell. *Tissue and Cell* **4**, 457–468.

SMYTH, J. D. (1976). *Introduction to Animal Parasitology*. 2nd Edition. London, Hodder and Stoughton.

SULLIVAN, L. C., DANIELS, C. B., PHILLIPS, I. D., ORGEIG, S. & WHITSETT, J. A. (1998). Conservation of surfactant protein A: evidence for a single origin for vertebrate pulmonary surfactant. *Journal of Molecular Evolution* **46**, 131–138.

THOMAS, G. & BÖCKELER, W. (1992a). Light and electron microscopical investigations on the feeding mechanism of *Reighardia sternae* (Pentastomida: Cephalobaenida). *Zoologischer Jahrbücher Anatomie* **122**, 1–12.

THOMAS, G. & BÖCKELER, W. (1992b). Light and electron microscopical investigations of the midgut epithelium of different Cephalobaenida (Pentastomida) during digestion. *Parasitological Research* **78**, 587–593.

TINSLEY, R. C. & EARLE, C. M. (1983). Invasion of vertebrate lungs by the polystomatid monogeneans *Pseudodiplorchis americanus* and *Neodiplorchis scaphiopodis*. *Parasitology* **86**, 501–517.

TURK, F. A. (1947). Studies of Acari IV. A review of the lung mites of snakes. *Parasitology* **38**, 17–26.

VON HAFFNER, K. (1972). Die Muskultur und Ortsbewegung der Stachellarven (= Wandernlarven, Nymphen) von *Neolinguatula nuttali* (Pentastomida). *Zoologischer Anzeiger* **189**, 235–244.

WALOSSEK, D. & MÜLLER, K. J. (1994). Pentastomid parasites from the Lower Palaeozoic of Sweden. *Transactions of the Royal Society of Edinburgh, Earth Sciences* **85**, 1–37.

WALOSSEK, D., REPETSKI, J. E. & MÜLLER, K. J. (1994). An exceptionally preserved parasitic arthropod,

Heymonsicambria taylori n. sp. (Arthropoda incertae sedis: Pentastomida), from Cambrian-Ordovician boundary beds of Newfoundland, Canada. *Canadian Journal of Earth Sciences* **31**, 1664–1671.

WILLIAMS, M. C. (1977). Conversion of lamellar body membranes into tubular myelin in alveoli of fetal rat lungs. *Journal of Cell Biology* **72**, 260–277.

WILSHER, M. L., HUGHES, D. A. & HASLAM, P. L. (1988). Immunoregulatory properties of pulmonary surfactant: effect of lung lining fluid on proliferation of human blood lymphocytes. *Thorax* **43**, 354–359.

WRIGHT, J. R. (1997). Immunoregulatory functions of surfactant. *Physiological Reviews* **77**, 931–962.

WRIGHT, J. R. & CLEMENTS, J. A. (1987). Metabolism and turnover of lung surfactant. *American Review of Respiratory Diseases* **135**, 426–444.

Do parasites live in extreme environments? Constructing hostile niches and living in them

C. COMBES* and S. MORAND

Laboratoire de Biologie Animale, UMR 5555 CNRS, Centre de Biologie et d'Ecologie Tropicale et Méditerranéenne, Université de Perpignan, Avenue de Villeneuve, F-66860 Perpignan Cedex, France

SUMMARY

We develop the hypothesis that parasites do not invade extreme environments, i.e. hostile hosts, but rather 'create' them. We argue that parasites may have driven the evolution of the constitutive and adaptive immune system. This leads to several implications. First, parasites respond to 'genes to kill' by 'genes to survive' and this triggers an indefinite selection of measures and counter-measures. Second, these coevolutionary arms races may lead to local adaptation, in which parasite populations perform better on local hosts. Third, the evolution of the immune system, whose responses are predictable, may allow parasites to specialize, to evade and even to manipulate. Finally we show that the correlations between the increase in the antibody repertoire, the expansion of MHC loci and parasite pressures support our hypothesis that both host complexity and parasite pressures can be invoked to explain the diversity of antibodies, T-receptors and MHC molecules.

Key words: immune system, host niche, host specificity, MHC.

INTRODUCTION

When parasites exploit organisms which live in extreme environments (for instance at the most extreme temperatures which are still compatible with life), their free-living stages are exposed to extreme conditions. This is one aspect of the adaptation of parasites to extreme environments. A different question is: can the environment where parasites live (i.e. the host) be qualified as extreme or not?

In recent years, it has been increasingly perceived that organisms modify their environments and that these changes can, in their turn, have consequences for selective pressures exerted by these environments. These changes apply to other organisms in the same ecosystem and to the organisms responsible for the environmental alteration itself. The way an organism modifies natural selection for other organisms sharing its ecosystem relates to the concept of ecosystem engineers developed by Jones, Lawton & Shachak (1994, 1997). The way an organism modifies natural selection for itself is closer to the concept of niche construction developed by Odling-Smee, Laland & Feldmann (1996), and by Laland, Odling-Smee & Feldmann (1996, 2000).

Parasites have already been considered as ecosystem engineers. For instance, Thomas *et al.* (1998) and Poulin (1999) have shown that the behaviour of the cockle *Austrovenus stutchburyi* was altered by metacercariae of the trematode *Curtuteria australis* in such a way that parasitized individuals were unable to bury themselves and remained exposed at the surface of the sediment. In the area under study (shores of New Zealand), cockles are one of the rare hard surfaces where an intertidal community of benthic invertebrates (limpets, anemones, bryozoans, barnacles, etc.) can establish. When a cockle was infected by *C. australis*, limpets (*Notoacmea helmsi*) were significantly more frequent and anemones (*Anthopleura aureoradiata*) less frequent, in relation to the depth at which the substrate provided by the cockle was available. The trematode was thus an ecosystem engineer since it modified the quality of the environment for organisms not directly involved in the parasite–host association.

PARASITES CONSTRUCT HOSTILE NICHES

Regarding the influence that an organism may have on the selective pressures exerted on itself (niche construction), the best possible example is provided by parasite-host relationships themselves because 'living in a living environment' always provokes a dramatic change in the quality of the niche. A living organism which would otherwise provide the parasite with a peaceful habitat and an indefinite source of energy soon becomes a killing machine possessing an impressive battery of weapons (phagocytic cells, cytotoxic cells, antibodies, enzymes, etc.) (Frost, 1999).

Parasites do not invade extreme environments. They 'create' them as we may argue that parasites

* Corresponding author. Tel: 33 4 68 66 21 81; Fax: 33 4 68 66 22 81. E-mail: combes@univ-perp.fr

have driven the evolution of the constitutive and adaptive immune system.[1]

This has several important evolutionary consequences. One is that parasites respond to 'genes to kill' by 'genes to survive' (Combes, 1995). This triggers an indefinite selection of measures and counter-measures which are probably one of the main supporting examples of the Red Queen hypothesis (Van Valen, 1973). In their turn, such arms races provide one of the most likely explanations for maintenance of sex (see for instance Hamilton, Axelrod & Tanese, 1990; Ladle, 1992) and for sexual selection (the quest for 'good genes' which constitutes the basis of Hamilton and Zuk's 1982 hypothesis).

Second, these coevolutionary arms races and the short generation times of parasites as compared to those of their hosts, may lead to local adaptation, in which parasite populations perform better on local hosts. Indeed, several mathematical models (Gandon et al. 1996; Morand, Manning & Woolhouse, 1996) and empirical studies (Lively, 1989, 1999; Xia, Jourdane & Combes, 1998; Moné, Mouahid & Morand, 2000) have shown that parasites are more compatible or better infective to their local hosts, i.e. in sympatric conditions, than to geographically distant hosts, i.e. in allopatric conditions (see Kaltz & Shykoff, 1998). Although the underlying processes and mechanisms are still poorly known, they represent one of the aspects of the survivorship of parasites in the hostile niche.

HOSTILE NICHE AND HOST SPECIFICITY

Bristow (1988) remarked that, in predator-prey systems, specialization arises when a particular defence mechanism has been selected in prey, for instance a thick carapace, the secretion of toxic substances, or an homochromy with the milieu. When such adaptations are selected in prey, counter-adaptations are usually selected in only a small number of predators and allow them to specialize on the resource. Combes & Théron (2000) suggested that specialization in parasites could be the result of a similar process: parasites would be specific (in general) because adaptations necessary to survive in the hostile environments they have 'created' in hosts are extremely costly. Since immunity is achieved through different pathways in different host species (in the same way that different plant species produce different compounds as an adaptation to combat herbivorous insects), parasites specialize to survive in a limited number of these hostile environments. It does seem that the efficiency of evasion mechanisms of a parasite or parasitoid is better when it exploits few host species than when it exploits many (Dupas & Boscaro, 1999) which is an indirect argument for the above statement. 'Extreme' environments following infection in hosts have the same evolutionary consequences as plant defences and trigger the emergence of specialist parasites.

The immune response to a given parasite implies a rather precise cascade of specific and non-specific mediators within a defined temporal sequence (see Davey, 1990). This means that immune responses are predictable, which allows parasites to specialize, to evade and even to manipulate (see Frost, 1999).

Host specificity can thus be the result of building a hostile niche and living in it.[2]

PARASITES AND THEIR EVER-EVOLVING HOSTILE ENVIRONMENT

The sophistication of immune systems has obviously increased from 'lower' to 'higher' organisms. This is especially true when MHC molecules, T-receptors and the antibody repertoire of vertebrates are considered. This 'ever-evolving hostile environment' might have had two different causes, acting in parallel. (1) Through evolution, the increasing complexity of free-living organisms has provided pathogens with an increasing diversity of niches opened to colonization, the result of which was an increasing diversity of parasites. Klein (1991) argued that, in order to have an impact on MHC evolution, a parasite must coevolve over a long period of time, also emphasising which parasites may play this role. We may argue that it is not a parasite species nor a parasite group *per se* (say trypanosomes as exemplified by Klein, 1991) that is responsible for MHC evolution, but rather a community of parasite species, i.e. parasite species richness (but see Paterson, Wilson & Pemberton, 1998). By using parasite species richness as an indicator, Morand (2000) showed that helminth species richness is low in reptiles and amphibians (from 3·7 to 6·7 parasite species per host species) and high in birds (14·0) and mammals (12·0) (Poulin & Morand, 2000; Morand,

---

[1] Ectoparasites 'cumulate' classical pressures from the abiotic and biotic environments with pressures due to immunological responses of the hosts. For instance, ectoparasites living on the gills of fish are directly affected by external factors such as water temperature, water currents, activities of cleaners (Hart et al. 1990). Cleaner fishes in particular have a strong influence on their client's ectoparasite load (Grutter, 1996, 1999; Arnal & Côté, 1998). Endoparasites themselves are submitted to factors that can be qualified as extreme and do not originate in immunity, such as pH, enzyme activities, fluid movements, etc. (Sukhdeo, 1997).

[2] This does not mean that specialization always arises from the hostility of the niche. For instance specialization may occur from a 'lock-and-key' relationship between two anatomical features of the host and the parasite (see Morand et al. 2000 for a positive relationship between gopher hair-shaft diameter and louse head-groove width).

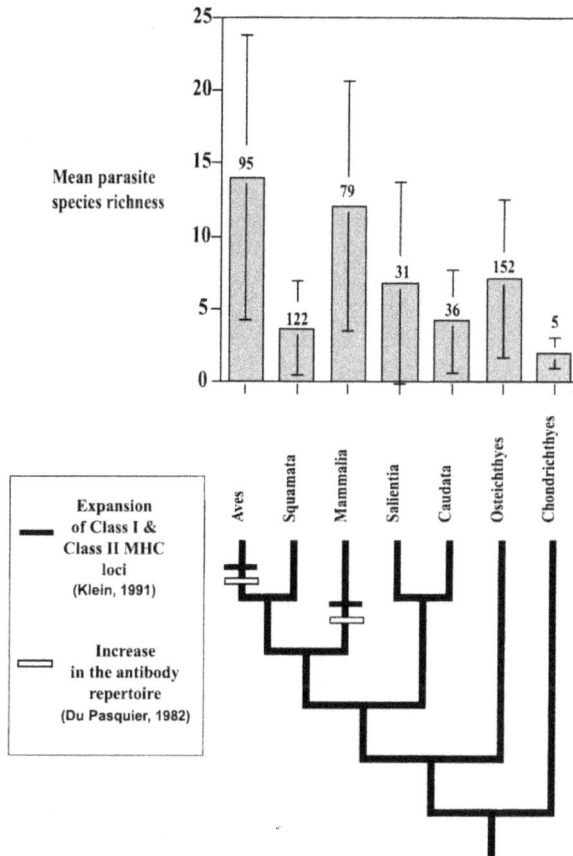

Fig. 1. Correlation between the increase of the antibody repertoire, expansion of MHC loci and parasite richness in vertebrates.

2000) (Fig. 1). Although the helminths are not the only pathogens that may have an effect, we may expect a co-variation in parasite diversity, in such a way that host species harbouring a high species richness of helminths should also harbour a high diversity of protozoans, bacteria or viruses.

The high diversity of parasites was responsible for increasing pressures on immune systems, which responded by more and more sophisticated weapons, especially in vertebrates (Klein, 1991). Lower vertebrates possess fewer specific antibodies than birds and mammals, respectively fewer than 500000 as compared with between 10^7 and 10^9 (Du Pasquier, 1982; see Frost, 1999). Similarly, the emergence of the MHC in the lower vertebrates was followed by dramatic expansion and duplication of MHC genes in birds and mammals (Klein, 1991). MHC genes are not neutral and their persistence and the level of their polymorphism can be explained by balancing selection caused by parasites and pathogens (Takahata, 1990). Other comparative studies should be carried out both inter- and intra-specifically in order to confirm the hypothesis of a parasite-driven force. In response to the evolving immune systems (a 'Red Queen' process ...), pathogens acquired adaptations to escape recognition, using for example antigens similar to host self antigens, to provoke immunosuppression or to provoke non-specific lymphocyte polyclonal activation, which results in the dilution of the specific response against the parasite (Reina-San-Martin, Cosson & Minoprio, 2000).

(2) However, it is probable that pathogens were not alone in being responsible for the complexity of immune systems. We may hypothesize that antigenic diversity increased in proportion to structural complexity of organisms and that the development of the immune system also has been influenced by the necessity to become tolerant to an increasing number of self antigens.

In the current state of knowledge, we think that the correlations between the increase in the antibody repertoire, the expansion of MHC loci and parasite pressures (Fig. 1) support the hypothesis that *both* host complexity and parasite pressures can be invoked to explain the diversity of antibodies, T-receptors and MHC molecules.

CONCLUSION

As a rule, extreme environments and/or hostile environments are defined on the basis of the difficulty of maintaining life and are thus characterized by a low diversity in term of species richness (see for instance Vernon, Vannier & Trehen, 1998). Contrary to this, living in a hostile niche seems to promote both specialization and diversification in parasites. Poulin (1992) and Sasal, Desdevises & Morand (1998) found that high host specificity is correlated with high species diversity in fish parasites.

Finally, it must be stressed that the notion of the extreme environment is somewhat anthropomorphic. For an animal adapted to live in a desert (usually referred to as an extreme environment), it can be lethal to be transported to a temperate climate (supposed to be not extreme). Once an organism is adapted to an environment, this is the environment where its fitness is best, whether we qualify it as extreme or not. It is the same for parasites. It might be concluded that it is more comfortable to live inside a polar bear than on the pack ice. The polar bear does live in an extreme environment. Its parasites do not.

REFERENCES

ARNAL, C. & CÔTÉ, I. M. (1998). Interactions between cleaning gobies and territorial damselfish on coral reefs. *Animal Behaviour* **55**, 1429–1442.
BRISTOW, C. M. (1988). What makes a predator specialize? *Trends in Ecology and Evolution* **3**, 1–2.
COMBES, C. (1995). *Interactions Durables. Ecologie et Evolution du Parasitisme*. Paris, Masson.
COMBES, C. & THÉRON, A. (2000). Metazoan parasites and resource heterogeneity: constraints and benefits. *International Journal for Parasitology* **30**, 299–304.
DAVEY, B. (1990). *Immunology. A Foundation Text*. New Jersey, Prentice Hall.
DU PASQUIER, L. (1982). Antibody diversity in lower

vertebrates – why is it so restricted? *Nature* **296**, 311–313.

DUPAS, S & BOSCARO, M. (1999). Geographic variation and evolution of immunosuppressive genes in a *Drosophila* parasitoid. *Ecography* **22**, 284–291.

FROST, S. D. W. (1999). The immune system as an inducible defense. In *The Ecology and Evolution of Inducible Defenses* (ed. Tollrian, R. & Harvell, C. D.), pp. 104–126. Princeton, Princeton University Press.

GANDON, S., CAPOWIEZ, Y., DUBOIS, Y., MICHALAKIS, Y. & OLIVIERI, I. (1996). Local adaptation and gene-for-gene coevolution in a metapopulation model. *Proceedings of the Royal Society of London, Series B Biological Sciences* **263**, 1003–1009.

GRUTTER, A. S. (1996). Parasite removal rates by the wrasse *Labroides dimidiatus*. *Marine Ecology Progress Series* **130**, 61–70.

GRUTTER, A. S. (1999). Cleaner fish do clean. *Nature* **398**, 672–673.

HAMILTON, W. D., AXELROD, R. & TANESE, R. (1990). Sexual reproduction as an adaptation to resist parasites (a review). *Proceedings of the National Academy of Sciences, USA* **87**, 3566–3573.

HAMILTON, W. D. & ZUK, M. (1982). Heritable true fitness and bright birds: a role for parasites? *Science* **218**, 384–386.

HART, B. L., HART, L. A. & MOORING, M. S. (1990). Differential foraging of oxpeckers on impala in comparison with sympatric antelope species. *African Journal of Ecology* **28**, 240–249.

JONES, C. G., LAWTON, J. H. & SHACHAK, M. (1994). Organisms as ecosystem engineers. *Oikos* **69**, 373–386.

JONES, C. G., LAWTON, J. H. & SHACHAK, M. (1997). Positive and negative effects of organisms as physical ecosystem engineers. *Ecology* **78**, 1946–1957.

KALTZ, O. & SHYKOFF, J. A. (1998). Local adaptation in host–parasite systems. *Heredity* **81**, 361–370.

KLEIN, J. (1991). Of HLA, Tryps, and selection: an essay on coevolution of MHC and parasites. *Human Immunology* **30**, 247–258.

LADLE, R. J. (1992). Parasites and sex: catching the Red Queen. *Trends in Ecology and Evolution* **7**, 405–408.

LALAND, K. N., ODLING-SMEE, F. J. & FELDMANN, M. W. (1996). On the evolutionary consequences of niche construction. *Journal of Evolutionary Biology* **9**, 293–316.

LALAND, K. N., ODLING-SMEE, F. J. & FELDMANN, M. W. (2000). Niche construction, biological evolution, and cultural change. *Behavioral and Brain Sciences* **23**, 131–175.

LIVELY, C. M. (1989). Adaptation by a parasitic trematode to local populations of its snail host. *Evolution* **43**, 1663–1671.

LIVELY, C. M. (1999). Migration, virulence, and the geographic mosaic of adaptation by parasites. *The American Naturalist* **153**, S34–S47.

MONÉ, H., MOUAHID, G. & MORAND, S. (2000). On biogeographical history of *Schistosoma bovis* Sonsino, 1876 in the light of both intermediate host spectrum and compatibility in the mollusc–parasite association. *Advances in Parasitology* **44**, 99–138.

MORAND, S. (2000). Wormy world: comparative tests of theoretical hypotheses on parasite species richness. In *Evolutionary Biology of Host–Parasite Relationships: Reality Meets Models* (ed. Poulin, R., Skorping, A. & Morand, S.). Elsevier (in press).

MORAND, S, HAFNER, M. S, PAGE, R. D. M & REED, D. L. (2000). Comparative body size relationships in Pocket Gophers and their Chewing Lice. *Biological Journal of the Linnean Society* **70**, 239–249.

MORAND, S., MANNING, S. D. & WOOLHOUSE, M. E. J. (1996). Parasite-host coevolution and geographic patterns of parasite infectivity and host susceptibility. *Proceedings of the Royal Society of London, Series B Biological Sciences* **263**, 119–128.

ODLING-SMEE, F. J., LALAND, K. N. & FELDMANN, M. W. (1996). Niche construction. *American Naturalist* **147**, 641–648.

PATERSON, S., WILSON, K. & PEMBERTON, J. M. (1998). Major histocompatibility complex variation associated with juvenile survival and parasite resistance in a large unmanaged ungulate population (*Ovis aries* L.). *Proceedings of the National Academy of Sciences, USA* **95**, 3714–3719.

POULIN, R. (1992). Determinants of host-specificity in parasites of freshwater fishes. *International Journal for Parasitology* **22**, 753–758.

POULIN, R. (1999). The functional importance of parasites in animal communities: many roles at many levels? *International Journal for Parasitology* **29**, 903–914.

POULIN, R. & MORAND, S. (2000). The diversity of parasites. *Quarterly Review of Biology* (in press).

REINA-SAN-MARTIN, B., COSSON, A. & MINOPRIO, P. (2000). Lymphocyte polyclonal activation: a pitfall for vaccination design against infectious agents. *Parasitology Today* **16**, 62–67.

SASAL, P., DESDEVISES, Y. & MORAND, S. (1998). Host-specialization and species diversity in fish parasites: phylogenetic conservatism? *Ecography* **21**, 639–645.

SUKHDEO, M. V. K. (1997). Earth's third environment: the worm's eye view. *BioScience* **47**, 141–149.

TAKAHATA, N. (1990). A simple genealogical structure of strongly allelic lines and trans-species evolution of polymorphism. *Proceedings of the National Academy of Sciences, USA* **87**, 2419–2423.

THOMAS, F., RENAUD, F., DE MEEÜS, T. & POULIN, R. (1998). Manipulation of host behaviour by parasites: ecosystem engineering in the intertidal zone? *Proceedings of the Royal Society of London, Series B Biological Sciences* **265**, 1091–1096.

VAN VALEN, L. (1973). A new evolutionary law. *Evolutionary Theory* **1**, 1–30.

VERNON, P., VANNIER, G. & TREHEN, P. (1998). A comparative approach to the entomological diversity of polar regions. *Acta Oecologica* **19**, 303–308.

XIA, M., JOURDANE, J. & COMBES, C. (1998). Local adaptation of *Schistosoma japonicum* in its snail host demonstrated by transplantation of sporocysts. *ICOPA IX, Monduzzi Editore* (ed. Tada, I., Kojima, S. & Tsuji, M.), pp. 573–576.

Analysis of parasite host-switching: limitations on the use of phylogenies

J. A. JACKSON*

School of Biological Sciences, University of Bristol, Bristol BS8 1UG, UK

SUMMARY

Even the most generalist parasites usually occur in only a subset of potential host species, a tendency which reflects overriding environmental constraints on their distributions in nature. The periodic shifting of these limitations represented by host-switches may have been an important process in the evolution of many host-parasite assemblages. To study such events, however, it must first be established where and when they have occurred. Past host-switches within a group of parasites are usually inferred from a comparison of the parasite phylogeny with that of the hosts. Congruence between the phylogenies is often attributed to a history of association by descent with cospeciation, and incongruence to host-switching or extinction in 'duplicated' parasite lineages (which diverged without a corresponding branching of the host tree). The inference of host-switching from incongruent patterns is discussed. Difficulties arise because incongruence can frequently be explained by different combinations of biologically distinct events whose relative probabilities are uncertain. Also, the models of host–parasite relationships implicit in historical reconstructions may often not allow for plausible sources of incongruence other than host-switching or duplication/extinction, or for the possibility that colonization could, in some circumstances, be disguised by 'false' congruence.

Key words: Host-switching, phylogenies, cospeciation.

INTRODUCTION

The factors which contribute to host-specificity, or the tendency to infect a restricted group of hosts (Poulin, 1998), represent a key constraint on the occurrence of parasitic organisms in nature. As outlined by Combes (1991) this constraint may have two aspects. One is the possession by the parasite of adaptations that allow it to deal with the morphological, physiological and immunological characteristics of particular hosts only. The other is the accessibility of such compatible hosts in the environment. These limitations may not be static. Host-specificity can change over evolutionary time, with new host lineages being acquired and others lost. Such host-shifts could arise either because environmental changes bring parasites into contact with new host organisms to which they are pre-adapted, or because genetic changes in host or parasite lineages alter the way different hosts are utilized. It has been suggested that, due to the cost of host-specific adaptations, there may be a trade-off between the ability to exploit one host efficiently and the ability to exploit others (Poulin, 1998). Host-switching might also be linked to speciation, given that parasite populations dispersing to a new host could become isolated and/or be exposed to novel selection pressures. Changes of host by parasites are, therefore, events of intrinsic evolutionary interest and may have been important determinants of the distributions of present-day parasite groups. However, the factors involved in host-switching can only be studied by comparing lineages that have undergone a history of colonization with those which have not. For this to be possible, instances of host-switching must first be identified. As events that have occurred in the past, colonizations of host lineages by parasites are detected by some form of historical reconstruction. Because parasites have no fossil record, this reconstruction relies primarily on present-day host–parasite distributions and the phylogenies of the host and parasite groups involved. Although parasitologists have long been interested in such analyses (Klassen, 1992), numerical methods have only been developed in the last two decades (Brooks, 1981, 1987, 1988, 1990; Brooks & Bandoni, 1988; Page, 1990a, 1991, 1993, 1994, 1996; Page & Charleston, 1998; Ronquist & Nylin, 1990; Ronquist, 1995, 1998; Siddall, 1996; Charleston, 1998). The motivation for such studies is clear (Brooks, 1988; Paterson & Gray, 1997; Poulin, 1998). Knowledge of the history of a host–parasite association would allow a wide range of inferences about the way in which contemporary characteristics of that association evolved, or the way in which historical events themselves may have been influenced by biological characteristics of the hosts and parasites. It has become increasingly apparent (Brooks, 1980; Poulin 1998; Vickery & Poulin, 1998) that historical host associations may be important determinants of parasite communities, which have often only been

* Tel. +0117 928 7478. Fax: +0117 925 7374.
E-mail: Joe.Jackson@bristol.ac.uk

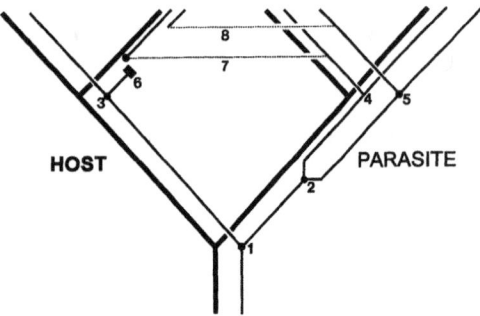

Fig. 1. Historical events in a hypothetical host–parasite association. Parasite evolutionary sequence superimposed on host phylogeny (bold) and to its right, with parasite speciation events indicated by closed circles. Host–parasite cospeciations (1, 3, 5), duplication (= intrahost speciation) (2), host speciation without corresponding parasite speciation (4), parasite extinction (= sorting event) (6), host-switch with speciation of dispersing population (7), host-switch without parasite speciation (8).

considered in the context of present-day ecological factors. Reconstruction of past host colonizations might also provide information on the evolution of important pathogens. For example, analyses of the phylogenies of immune deficiency viruses and their primate hosts have led to contradictory hypotheses on the origins of HIV infections in humans (Mindell, Shultz & Ewald, 1995; Siddall, 1997).

Both intuitive and numerical reconstructions assume that congruent host-parasite trees result from a continuous association between host and parasite lineages (with parasites being inherited by hosts from their ancestors), where parasite speciation events mirror those in the host tree. Incongruent trees are assumed to be the result of host-switching or certain patterns of parasite speciation and extinction. It has, to some extent, been recognized that limitations exist on how much historical information is recoverable by these analyses (e.g. Ronquist, 1995). However, recently developed numerical techniques are being increasingly used and advocated as rigorous methods to investigate historical host-parasite associations (e.g. Brooks, 1988, 1990; Page, Clayton & Paterson, 1996; Paterson & Gray, 1997; Siddall, 1997; Page & Charleston, 1998). Some systems of associating organisms, for example certain bacterial endosymbionts and their invertebrate hosts (Peek et al. 1998; Clark et al. 2000), have been demonstrated to show highly congruent phylogenies. Such instances might reasonably be considered strong evidence for a dominant history of association by descent and cospeciation. However, it is almost axiomatic that host and parasite phylogenies will usually show some degree of incongruence (Hafner & Nadler, 1990; Brooks & McLennan, 1991). While not questioning that a high degree of congruence is likely to be a signal of cospeciation (Hafner & Nadler, 1990), this study aims to examine critically the basis upon which reconstruction methods differentiate host-switching from other historical processes.

TERMINOLOGY

Various terminology has been applied to historical processes in host–parasite assemblages (Brooks, 1979; Brooks & McLennan, 1991; Page, 1993; Ronquist, 1995; Hoberg, Brooks & Siegel-Causey, 1997; Paterson & Gray, 1997; Page & Charleston, 1998). In the following account (see also Fig. 1) the key terms are defined as follows. *Association by descent* (Brooks & McLennan, 1991): where host lineages inherit parasites from their ancestors. *Cospeciation* (Brooks, 1979): where parasite speciation is coincidentally linked to host speciation, with each daughter host species coming to be infected by one of the daughter parasite species. *Duplication* (Page, 1993, 1994) (= 'intrahost speciation'): where a parasite lineage speciates without a corresponding speciation in the host lineage. *Host-switch*: where a parasite invades a new host lineage, leading to *association by colonization* (Brooks & McLennan, 1991). Host-switches may or may not be associated with parasite speciation. *Extinction*: where a parasite lineage dies out from a particular host lineage (irrespective of whether it survives in other lineages). Page (1993, 1994) used 'sorting event' to include extinctions, loss of a parasite lineage from a host population during the early stages of speciation (= 'missing the boat'), or sampling error. Here sampling error will be disregarded, whilst 'missing the boat' is considered a special case of extinction, distinguished by its occurrence in a small founder population (a biologically equivalent event might occur in a single host lineage undergoing a population bottleneck).

CONGRUENCE AND HISTORY

Reconstructions of the history of host–parasite assemblages are based on the degree of congruence or incongruence of the host and parasite phylogenies. It is assumed that, under a restricted set of circumstances, a parasite undergoing association by descent with a host group will show a phylogeny congruent with that of its hosts (Fig. 2). Departures from congruence might be explained by host-switches. However, a major problem is that patterns of parasite speciation and extinction can also produce incongruence ('false' incongruence) despite a history of association by descent (see Fig. 2). The nature of these two sources of incongruence (host-switching or speciation/extinction), as they relate to an analysis of historical host–parasite relationships, are now considered.

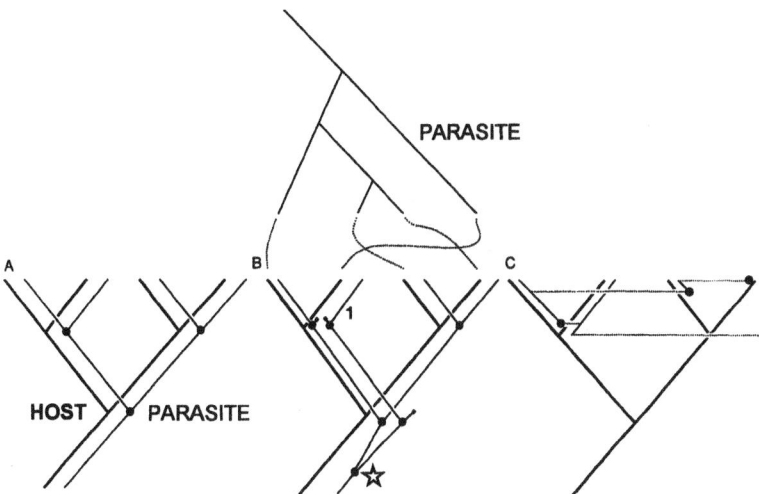

Fig. 2. The effects of cospeciation, duplication and parasite extinction on the congruence of host and parasite phylogenies. A–C. Parasite evolutionary sequence superimposed on host phylogeny (bold) and to its right. Parasite speciation events indicated by closed circles and extinctions by closed squares. A. Congruence due to cospeciation: parasites speciate contemporaneously with their host, with one daughter parasite species being inherited by each daughter host species, producing congruent phylogenies. B. Incongruence produced by one parasite duplication (☆) and three extinctions; parasite phylogeny also shown above with host–parasite associations indicated by stippled lines. C. Colonization sequence producing identical parasite phylogeny to scenario B (colonizations indicated by horizontal stippled lines). (Note ages of nodes. In scenario A (cospeciation) host and parasite nodes are contemporaneous. In scenario B (duplication/extinction) the single duplication (☆) is more ancient than any nodes in the host tree. In scenario C (colonization) speciations of parasite lineages switching between host lineages are always younger than the divergence of the hosts. Note, also, that if lineage 1 in scenario B were to switch to any other host and become extinct on its existing host, then it would still be derived from the ancient node (☆): in some combinations the roles of host-switching, duplication and extinction may not, then, be distinguished by comparison of host and parasite node ages.

Colonization, speciation and extinction

Host-switching. Factors that promote host-switches by parasite lineages are not well known but a number of reasonable speculations may be advanced. The presence in the immediate environment of organisms similar to an existing host in terms of ecology, morphology, physiology and immune responses might be important in reducing the 'adaptive jumps' to new hosts. While transfers of parasites between ecologically similar but phylogenetically distant hosts are well documented, it is likely that, in general, similarity in such characteristics between existing and potential hosts would often depend on their phylogenetic relatedness. This is supported by studies (e.g. Reed & Hafner, 1997) showing that host relatedness determines the level of parasite performance on unnatural compared to natural hosts. The probable link between host phylogeny and parasite host-switches may mean that colonization is more prevalent in parasites of diverse host groups. Poulin (1992), for example, has shown that host range is positively correlated with host group diversity in fish parasites. A further possibility is that the probability of colonization may be affected by potential competitive interactions between established and invading parasite lineages (Barker, 1991, 1996).

At the start of any analysis, the probability of host-switching is unknown and can only be inferred from the degree of incongruence in the host and parasite phylogenies (see below), taking other sources of incongruence into account. Also, given the variety of factors that could influence host-switches, the likelihood of these events might be expected to vary greatly in time and space. For example, parasite transfers may be common in a young, actively radiating host taxon (where many closely related host species are present in the environment), but become less frequent in older taxa composed of relatively ancient, divergent lineages with low speciation rates.

Speciation and extinction. As parasites lack a fossil record their historical speciation and extinction rates are not easy to estimate, the number of species at the present time being determined by the relative rates of speciation and extinction in the past. Identical present-day diversity could, therefore, occur in parasite clades with a rapid or slow turnover of species. It is a major starting limitation of any historical reconstruction that the parasite speciation and extinction rates are unknown or can only be estimated with a high degree of uncertainty. Nee *et al.* (1994*a, b*) proposed modelling methods to derive past (and future) speciation and extinction rates from molecular phylogenetic trees. These are likely to be highly sensitive to starting assumptions about the

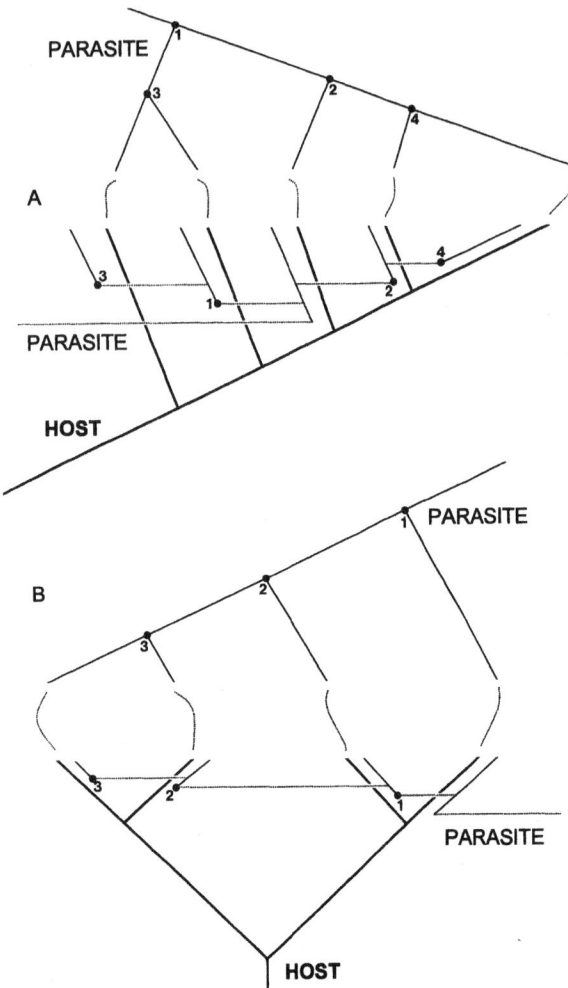

Fig. 3. Sequential colonization: where parasites switch to the nearest unoccupied relatives of their current host. A–B. Host trees shown below (bold), with parasite colonization sequence to one side. Resulting parasite trees shown above with host associations indicated by vertical stippled lines. Closed circles represent parasite speciations (numbering identifies individual nodes) and horizontal stippled lines host-switches. A. Colonization in a pectinate host phylogeny. Switching from the first-colonized host lineage onto the host terminal sister lineages produces a section of parasite phylogeny congruent with the host tree. Colonizations of host lineages more ancient than the one initially invaded result in incongruence. B. Colonization in a symmetrical host tree: host-switches produce an incongruent phylogeny, except for the final switch between host sister lineages.

relationship of speciation and extinction rates over time and to other model characteristics (Nee *et al.* 1994*a, b*).

Host–parasite systems with a high parasite species turnover may have a predisposition towards false incongruence. Even if overall speciation and extinction rates could be estimated from molecular phylogenies however, parasite speciation is of three different types. Cospeciation, duplication and speciation associated with host-switching all have differ-

ing consequences for incongruence between host and parasite phylogenies and it is not clear how their relative probabilities could be estimated.

Other significant historical processes in host–parasite associations

Historical analyses usually only consider congruence as a result of association by descent and incongruence as a result of host-switching or parasite speciation/ extinction patterns (although some authors have considered scenarios in which these assumptions might not hold, e.g. Brooks, 1987, 1988, 1990; Page, 1994). In the following section, the potential significance of other sources of 'false' incongruence (incongruence despite a history of association by descent) and the potential for 'false' congruence (congruence despite a history of host-switching) are considered.

Colonization and phylogeny. False congruence occurs where a history of parasite colonization (Fig. 3) fortuitously produces a parasite phylogeny congruent with that of the hosts. Such scenarios have been considered relatively unlikely on the assumption that colonization is a random process and would only rarely produce congruent patterns. Brooks (1987) suggested spurious congruence might be most likely where a small number of parasites occur in a small number of distantly related hosts. However, as discussed above, the probabilities of host-switches may often be influenced by the phylogenetic relatedness between existing and potential hosts. If colonization occurred in a sequence, with switches occurring from each host to its closest unoccupied relatives, then terminal regions of the parasite phylogeny might show a higher degree of congruence than if host-switching were random (Fig. 3). In this type of scenario, outcomes would depend on host tree topology and the position of initial colonization (Fig. 3). Sequential colonization (subsequent to divergence of the host lineages) in symmetrical host trees would always produce parasite trees that are incongruent. This is except for the final pair of hosts occupied by each invading parasite lineage, if these are derived from a common ancestor from which no other parasitized lineages arise. In the case of pectinate host phylogenies, an initial invasion of a derived host lineage by the parasite would produce a colonization sequence leading to complete incongruence. However, colonization of terminal sister host lineages from a third, basal lineage could result in a congruent section of parasite phylogeny (see Fig. 3). It is perhaps significant that parasite switching between a source host and its unoccupied sister lineage is never incongruent (Fig. 4) and might be common, given the close host genetic relationship (see above). The possibility of 'pseudocospeciation'

Inference of host-switching

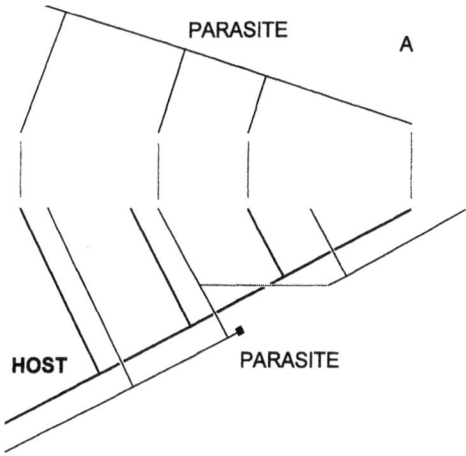

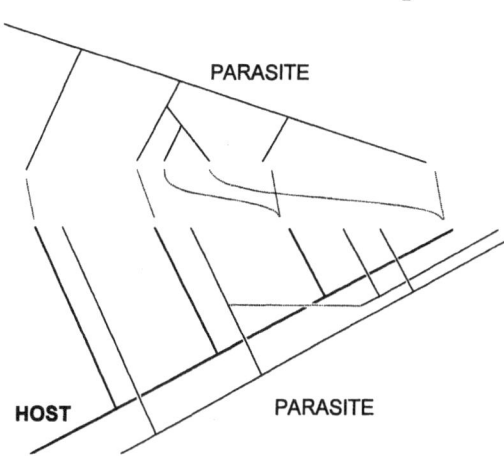

Fig. 4. Detectability of host-switches. A–B. Host trees shown below (bold), with parasite speciation/colonization sequence to one side. Resulting parasite trees shown above with host associations indicated by vertical stippled lines. All nodes represent speciation events. A. A parasite clade colonizes (horizontal stippled line) the sister lineage of its original host which is unoccupied due to an earlier extinction. The resulting host and parasite phylogenies are congruent and could be interpreted as evidence of full cospeciation. B. The same scheme without extinction: the parasite phylogeny is now incongruent with that of the host. In an analysis this incongruence might be explained by a host-switch or duplication and extinction events. Note: host-switches onto the unoccupied sister lineage of a source host are never incongruent. If colonization is influenced by host relatedness and potential competitive interactions with other parasites (see text), this class of host-switch might be relatively common.

by this mechanism has previously been suggested by Hafner & Nadler (1988, 1990). Host-switching may also be associated with the extinction of established parasite lineages (Barker, 1991, 1996). For example, colonization of a host might be more likely if its own parasites had gone extinct due to some stochastic process (leaving an 'empty niche' available), or a colonizing parasite might competitively exclude the previous incumbents. It seems possible, then, that the combination of these two putative processes, the

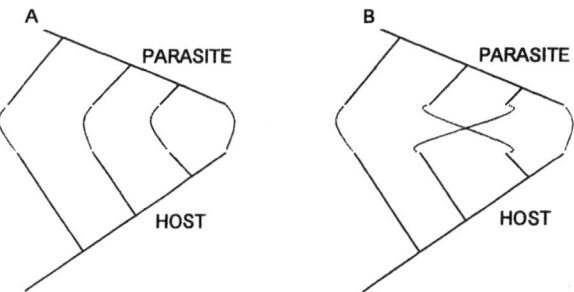

Fig. 5. Conversion of incongruent parasite trees by host extinction (parasite trees above, host trees below; host associations indicated by stippled lines; all nodes represent speciation events). A. Congruent trees: any combination of host extinctions would leave the remaining trees congruent (the same applies to any fully congruent host–parasite phylogeny). B. Incongruent trees: extinction of any one of the terminal three host branches would render the host and parasite trees congruent.

tendency for host-switches to occur between the most closely related hosts (sister lineages) and of colonization to be correlated with extinction of established parasite lineages, could produce a significant degree of false congruence (see Fig. 4).

Host extinction (with corresponding extinction of parasite lines) might also have a ratchet-like tendency to maximize congruence over time. Such events could remove whole sections of the parasite tree, whether incongruent or congruent. However, while local extinctions within a congruent section of host and parasite tree cannot make the residual trees incongruent, they could remove branches from incongruent trees to make these congruent (Fig. 5).

Another scenario in which colonization might produce congruent host and parasite phylogenies is where a parasite lineage colonizes a group of hosts and then speciates in a time sequence influenced by the phylogenetic relatedness of host species (Fig. 6) (effectively, the probability of parasite speciation at any one time decreases with increasing phylogenetic relatedness of hosts).

These considerations call into question a basic assumption of historical reconstruction of host–parasite relationships: that in a host and parasite associated by colonization, no more phylogenetic congruence is expected than if the parasite distribution and phylogeny were independent of the host phylogeny. It is worth noting that, given the vacuum of knowledge concerning what processes are important in past host–parasite relationships, any decision to ignore false congruence as being 'unlikely' is an arbitrary one.

Asynchronous cospeciation. It has previously been recognized that, in a cospeciating system, corresponding speciation events in the host and parasite phylogenies are not necessarily contemporaneous (Hafner & Nadler, 1990; Hafner & Page, 1995). An

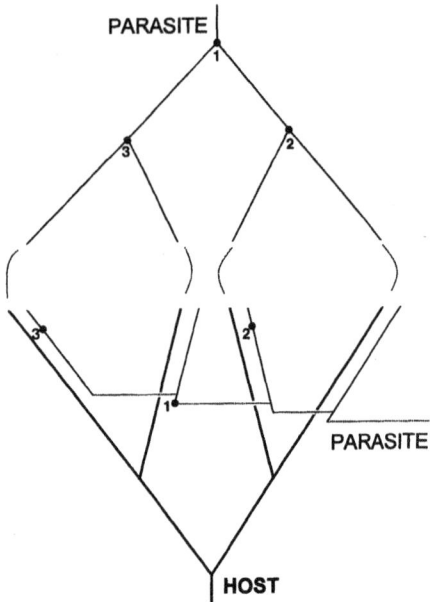

Fig. 6. Congruent host and parasite phylogenies resulting where the speciation probability of host-switching parasite lineages per unit time depends on the genetic relatedness of the colonized hosts. Below: parasite colonization and speciation sequence shown to right of host tree (bold). Above: resulting parasite phylogeny, with host associations indicated by vertical stippled lines. Closed circles indicate parasite speciations (numbering identifies individual nodes) and horizontal stippled lines colonizations. A hypothetical parasite lineage invades members of a host clade during a short episode of colonization: individual sublineages speciate immediately when switching between relatively distantly related hosts and with a delay when switching between closely related hosts.

underlying assumption in historical reconstructions is that where parasite lineages isolated in diverging host lineages speciate, the speciations will be time-ordered with those of the host. However, it is possible to consider biologically plausible scenarios where this might not be the case. For example (Fig. 7): a single parasite species occurs in a host lineage, which radiates into several lineages. Initially the parasite does not co-speciate, retaining some degree of gene flow between populations on the different hosts. As these host clades diverge phenotypically and ecologically at different rates, their parasite populations become isolated and speciate in a different order from their hosts. Under some sets of conditions, association by descent, with pairwise (but asynchronous) host–parasite cospeciation, could therefore produce incongruent host and parasite phylogenies.

Widespread parasites. Speciation within a host lineage that occurs without speciation in its parasites (Fig. 8) can result in 'widespread parasites' (Brooks, 1990; Page, 1994, see also Page, 1990b) that are

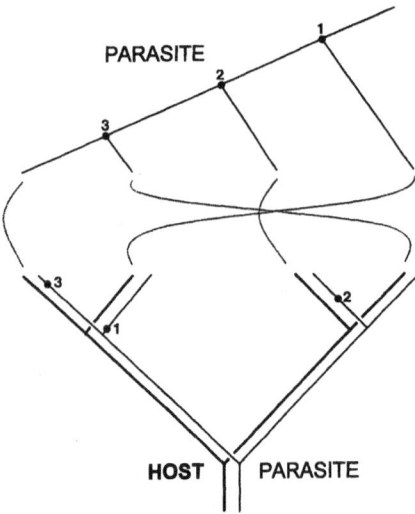

Fig. 7. Asynchronous cospeciation. Below: host tree (bold) with speciation sequence of an associated parasite lineage shown to right (parasite speciation events indicated by closed circles). Above: resulting parasite phylogeny (numbering identifies individual parasite nodes and host associations are shown by stippled lines). Each speciating host lineage inherits the ancestral parasite lineage (association by descent). Parasite populations in individual hosts then speciate in random order, producing incongruent host and parasite phylogenies.

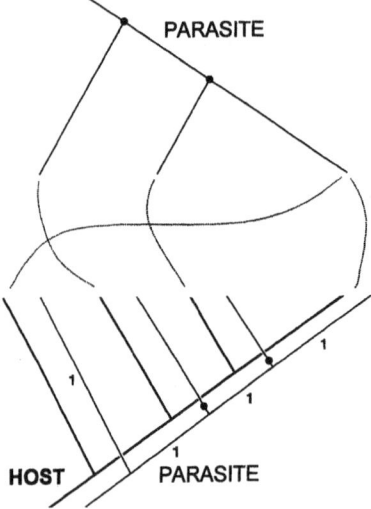

Fig. 8. Widespread parasite species. Below: parasite speciation sequence superimposed on host phylogeny (bold) and to its right. Parasite speciation events indicated by closed circles. Above: resulting parasite phylogeny (host associations shown by stippled lines). Host speciation where the parasite does not cospeciate (i.e. at the basal host node) results in the same parasite taxon being present in basal and derived host lineages (i.e. lineage 1 retains a common genetic identity). Such a distribution could also be explained by host-switching or duplication/extinction. Note, also, that the nodes in the parasite tree would have the same timing as in the present scheme if lineage 1 had colonized the basal host lineage subsequent to the two speciations.

Inference of host-switching

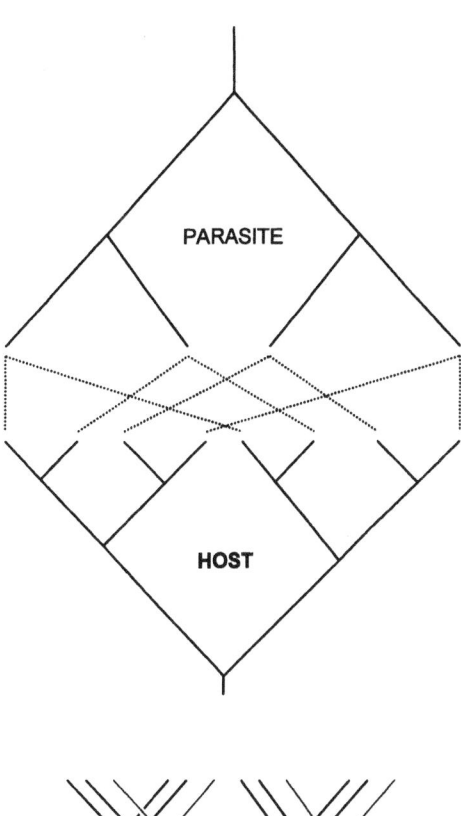

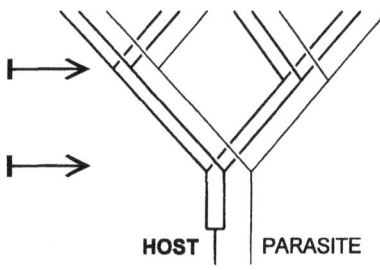

Fig. 9. Cospeciation with host assemblages. Below: a hypothetical scenario where a single parasite initially occurs in host sister lineages. Both host (bold) and parasite lineages (right) then undergo coincident rounds of speciation due to prevailing ecological/geographical factors (arrows) (e.g. habitat fragmentations which successively isolate descendants of both sister lineages and the parasite). Above: resulting host–parasite phylogenies are partially incongruent (parasite phylogeny shown above host phylogeny; all nodes in phylogenies represent speciation events; host-parasite associations indicated by stippled lines).

inherited by unrelated host species. For example, in Fig. 8 the first branching of the host tree is not mirrored by the parasite. Subsequently two cospeciation events occur with two parasite lineages diverging from another, which retains gene flow and genetic identity with populations in the basal host line. This scheme results in one parasite species being present in both the basal and a derived host, a distribution which might equally be explained by host-switching or duplication/extinction events. In effect, scenarios of this type are a special case of the asynchronous cospeciation described above. Rather than parasite speciation responding to host speciation with a variable time delay, it occurs either contemporaneously or not at all.

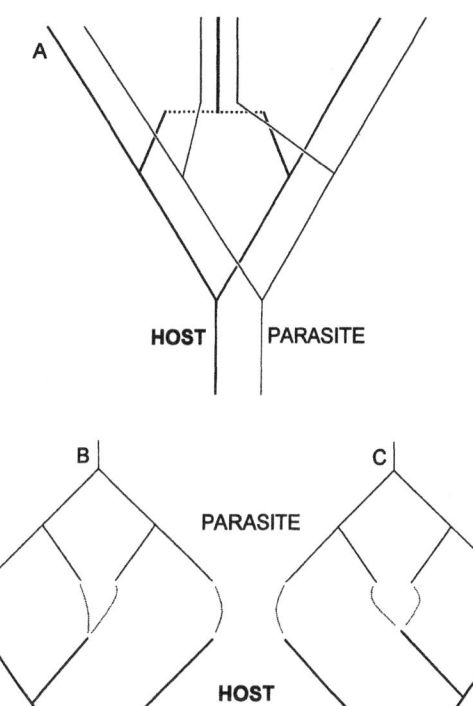

Fig. 10. Effects of coalescent host lineages on historical reconstructions. A. A symmetrical host tree (bold) undergoing coalescence of two non-sister lineages; the parasites (tree shown to right) have previously cospeciated with the hosts and the coalescent host lineage inherits taxa from both of its 'parents'. B–C. If the lineage coalescence went unrecognised, then phylogenetic analysis of the hosts would produce two possible trees. These are shown in B and C (host trees shown below in bold, parasite trees above; host–parasite associations indicated by stippled lines). Each host tree would support a different, spurious, historical scenario.

Cospeciation with host assemblages. A number of historical processes (host-switching, extinction, independent host speciation) may lead to individual parasite species being present in different, unrelated host lineages. Indeed host distributions of this nature are indisputably common in the present day. If parasites had a tendency to coevolve with such non-monophyletic host assemblages, rather than individual lineages, this would have the potential to produce complex, mixed patterns of congruence and incongruence without further host-switching or duplication/extinction. Fig. 9 shows an example where a parasite clade tracks assemblages of non-monophyletic hosts as these speciate in response to prevailing geographical/ecological isolating barriers. In such a case subsequent analysis might suggest a pattern of partial incongruence, even though host-switching and parasite extinctions had not occurred (Fig. 9).

Reticulate evolution. While the possibility of coalescence between lineages is often ignored in historical phylogenetic studies (but see Brooks, 1988), interspecies hybridization and gene introgression occurs in many modern-day animal and plant taxa. Speciation involving polyploidization of hybrids (allopolyploidization) is a significant process in plants, and can also occur in certain animal groups (Kobel, 1996; Chenuil, Galtier & Berrebi, 1999). Hybrid host lineages could potentially inherit all, or a subset, of the parasites from their parental lineages (e.g. Tinsley & Jackson, 1998; Jackson, Tinsley & Kigoolo, 1998). Similarly, hybrid parasites might remain infecting either all, or a subset, of the hosts of their parental lineages. In such situations phylogenetic analysis of the hybrids (if hybridization had occurred between non-sister groups) and their close relatives would produce spurious cladistic reconstructions, which could be congruent or incongruent with those of the associating host/parasite, but which would result in misleading historical reconstructions (Fig. 10). The same principle might apply to fully coalesced lineages as to those which underwent a degree of gene introgression and then diverged (here spurious phylogenies could be constructed if morphological or molecular characters included those affected by introgressed genes). Unfortunately, hybridization and lineage coalescence would be expected to occur sporadically in time and not be easily detectable after the event.

HISTORICAL RECONSTRUCTION

Methodologies

A number of different approaches to the historical reconstruction of host-parasite relationships have been adopted. At their simplest, such analyses are limited to an intuitive identification of possible evolutionary scenarios from a visual examination of host–parasite phylogenies (Brooks, 1979). This has the disadvantage of lacking an explicit, repeatable and exhaustive methodology. Brooks Parsimony Analysis (BPA) (Brooks, 1981, 1987, 1988, 1990) was the first numerical technique to be proposed. It involves mapping of the parasite tree, which is coded as a multistate transformation series, onto the host phylogeny. Parallelism might be interpreted as host-switches or either host-switches or speciation/extinction (Brooks, 1988; Paterson & Gray, 1997). BPA has been criticised (Page, 1993; Paterson & Gray, 1997) for producing spurious reconstructions in some circumstances, and for overemphasizing the importance of host-switching. Brooks (1988), however, seems to emphasize that BPA is properly used as a relatively neutral tool to examine instances of incongruence. These may be attributed to host-switching or other historical processes on intuitive grounds, or on the basis of biogeographic and ecological evidence. Other methods were developed from reconciliation analysis (Page, 1993), a technique originally used in vicariance biogeography (Nelson & Platnick, 1981). Here a parasite tree is reconciled with the host tree in a way that maximizes the number of cospeciations (accounting for incongruence with hypothetical speciations and extinctions). Host-switching is not recognized and can only be inferred by *ad hoc* criteria (Page, 1993). A further development of reconciliation-based methods, TREEMAP analysis (Page, 1994), incorporates searches for optimal reconstructions that do involve host-switches. This may postulate mutually incompatible host-switches, where one switch requires that the other had occurred between hosts not existing at the same time (Ronquist, 1995; Page & Charleston, 1998). More recently, analytical methods that solve the later problem have been proposed (Ronquist, 1995, 1998; Page & Charleston, 1998; Charleston, 1998). These attach relative costs to different events (such as host-switching or speciation), and then seek optimal reconstructions minimizing these costs.

Practical application of numerical methods

The range of host-parasite systems in which some form of numerical reconstruction has been attempted is now relatively varied (see, for example, Brooks, 1988; Brooks & O'Grady, 1989; Hafner & Page, 1995; Clayton, Price & Page, 1996; Hoberg *et al.* 1997; Page *et al.* 1998; Paterson & Gray, 1997; Siddall, 1997; Paterson & Poulin, 1999) and includes digenean, cestode, crustacean, insect and viral parasites/pathogens in vertebrates. The main reconstruction techniques applied have been BPA and reconciliation or TREEMAP analysis, while a few studies have compared both methods. Hoberg *et al.* (1997) analysed highly incongruent relationships in a cestode-bird system with BPA and reconciliation analysis. In contrast, Paterson & Gray (1997) applied BPA and TREEMAP analysis to a mainly congruent louse-bird system with a significant degree of incongruence. Both BPA and reconciliation-based methods indicated the general importance of colonization for the cestode-bird system (Hoberg *et al.* 1997) and of association by descent for the louse-bird system (Paterson & Gray, 1997). However, alternative reconstruction techniques implied quite different evolutionary scenarios for the same host-parasite systems. Hoberg *et al.*'s study found that BPA supported a history of recent colonization, whereas reconciliation analysis explained the host-parasite relationships by more ancient host-switching and cospeciation (with the postulation of higher numbers of speciation and extinction events). Paterson & Gray's TREEMAP analysis inferred ancient duplication of parasite lineages, with a relatively large number of subsequent extinction events. Their BPA suggested a general history of cospeciation, but differed in the way possible instances of host-switching were treated and, the authors noted, made less explicit assertions about the history of parasite lineages.

Different numerical methods may therefore vary markedly in the detail of the scenarios proposed and in the number, age and type of historical processes that are inferred.

Host switching versus extinction

In practice, either parasite host-switching or duplication/extinction is usually invoked to explain incongruent host and parasite phylogenies. However, it has been widely recognized that the patterns produced by these two processes are potentially confounded (Brooks, 1988; Page, 1993, 1994; Barker, 1996). Fig. 2 illustrates how different colonization or duplication/extinction scenarios may result in the same incongruent host–parasite phylogenies and associations. Within the context of BPA or reconciliation-based analyses (e.g. TREEMAP, Page, 1994), the aim is to produce an explanation of incongruence with a minimum number of plausible evolutionary events. The types of event that are implicated may reflect the analytical method used. They may also depend on interpretation by the analyst: one potential reconstruction might be selected over another on intuitive grounds, or on the basis of additional information (e.g. geological or biogeographical data). An initial reservation with such approaches is that it is unreasonable to expect that the evolution of host–parasite associations will always have proceeded through the minimum possible number of steps. Whilst a parsimony criterion may be suitable for phylogeny reconstructions from large sets of characters, it is arguably an unsuitable basis for estimating a one-off historical sequence from what is, effectively, a single complex character (the incongruence between two phylogenies). Further, the different events which could be used to explain incongruence result from fundamentally distinct biological processes (host-switches, speciation, extinction) (Ronquist, 1995) whose relative probabilities may be unknown. As a reconciliation method, TREEMAP analysis maximizes cospeciation and explains incongruence by parasite duplication/extinction. At the same time it allows searches for optimal host-switches that account for incongruence with reduced numbers of speciations and extinctions. The balance between assigning incongruent patterns to host-switching or to speciation/extinction (i.e. selection of particular reconstructions) must ultimately be determined by the investigator. BPA may, in some senses, be considered freer from *a priori* assumptions than reconciliation methods. It does not, as its basis, attempt to maximize one process (cospeciation) by the postulation of certain types of event (speciation/extinction). However, there would usually be a working assumption that congruence is due to cospeciation. Also, where instances of incongruence are identified (and no evidence from other sources is available) their cause can only be assigned by intuitive criteria. As discussed above, alternative numerical methods tend to attribute incongruence to deep parasite nodes with a large number of subsequent extinctions (e.g. TREEMAP analysis) or to more recent host-switching (e.g. possible interpretations of BPA). Given that departures from phylogenetic congruence arise from different classes of event whose relative probabilities are unknown, the information contained within incongruent patterns is inherently ambiguous. A real danger exists, therefore, that any arbitrary assumptions and/or biases underlying analytical procedures may be reflected directly in the scenarios which are inferred.

Some recent studies have attempted to address the problem that distinct historical processes might occur with different probabilities. Ronquist (1995, 1998) and Charleston (1998) have emphasized the assignment of costs (or weights) to particular events (e.g. host-switching, duplication or extinction). These costs relate to the relative probabilities of the events: so that higher costs would be attached to more improbable events and reconstructions would, in some way, minimize the overall cost of an observed pattern in terms of the different event types. However, even if such methodologies are applied, the problem remains as to how the relative costs may be estimated (Charleston, 1998). A possibility is that modelling techniques (Nee et al. 1994 a, b) could be used to determine speciation and extinction rates from molecular phylogenies (Page, 1996). Here, any uncertainty in the estimates would be compounded with that arising from the analysis of congruence/incongruence itself. Also (as discussed above), it may not be possible to distinguish the relative probabilities of different types of parasite speciation (cospeciation, duplication and speciation associated with host-switching), each of which is associated with a different historical process. Huelsenbeck, Rannala & Yang (1997) modelled a host-parasite association (the gophers and chewing lice of Hafner et al. 1994) after estimating host speciation and extinction rates with the methods of Nee et al. (1994 a, b). They assumed that parasites only speciated and became extinct alongside their hosts, and allowed only one extant parasite lineage per host. The actual host-switching rate was estimated by comparing the observed deviance of the host and parasite trees to that produced in simulations using different host-switching rates. However, this approach (and that of Huelsenbeck, Rannala & Larget, 2000) was undertaken at the expense of ignoring parasite duplication and extinction as a source of incongruence.

A further complication is that in cospeciating, non-symmetrical host and parasite trees, the phylo-

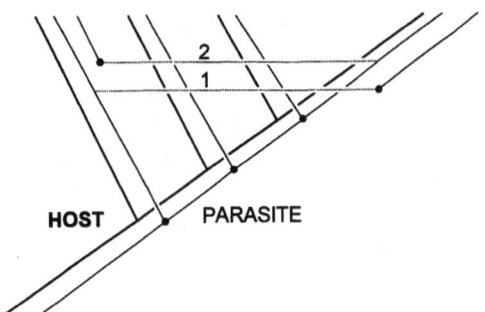

Fig. 11. Equally probable host-switches producing alternative scenarios with different 'costs' (i.e. numbers of events). Host tree (bold) with superimposed parasite tree (right) showing two hypothetical host-switches (stippled lines) between derived and basal hosts. Parasite speciation events indicated by closed circles. Alternative explanations for the resulting host–parasite phylogenies, which rule out host-switches, can be sought by reconciling the parasite trees with those of the host (using the program TreeMap 1.0, Page, 1995, Division of Environmental and Evolutionary Biology, Institute of Biomedical and Life Sciences, University of Glasgow). (Cospeciation is maximized, ignoring the possibility of host-switching, and a reconstruction is produced with the least number of duplications and extinctions.) Reconciliation postulates 3 cospeciations, 1 duplication and 3 extinctions for the scenario involving host-switch (1), and 1 cospeciation, 3 duplications and 8 extinctions for the reciprocal host switch.

genetic position of the parasite (i.e. basal or derived) may determine the degree of incongruence produced by a host-switch. In the example shown in Fig. 11, two host-switch scenarios are compared in terms of the minimum number of speciation and extinction events required to produce an association by descent explanation. Under these criteria, transfer of the basal parasite lineage to one of the most derived hosts requires 3 cospeciations, 1 duplication and 3 extinctions (7 events). The reciprocal switch (i.e. by one of the most derived parasite lineages to the basal host) would require 1 cospeciation, 3 duplications and 8 extinctions (12 events). Amongst equally probable host-switches some, therefore, may be relatively 'cheap' and others 'expensive' in terms of how many events of other types are required to produce alternative explanations. As a result, reconstructions based on minimizing the total costs of different events should ideally be required to weight the cost of host-switches according to their phylogenetic context.

Assumed models of historical host–parasite associations

The above section deals with the limitations of inferring host-switching or speciation/extinction scenarios from incongruent host and parasite phylogenies. However, a further problem with reconstruction methods is that they depend on a restricted model of potential host–parasite association and that other plausible confounding processes are often not taken into account. In practice, only two sources of incongruence are usually considered: host-switching and parasite speciation and extinction (e.g. Page, 1993; Hoberg et al. 1997; Charleston, 1998). It is implicit that parasite cospeciation is time-ordered with host speciation (e.g. Brooks, 1979) and occurs with a monophyletic host lineage. Further, it is assumed that colonization produces congruent host and parasite trees no more often than if host-switches were random and occurred with equal probability between any pair of contemporaneous hosts in an analysis (i.e. the phylogeny and host associations of a colonizing parasite are expected to be independent of the host phylogeny). There is thus a tendency for congruent patterns to be attributed only to association by descent. As outlined above, many of these assumptions are questionable. A number of effects have been identified which might lead to spurious congruence between host and parasite phylogenies after a history of colonization. False incongruence can result from several processes (asynchronous cospeciation; parasite cospeciation with host assemblages; independent host speciation; reticulate evolution scenarios) whose effects may be indistinguishable from each other and from the consequences of host-switching and duplication/extinction.

Timing

The preceding discussions primarily relate to situations in which the ages of branching points within host and parasite phylogenies cannot be considered. Brooks (1987) suggested that instances of spurious congruence could be identified by evidence for the age of host–parasite relationships derived from historical geology. Comparison of the timing of nodes in both host and parasite trees using molecular techniques (Hafner & Nadler, 1990; Hafner & Page, 1995; Page, 1996; Page et al. 1998) might also eliminate many otherwise plausible scenarios from a historical reconstruction. The potential difficulties of inferring lineage age where there might have been local variations in evolutionary rate (perhaps associated with host-switching in the parasite) are beyond the scope of this review. However, if relative branching times can be established this may allow some colonization and association by descent scenarios to be distinguished by the ages of corresponding nodes in the host and parasite phylogenies. Synchronous divergence might be interpreted as evidence for cospeciation (Page et al. 1998). Host-switches, on the other hand, would often be associated with more recent parasite nodes, and duplication/extinction scenarios with more ancient parasite nodes (Page et al. 1998; see also Fig. 2). Procedures for assessing the timing of nodes in host and parasite trees have been considered by Hafner & Nadler (1990), Hafner & Page (1995), Page (1991, 1996) and Huelsenbeck *et al.* (1997). However, even if a temporal comparison of branching events can be achieved, some potential problems remain. For example, colonizations by parasites against a background of duplication and extinction, might, in some circumstances, be associated with nodes more ancient than the corresponding host node (see Fig. 2). Knowledge of node timings cannot be used to distinguish incongruence due to host-switching from false incongruence due to some of the processes described above (e.g. asynchronous cospeciation, coalescent evolution scenarios). It would also not resolve some of the ambiguities which might arise from the presence of 'widespread' parasites (see Fig. 8).

CONCLUSIONS

Numerical reconstructions of historical host-parasite associations usually allow for up to four types of event in parasite lineages: host-switching, cospeciation, duplication and extinction. They assume that association by descent occurs with monophyletic host lineages and that, during episodes of cospeciation, parasite speciations take place in the same temporal sequence as those of their hosts. In practice, host-switching or parasite duplication and extinction are usually invoked to explain incongruence, and significant congruence is attributed to cospeciation. This model may be unrealistically narrow for several reasons: (1) Some effects, including the influence of host phylogenetic relationships on parasite colonization and speciation, may produce spurious congruence. (2) Incongruence may occur despite association by descent if parasite lineages speciate in a different temporal sequence to their hosts (i.e. asynchronous cospeciation). (3) Host speciation without corresponding parasite speciation may result in the same parasite species being inherited by distantly related hosts. (4) The cospeciation of parasite lineages with non-monophyletic host assemblages (without host-switching or duplication/extinction) could be a common source of mixed congruent/incongruent patterns, given the frequency of such associations in nature. (5) Historical coalescence of host or parasite lineages, if undetected, would result in spurious reconstructions.

Even if a limited model of host–parasite historical association is accepted, and incongruence can only be produced by host-switching or duplications and extinctions, these processes are confounded. Numerical methods, which, within the context of the technique applied, aim to produce a reconstruction involving the least number of plausible evolutionary events, do not deal with the problem that different events (e.g. host-switching, speciation associated with host-switching, cospeciation, speciation within a host lineage, or extinction) would occur with different probabilities. The advantage of methods seeking to assign relative 'costs' to distinct events (which are then minimized in a historical reconstruction) is limited by the difficulty of estimating such quantities. Comparison of node timings in both parasite and host trees might distinguish between different historical scenarios involving cospeciation, colonization or duplication/extinction, but some sources of incongruence would still be potentially confounded with others.

The inference of parasite host-switches from incongruence between host and parasite phylogenies, then, is seriously compromised by alternative explanations for incongruence and by the possibility that colonization does not always produce incongruence.

ACKNOWLEDGEMENTS

I am very grateful to Professor R. C. Tinsley and to three referees for suggestions which improved the original manuscript and for support from BBSRC research grant 7/S12169.

REFERENCES

BARKER, S. C. (1991). Evolution of host-parasite associations among species of lice and rockwallabies: coevolution? *International Journal for Parasitology* **21**, 497–501.

BARKER, S. C. (1996). Lice, cospeciation and parasitism. *International Journal for Parasitology* **26**, 219–222.

BROOKS, D. R. (1979). Testing the context and extent of host-parasite coevolution. *Systematic Zoology* **28**, 299–307.

BROOKS, D. R. (1980). Allopatric speciation and non-interactive parasite community structure. *Systematic Zoology* **29**, 192–203.

BROOKS, D. R. (1981). Hennig's parasitological method: a proposed solution. *Systematic Zoology* **30**, 229–249.

BROOKS, D. R. (1987). Analysis of host–parasite coevolution. *International Journal for Parasitology* **17**, 291–297.

BROOKS, D. R. (1988). Macroevolutionary comparisons of host and parasite phylogenies. *Annual Review of Ecology and Systematics* **19**, 235–259.

BROOKS, D. R. (1990). Parsimony analysis in historical biogeography and coevolution – methodological and theoretical update. *Systematic Zoology* **39**, 14–30.

BROOKS, D. R. & BANDONI, S. M. (1988). Coevolution and relicts. *Systematic Zoology* **37**, 19–33.

BROOKS, D. R. & MCLENNAN, D. A. (1991). *Phylogeny, Ecology and Behaviour*. Chicago, Chicago University Press.

BROOKS, D. R. & O'GRADY, R. T. (1989). Crocodilians and their helminth parasites: macroevolutionary considerations. *American Zoologist* **29**, 873–883.

CHARLESTON, M. A. (1998). Jungles: a new solution to the host/parasite phylogeny reconciliation problem. *Mathematical Biosciences* **149**, 191–223.

CHENUIL, A., GALTIER, N. & BERREBI, P. (1999). A test of the hypothesis of an autopolyploid vs. allopolyploid origin for a tetraploid lineage: application to the genus *Barbus* (Cyprinidae). *Heredity* **82**, 373–380.

CLARK, M. A., MORAN, N. A., BAUMANN, P. & WERNEGREEN, J. J. (2000). Cospeciation between bacterial endosymbionts (*Buchnera*) and a recent radiation of aphids (*Uroleucon*) and pitfalls of testing for phylogenetic congruence. *Evolution* **54**, 517–525.

CLAYTON, D. H., PRICE, R. D. & PAGE, R. D. M. (1996). Revision of *Dennyus* (*Collodennyus*) lice (Phthiraptera: Menoponidae) from swiftlets, with descriptions of new taxa and a comparison of host–parasite relationships. *Systematic Entomology* **21**, 179–204.

COMBES, C. (1991). Evolution of parasite life cycles. In *Parasite–Host Associations: Coexistence or Conflict?* (ed. Toft, C. A., Aeschlimann, A. & Bolis, L.), pp. 62–82. Oxford, Oxford University Press.

HAFNER, M. S., HAFNER, P. D., SUDMAN, F. X., VILLABLANCA, X., SPRADLING, T. A., DEMASTES, J. W. & NADLER, S. A. (1994). Disparate rates of molecular evolution in cospeciating hosts and parasites. *Science* **265**, 1087–1090.

HAFNER, M. S. & NADLER, S. A. (1988). Phylogenetic trees support the coevolution of parasites and their hosts. *Nature* **332**, 258–259.

HAFNER, M. S. & NADLER, S. A. (1990). Cospeciation in host–parasite assemblages: comparative analysis of rates of evolution and timing of cospeciation. *Systematic Zoology* **39**, 192–204.

HAFNER, M. S. & PAGE, R. D. M. (1995). Molecular phylogenies and host–parasite cospeciation: gophers and lice as a model. *Philosophical Transactions of The Royal Society of London, B* **349**, 77–83.

HOBERG, E. P., BROOKS, D. R. & SIEGEL-CAUSEY, D. (1997). Host–parasite co-speciation: history, principles and prospects. In *Host–Parasite Evolution: General Principles and Avian Models* (ed. Clayton, D. H. & Moore, J.), pp. 212–235. Oxford, Oxford University Press.

HUELSENBECK, J. P., RANNALA, B. & LARGET, B. (2000). A Bayesian framework for the analysis of cospeciation. *Evolution* **54**, 352–364.

HUELSENBECK, J. P., RANNALA, B. & YANG, Z. H. (1997). Statistical tests of host–parasite cospeciation. *Evolution* **51**, 410–419.

JACKSON, J. A., TINSLEY, R. C. & KIGOOLO, S. (1998). Polyploidy and parasitic infection in *Xenopus* species from western Uganda. *Herpetological Journal* **8**, 19–22.

KLASSEN, G. J. (1992). Coevolution: a history of the macroevolutionary approach to studying host–parasite associations. *Journal of Parasitology* **78**, 573–587.

KOBEL, H. R. (1996). Allopolyploid speciation. In *The Biology of* Xenopus (ed. Tinsley, R. C. & Kobel, H. R.), pp. 391–401. Oxford, Oxford University Press.

MINDELL, D. P., SHULTZ, J. W. & EWALD, P. W. (1995). The AIDS pandemic is new, but is HIV new? *Systematic Biology* **44**, 77–92.

NEE, S., HOLMES, E. C., MAY, R. M. & HARVEY, P. H. (1994a). Extinction rates can be estimated from molecular phylogenies. *Philosophical Transactions of The Royal Society of London, B* **344**, 77–82.

NEE, S., MAY, R. M. & HARVEY, P. H. (1994b). The reconstructed evolutionary process. *Philosophical Transactions of The Royal Society of London, B* **344**, 305–311.

NELSON, G. & PLATNICK, N. (1981). *Systematics and Biogeography: Cladistics and Vicariance.* New York, Columbia University Press.

PAGE, R. D. M. (1990a). Temporal congruence and cladistic analysis of biogeography and cospeciation. *Systematic Zoology* **39**, 205–226.

PAGE, R. D. M. (1990b). Component analysis: a valiant failure? *Cladistics* **6**, 119–136.

PAGE, R. D. M. (1991). Clocks, clades, and cospeciation – comparing rates of evolution and timing of cospeciation events in host–parasite assemblages. *Systematic Zoology* **40**, 188–198.

PAGE, R. D. M. (1993). Parasites, phylogeny and cospeciation. *International Journal for Parasitology* **23**, 499–506.

PAGE, R. D. M. (1994). Parallel phylogenies: reconstructing the history of host–parasite assemblages. *Cladistics* **10**, 155–173.

PAGE, R. D. M. (1996). Temporal congruence revisited: comparison of mitochondrial DNA sequence divergence in cospeciating pocket gophers and their chewing lice. *Systematic Biology* **45**, 151–167.

PAGE, R. D. M. & CHARLESTON, M. A. (1998). Trees within trees: phylogeny and historical associations. *Trends in Ecology and Evolution* **13**, 356–359.

PAGE, R. D. M., CLAYTON, D. H. & PATERSON, A. M. (1996). Lice and cospeciation: a response to Barker. *International Journal for Parasitology* **26**, 213–218.

PAGE, R. D. M., LEE, P. L. M., BECHER, S. A., GRIFFITHS, R. & CLAYTON, D. H. (1998). A different tempo of mitochondrial DNA evolution in birds and their parasitic lice. *Molecular Phylogenetics and Evolution* **9**, 276–293.

PATERSON, A. M. & GRAY, R. D. (1997). Host–parasite co-speciation, host switching, and missing the boat. In *Host–Parasite Evolution: General Principles and Avian Models* (ed. Clayton, D. H. & Moore, J.), pp. 236–250. Oxford, Oxford University Press.

PATERSON, A. M. & POULIN, R. (1999). Have chondracanthid copepods co-speciated with their teleost hosts? *Systematic Parasitology* **44**, 79–85.

PEEK, A. S., FELDMAN, R. A., LUTZ, R. A. & VRIJENHOEK, R. C. (1998). Cospeciation of chemoautotrophic bacteria and deep sea clams. *Proceedings of the National Academy of Sciences, USA* **95**, 9962–9966.

POULIN, R. (1992). Determinants of host-specificity in parasites of freshwater fish. *International Journal for Parasitology* **22**, 753–758.

POULIN, R. (1998). *Evolutionary Ecology of Parasites.* London, Chapman & Hall.

REED, D. L. & HAFNER, M. S. (1997). Host specificity of chewing lice on pocket gophers: a potential mechanism for cospeciation. *Journal of Mammalogy* **78**, 655–660.

RONQUIST, F. (1995). Reconstructing the history of host–parasite associations using generalised parsimony. *Cladistics* **11**, 73–89.

RONQUIST, F. (1998). Three-dimensional cost-matrix optimization and maximum cospeciation. *Cladistics* **14**, 167–172.

RONQUIST, F. & NYLIN, S. (1990). Process and pattern in the evolution of species associations. *Systematic Zoology* **39**, 323–344.

SIDDALL, M. E. (1996). Phylogenetic covariance probability: confidence and historical associations. *Systematic Biology* **45**, 48–66.

SIDDALL, M. E. (1997). The AIDS pandemic is new, but is HIV not new? *Cladistics* **13**, 267–273.

TINSLEY, R. C. & JACKSON, J. A. (1998). Correlation of parasite speciation and specificity with host evolutionary relationships. *International Journal for Parasitology* **28**, 1573–1582.

VICKERY, W. L. & POULIN, R. (1998). Parasite extinction and colonisation and the evolution of parasite communities: a simulation study. *International Journal for Parasitology* **28**, 727–737.

Digenean parasites of deep-sea teleosts: a review and case studies of intrageneric phylogenies

R. A. BRAY[1]*, D. T. J. LITTLEWOOD[1,2], E. A. HERNIOU[1], B. WILLIAMS[1] and R. E. HENDERSON[3]

[1] *Department of Zoology, The Natural History Museum, Cromwell Road, London SW7 5BD, UK*
[2] *Division of Life Sciences, Franklin Wilkins Building, King's College London, 150 Stamford Street, London SE1 8WA, UK*
[3] *Department of Zoology, University of Aberdeen, Tillydrone Avenue, Aberdeen AB9 2TN, UK*
(Current address: International Centre for Island Technology (ICIT), Stromness, Orkney, UK)

SUMMARY

Studies on the digenean parasites of deep-sea (> 200 m depth) teleosts are reviewed and two case study generic phylogenies are presented based on LSU rDNA and ND1 mtDNA sequences. The phylogeny of the lepocreadiid genus *Lepidapedon*, the most common deep-sea digenean genus, is not clearly resolved as the two gene trees are not compatible. It can be inferred, however, that the genus has radiated in the deeper waters off the continental shelf, mainly in fishes of the gadiform family Macrouridae. *Steringophorus*, a fellodistomid genus, is better resolved. In this case a deep-sea radiation is also indicated, but the pattern of host-specificity is not clear, with evidence of much host-switching. Results of studies of the parasites of the macrourid fish *Coryphaenoides* (*Nematonurus*) *armatus* from various depths have reinforced recent views on the lack of zoned depth-related communities in the deep-sea. The diversity of deep-sea digeneans is relatively low with only 18 families (of about 60) reported. Little, or nothing, is known from most deep-sea areas and nothing from trenches and mid-ocean ridge systems.

Key words: Digenea, deep-sea, abyss, molecular phylogeny, *Lepidapedon*, *Steringophorus*, *Coryphaenoides*.

'Ninety per cent of the two-thirds of the surface of the earth covered by sea lies beyond the shallow margins of the continents; and most lies under 2 km or more of water. We may therefore, with some justification, speak of the deep-sea bottom as constituting the most typical environment, and its inhabitants as the typical life-forms, of the solid face of the planet' (Gage & Tyler, 1991).

INTRODUCTION

On the evening of Friday 5 March 1880 H. N. Moseley gave a lecture at the Royal Institution in London describing the findings of the *Challenger* expedition, which had recently returned from its three-year journey around the world. The *Challenger* cruise is widely recognized as the most significant early contribution to deep-sea biology. Moseley, one of the naturalists on the cruise, is reported as stating 'The unhappy deep-sea animals have not escaped their parasites in their cold and gloomy retreat' (Moseley, 1880). Two points stand out from his statement. Firstly, it was recognized that deep-sea animals were parasitized. Secondly, he spoke of the deep-sea as a 'retreat', a view that the *Challenger*

* Corresponding author: Dr R. A. Bray, The Department of Zoology, The Natural History Museum, Cromwell Road, London SW7 5BD, UK. Tel: 020 7942 5752; fax: 020 7942 5151; e-mail: R.Bray@nhm.ac.uk

cruise did much to alter. The deep-sea does, however, present its inhabitants with environmental conditions which are distinct, and from an anthropomorphic point of view, extreme. Our epigram indicates that, although relatively extreme, the deep-sea environment is not extraordinary. 'It may in general be a more 'benign' environment... than most parts of the planet' (Grassle, quoted in Kunzig, 1999).

Many species of teleost fishes are found in the deep-sea and it is apparent that many harbour trematode parasites of the subclass Digenea. These are not, as yet, well known, but it has become clear that these digeneans are generally distinct from those recovered in shallow waters. The nature of this distinction is less clear, in that it is not known whether there are distinct lineages of digeneans which have radiated in the deep-sea, or whether these adaptable animals have continuously invaded the deep-sea from other environments. Evidence from recent phylogenetic studies (Cribb et al. in press) indicates that none of the common deep-sea fish digeneans is likely to be basal for the group, indicating that the group is unlikely to have arisen in deep-water. To attempt to give some indications of the mechanism and frequency of the invasion of the deep-sea, and from this to estimate the ease with which these organisms are able to adapt to such conditions as high pressure and low temperature, we have made inferences relating to the phylogeny of two digenean genera commonly found in deep-sea

Table 1. Definitions of bathymetric adjectives

Term	Pertaining to
Neritic	The shallow waters over the continental shelf
Epipelagic	The pelagic zone above about 200 m depth
Bathyal	The sea-floor between the shelf-break (usually about 200 m) and about 3000 m depth
Bathypelagic	The pelagic zone between 1000 m and about 3000 m depth
Benthopelagic	The pelagic zone below about 100 m above the deep sea floor, i.e. a narrow zone above the sea-floor
Demersal	The benthic zone and the pelagic zone close to the sea-floor
Mesopelagic	The pelagic zone from between about 200 m and 1000 m depth
Abyssopelagic	The pelagic zone from between about 3000 m depth and about 100 m off the sea-floor
Abyssal	The sea-floor below about 3000 m depth (Angel, 1997), but not including deep trenches, i.e. the bottom of the continental rise and the abyssal plain
Hadal	Deep trenches below about 6000 m (cf. Hades)

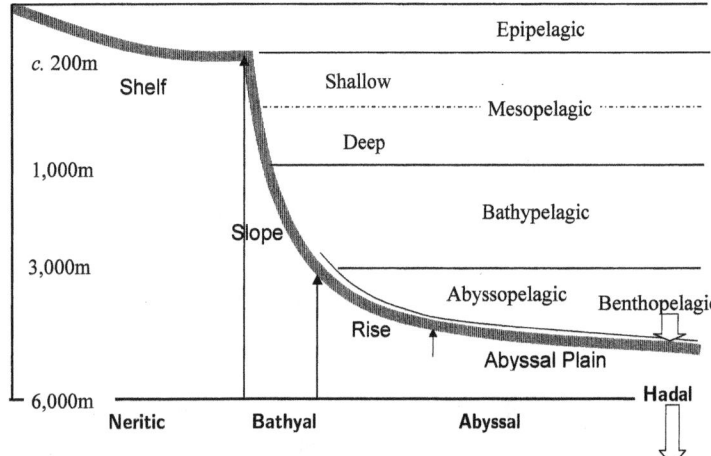

Fig. 1. The nomenclature of the deep-sea, see also Table 1.

teleosts. Phylogenies were estimated using partial nuclear large subunit (28S) and partial mitochondrial nicotinamide adenine dinucleotide dehydrogenase subunit 1 (ND1) gene sequences. In addition, we have reviewed current knowledge of deep-sea digeneans in an attempt to give an outline of some of the adaptations they have acquired to survive in this environment, where suitable hosts, both intermediate and definitive, may be sparsely distributed.

The deep-sea

We consider the 'deep-sea' to be those regions off the continental shelf, that is the area beginning at the 'shelf break'. In many areas this is at c. 200 m, but may be as deep as c. 500 m in the Antarctic due to the weight of the ice-cap (Gage & Tyler, 1991). In this sense the deep-sea incorporates the benthic and benthopelagic areas of the sea off the continental shelf. This includes the continental slope, continental rise and abyssal plain and would have included the trenches and mid-ocean ridges if any data had been available. The nomenclature of the parts of the deep sea is described in Table 1 and shown in Fig. 1. The continental shelves are the shallow parts of the ocean; the region where light is available for attached plants and the region where most demersal fisheries are carried on. The edge of the shelf at the 'shelf break' gives way to the relatively steeply shelving continental slope, which in turn becomes the less steeply shelving continental rise. At the base of the rise the sea bottom flattens out to form the abyssal plain (around 5000 m deep in the NE Atlantic, about 4000–6000 m elsewhere). Trenches, which occur near tectonic subduction zones, may be much deeper. Submarine canyons may dissect the continental slope and sea-mounts may arise from the slope, rise or abyssal plain. The abyssal plain covers most of the deep-sea floor and, like the continental shelf, has a gradient of less than 1 in 1000, whereas the slope gradient may be greater than 1 in 40. A high proportion (53·5%) of the earth's surface is covered by the deep-sea below about 3279 m, the mean depth of the ocean. Clearly our epigram is not an exaggeration. Approximately 79% of the volume of the marine component of the biosphere lies below 1000 m (Somero, 1992). Whilst in topographical terms the deep-sea starts at the edge of the continental shelf, in hydrographic terms it starts at the thermocline, above which the temperature drops rapidly from surface temperature to below 4 °C, but below which the temperature drops only gradually. In fact the thermocline can vary greatly in depth, vary with season and latitude, or be interrupted by up-welling or sinking of water bodies. Most cold deep-water is fed to the rest of the globe from the Antarctic and the northeastern Atlantic, as there is little contact between Arctic waters and both the north Pacific and the northwestern Atlantic, and none with the Indian Ocean.

For many years it was thought that the deep-sea fauna evolved from the shallow water fauna and had sought refuge and exploited new niches by invading the deeper areas of the ocean (Woodward, 1898). Recoveries in the 19th century of such apparently ancient forms as stalked crinoids, glass-sponges and *Spirula* fed this belief, but the discovery of a highly complex and rich fauna by the *Challenger* expedition cast doubt on this assumption (Broad, 1997). Now the deep-sea itself is seen as a major environment rather than a refuge.

Along with gradients in abiotic factors (see Table 2), there are bathymetric gradients of organisms in the deep-sea. The depth over which a species ranges varies, of course, with eurybathic species ranging over a wide vertical distribution and stenobathic species over a narrow depth range. It is no longer tenable to consider the fauna to consist of zoned communities, rather each species occurs over its own bathymetric range (Haedrich, 1997). Ekman (1953) showed that it was not possible to define a generally applicable upper limit to the deep-sea fauna. Around the Antarctic continent the shelf is depressed by ice cover and there is no distinction between the shelf and deeper faunas.

Weitzman (1997) listed 157 fish families known to include deep-sea species or species which occasionally penetrate below 500 m. This is about 30% of known fish families, and includes mainly relatively 'primitive' teleosts. 'Derived' orders, such as the Perciformes, with over 150 families, have relatively few deep-sea representatives. In north Atlantic temperate waters, at least, the order Gadiformes includes a high proportion of the more prominent benthopelagic species, with the families Macrouridae and Moridae particularly common. In other parts of the world ophidiids, alepocephalids or scorpaeniforms predominate (Merrett & Haedrich, 1997). It has been reckoned that 'bigger fish live in deeper water', and this so-called 'bigger-deeper' trend has been referred to as 'Heincke's Law' (Haedrich, 1997). This interspecific relationship has been shown in general to be an artefact resulting from net avoidance by larger fishes in shallower water, but the relationship may hold within species, including in *Coryphaenoides* (*Nematonurus*) *armatus*, the host discussed in more detail in this paper.

Most collecting of deep-sea marine parasites has been from benthic or benthopelagic fishes. This is partly due to the relative ease of collecting by benthic trawl, and partly due to the unrewarding results from deep-pelagic sampling. Although conflicting in detail, the surveys of Collard (1970) of mesopelagic fishes and of Gartner & Zwerner (1989) of meso- and bathypelagic fishes, agree in the recovery of low numbers of digeneans (and indeed other helminth parasites). Gartner & Zwerner (1989) concluded that this was a reflection of the 'lower overall energy of, and reduced probability of host-to-host transfer in, the deep-sea pelagic ecosystem'. Most of our discussion below will be on benthic and benthopelagic ecosystems.

The collection of samples from the deep-sea is not easy, cheap nor, in many cases, comfortable. Most collecting of deep-sea demersal fishes has been by trawl, although some have been taken in traps and epibenthic sleds (see Gage & Tyler, 1991, Chapter 3 for descriptions of the equipment used in deep-sea collecting). Most of the original material discussed in this paper was collected in a semi-balloon otter trawl (OTSB) from the NERC research vessels *Challenger* and *Discovery*. Teleost fishes brought up from depth usually have the stomach everted through the mouth due to pressure changes and sometimes the rectum is everted through the anus. This may, presumably, lead to loss of parasites. The fish are dead on arrival on deck, but if their guts are dissected quickly, the parasites will usually be found alive (or at least moving). It is usually necessary to examine the guts or a suspension of gut contents under a binocular microscope, an activity that is not always straightforward at sea.

Digeneans in the deep-sea

The first record of a deep-sea digenean is probably that of Wagener (1852) who reported a metacercaria encysted in the gills of a macrourid fish (Table 3). As mentioned above, parasites were recognized in the collections made during the voyage of the H.M.S. *Challenger* between 1873 and 1876 (Linklater, 1972). Most were studied by von Linstow (1888), who reported no digeneans from deep-sea fishes and stated that there is 'in the wide ocean but a slight chance that infected intermediate hosts become the prey of the final victims'. Nevertheless, it appears that this same expedition supplied the worm described by Bell (1887). Virtually no further studies were undertaken during the nineteenth and early twentieth centuries, and systematic study of deep-sea digeneans can be reckoned to have started with Manter's (1934) study of moderately deep forms in Floridian waters (for details of this and other 'landmark' papers on deep-sea digeneans see Table 3). In the second half of the twentieth century many more studies have reported deep-sea digeneans, but few have set out to study them systematically. The

Table 2. Conditions in the deep-sea

Light	Surface incident light cannot sustain attached seaweeds below 200 m and is not detected below 1000 m (Tyler, 1995). Below 1000 m light is restricted to the dim glow emitted from black smoker vents at mid-oceanic ridges and bioluminescence.
Pressure	Terrestrial habitats at a pressure of 1 atmosphere [atm = 0·101 megapascal (Mpa)] or less account for less than 1% of the biosphere (Somero, 1992). Pressure increases 1 atm for every 10 m of depth (Tyler, 1995). The oceans have an average depth of about 3800 m and an average pressure of 381 atm (38·5 Mpa) and animal and bacterial life occurs even at pressures of 1100 atm. Adaptations to pressure appear to be an important factor in replacement of species over bathmetric gradients, and obligate barophiles may require high pressure for survival. The constraints of pressure physiology have been invoked as reasons for the decrease in diversity with depth. The region below 6000 m ($\sim$ 60·7 Mpa) forms an apparent boundary both faunistically and topographically, e.g. several higher taxa, such as decapod crustaceans, anemones and echinoids do not occur below 6000 m. Apart from a record of an unnamed digenean (also one monogenean, two cestodes, two nematodes and two copepods) from *Macrourus acrolepis*, captured between 2500 and 7000 m in the Pacific Ocean, we have no information on digeneans from deeper than the abyssal plain in the NE Atlantic ($\sim$ 5000 m) (Gusev, 1957). The tertiary structure of enzymes and structural proteins is altered, e.g. adaptation of fish enzymes to high pressure is associated with the acquisition of increased resistance to denaturation (Somero, 1992), but the 'amino acid sequence changes that are involved in adapting proteins for function at high pressure remain to be elucidated' (Somero, 1998, p. 53). Nevertheless, pressure effects on water density around proteins, ions or substrates are more likely to affect processes than changes in the volume of macromolecules themselves. The involvement of complex assemblages of macromolecules in transcription and translation result in these processes being highly sensitive to pressure, in bacteria at least (Somero, 1992). It is clear, therefore, that the physiological effects of pressure have major impacts on the evolutionary distinctness of forms at different depth. In other words, we would not expect one species to spread over a large bathymetric range, particularly into shallow water. In fact, however, species may appear to be adapted to a wide range, e.g. the vertebrate *Coryphaenoides* (*Nematonurus*) *armatus* and the digenean *Gonocerca phycidis*. *G. phycidis* is reported in macrourids from just below 200 m to abyssal depths (4850 m) (Bray, 1995a), but is replaced by the common species *Derogenes varicus* in shallow water. *G. phycidis* is also common in Antarctic nototheniids (Zdzitowiecki, 1997a). Pressures at depth of about 500 m (perhaps even shallower) are great enough to cause selection for pressure resistance (Somero, 1992). The eurybathy of many deeper-water digeneans and their occurrence in Antarctic shelf waters indicates that pressure effects are not as important as temperature in their distribution.
O_2 concentration	Most open areas of the deep-sea bottom are oxygenated (Tyler, 1995) and any oxygen deficiency, which increases rapidly in the sediment, probably does not affect digeneans or their hosts materially.
Temperature	Temperature variation below the thermocline (800–1300 m depth in stable temperature regimes, deeper elsewhere, e.g. where Mediterranean outflow depresses 4 °C isotherm) is low. At the thermocline the temperature is just below 4 °C and decreases slowly below that. Fauna must, therefore, be adapted to cold, but only to a narrow range of temperatures. Somero (1998) reckoned, however, that 'adaptation to low temperatures does not appear to pre-adapt animals for life at depth'.
Salinity	The salinity at depth is more or less constant at 35‰ and is 'unlikely to have any ecological consequences for the deep-sea benthic fauna' (Tyler, 1995).
Suspended particulate matter	The concentration of suspended particles decreases with depth, but there are regions and levels, the so called nepheloid layers, where currents scour the sediment and the load is high. Eventually this material settles and may form a food source for benthic feeders.
Radiation	Cosmic radiation is absorbed by the surface layers, but marine sediments may contain radioactive elements such as radium and uranium. The effects of naturally occurring radiation on deep-sea organisms are not well understood.
Seabed topography	The major sea-bed topography (e.g. mid-ocean ridges, sea mounts, etc.) is getting fairly well known. Less familiar and less understood are the effects of micro- and meso-topography on the scale of millimetres and decimetres.
Seabed composition	Hard rock strata form the sea bottom on areas where the slope is too steep for sedimentation. The dominant substratum is, however, sediment, where mean grain size decreases with depth. Sediment is susceptible to modification by 'benthic storms', turbidity currents and slumps. These changes in bottom conditions can greatly affect the biota, causing drastic changes in its character. Sediment type contributes to the small-scale patchiness of the deep-sea bed.

Table 2 (cont.)

Energy availability	The deep-sea bottom relies on food (energy) which descends from illuminated surface waters, except in the regions close to hydrothermal vents and cold seeps. Globally, phytoplankton is the primary producer, contributing to the deep-sea energy budget either directly via phytodetritus or indirectly by animal-remains. The pulse of phytodetritus to the deep-sea is now known to reflect surface seasonality, imposing seasonality on deep-sea communities. Food falls such as large vertebrates form a focus for relatively large scavenging animals, such as amphipods and fishes. Even terrestrial plant matter may form a significant source of energy for deep-sea organisms.
Currents	The deep-sea is not generally tranquil not quiescent (Tyler, 1995) in that there are currents and eddies, some moving large amounts of water around at depth. It is known that scavenging organisms locate their food downstream in currents. Filter-feeding organisms rely on currents to pass suspended matter past them.

Table 3. *Some landmark papers in the investigation of deep-sea digeneans*

Wagener (1852)	First record of digenean: the metacercaria of *Stephanostomum hystrix* (Dujardin, 1845) encysted in the gills of *Lepidoleprus trachyrhynchus* [now *Trachyrincus scabrus*].
Bell (1887)	Description of a worm collected during the *Challenger* expedition, as *Distomum halosauri* (now known as the gorgoderid *Degeneria halosauri*, see Campbell, 1977) from *Halosauropsis macrochir*, likely to have come from the Marion Islands at a depth of 2515 m (Harrison, 1966).
Linton (1898)	Description of the deep-sea hemiurid *Distomum laeve* (now *Dissosaccus laevis*) from *Macrourus* (now *Nezumia*) *bairdii* off Massachusetts.
Odhner (1911)	Description of the deep-sea zoogonid *Proctophantastes abyssorum* from *Coryphaenoides rupestris* off Trondheim, Norway.
Manter (1934)	Reported on digeneans collected down to 1064 m in Floridian waters. He noted the 'surprising' lack of overlap between the worms in shore fishes and those occurring at 91–110 m, and pointed out the narrow bathymetric distribution of some species.
Yamaguti (1938)	Descriptions of digeneans from argentinids, epigonids, macrourids, ophidiids and synaphobranchids from Japan.
Manter (1954)	Descriptions of digeneans from a macrourid, New Zealand.
Gusev (1957)	First record of parasites, including unnamed digeneans, to 7000 m depth.
McCauley (1968)	Description of six *Lepidapedon* species from macrourids and morids off Oregon.
Collard (1970)	Study of mesopelagic fishes, finding low levels of digenean prevalence.
Noble (1973)	First review article on deep-sea parasitism (including digeneans).
Campbell *et al.* (1980)	First detailed study of bathymetric distribution of parasites. Found high levels of helminth parasitism (80%) (including digeneans) for benthic deep-sea fishes in the NW Atlantic.
Zubchenko (1981*a*)	Use of parasites (including digeneans) as biological tags in assessment of population structure of macrourid *Coryphaenoides rupestris*.
Zubchenko (1981*b*, 1984)	Ecological surveys of macrourid and alepocephalid parasites, with prevalence and intensity data.
Campbell (1983, 1990)	Two review articles synthesizing work to date.
Gartner & Zwerner (1989)	Study of meso- and bathypelagic fishes, finding low levels of digenean prevalence.
Bray & Gibson (1991)	First record of digenean apparently restricted to abyssal depths.
Lumb *et al.* (1993)	The first use of molecular techniques to study deep-sea digeneans.
Bray (1995*a*)	Checklist of digeneans of macrourids, with graphic illustration of known bathymetric ranges.

most notable exceptions are the studies of Noble and colleagues (e.g. Noble, 1973), Campbell and colleagues (e.g. Campbell, Haedrich & Monroe, 1980), Zubchenko (1981*a*, *b*, 1984) and Blend and colleagues (e.g. Blend, Dronen & Armstrong, 2000). As a result of these studies and including data from many other papers, we have developed a database listing nearly 840 records of just over 200 species (see more detailed discussion below).

Two of the most commonly encountered deep-sea genera were chosen for phylogenetic study. Eight species of the fellodistomid genus *Steringophorus*

Table 4. Details of the material used in the molecular phylogenies. * indicates use as outgroup to root the trees

Parasite	Host, locality, depth, date of collection, research vessel	GenBank/EMBL	
		ND1	28S rDNA
Fellodistomidae			
*Fellodistomum agnotum Nicoll, 1909	ex *Anarhichas lupus* (Perciformes: Anarhichadidae), North Sea (60° 55′ N, 00° 34′ E, depth 135 m, 13.v.1995). RV *Scotia*.	AJ405249	AJ405289
*Fellodistomum fellis (Olsson, 1868)	ex *Anarhichas lupus* (Perciformes: Anarhichadidae), North Sea (60° 55′ N, 00° 34′ E, depth 135 m, 13.v.1995). RV *Scotia*.	AJ405250	AJ405290
*Olsonium turneri Bray & Gibson, 1980	ex *Alepocephalus agassizi* (Salmoniformes: Alepocephalidae), Goban Spur (50° 02′ N, 12° 48′ W, depth 2441 m, 11.viii.1997). RRS *Challenger*.	AJ405248	AJ405288
Steringophorus blackeri Bray, 1973	ex *Xenodermichthys copei* (Salmoniformes: Alepocephalidae), Goban Spur (49° 35′ N, 11° 50′ W, depth 960 m, 14–15.viii.1992). RRS *Challenger*.	AJ405256	AJ405296
Steringophorus dorsolineatum (Reimer, 1985)	ex *Bathypterois dubius* (Aulopiformes: Chlorophthalmidae), Goban Spur (49° 32′ N, 12° 48′ W, depth 1541 m, 16.viii.1997). RRS *Challenger*.	AJ405251	AJ405291
Steringophorus furciger (Olsson, 1868)	ex *Limanda limanda* (Pleuronectiformes: Pleuronectidae), St Abbs (56° 05′ N, 02° 07′ W, depth 59 m, 23.v.1990). RV *Scotia*.	AJ405252	AJ405292
Steringophorus haedrichi Bray & Campbell, 1995	ex *Spectrunculus grandis* (Ophidiiformes: Ophidiidae), Goban Spur (49° 54′ N, 13° 03′ W, depth 2570 m, 8.vii.1994). RRS *Challenger*.	AJ405253	AJ405293
Steringophorus margolisi Bray, 1995	ex *Spectrunculus grandis* (Ophidiiformes: Ophidiidae), Rockall Trough (56° 44′ N, 09° 45′ W, depth 1745–1760 m, 19.xi.1991). RRS *Challenger*.	AJ405254	AJ405294
Steringophorus pritchardae (Campbell, 1975)	ex *Alepocephalus rostratus* (Salmoniformes: Alepocephalidae), Goban Spur (49° 30′ N, 13° 09′ W, 1625 m, 12.viii.1992). RRS *Challenger*.	AJ405255	AJ405295
Steringophorus sp.	ex *Cataetyx laticeps* (Ophidiiformes: Bythidiidae), Goban Spur (49° 33′ N, 13° 10′ W, depth 1654 m, 7.vii.1994). RRS *Challenger*.	AJ405259	AJ405299
Steringophorus thulini Bray & Gibson, 1980 – 1	ex *Coryphaenoides* (*Chalinura*) *mediterranea* (Gadiformes: Macrouridae), Rockall Trough (56° 44′ N, 09° 45′ W, depth 1745–1760 m, 19.xi.1991). RRS *Challenger*.	AJ405257	AJ405297
Steringophorus thulini Bray & Gibson, 1980 – 2	ex *Coryphaenoides* (*Chalinura*) *leptolepis* (Gadiformes: Macrouridae), Porcupine Abyssal Plain (50° 10′ N, 14° 53′ W, depth 4100 m, 5–6.viii.1992). RRS *Challenger*.	AJ405258	AJ405298

Lepocreadiidae

Neolepidapedon smithi Bray & Gibson, 1989	ex *Mora moro* (Gadiformes: Moridae), Goban Spur (49° 38′ N, 11° 49′ W, depth 985 m, 14.viii.1997). RRS *Challenger*.	AJ405284 AJ405270
Preptetos caballeroi Pritchard, 1960	ex *Naso vlamingii* (Perciformes: Acanthuridae), Heron Island, Queensland, Australia (23° 27′ S, 151° 55′ E).	AJ405269 AJ405273
Prodistomum priedei Bray & Merrett, 1998	ex *Epigonus telescopus* (Perciformes: Epigonidae,), Goban Spur (49° 33′ N, 11° 37′ W, depth 747 m, 14.vii.1997). RRS *Challenger*.	AJ405286 AJ405272
Profundivermis intercalarius Bray & Gibson, 1991	ex *Coryphaenoides* (*Nematonurus*) *armatus* (Gadiformes: Macrouridae), Porcupine Abyssal Plain (49° 50′ N, 14° 18′ W, depth 4143 m, 9–10.viii.1997). RRS *Challenger*.	AJ405285 AJ405271
Lepidapedon arlenae Bray & Gibson, 1995	ex *Trachyrincus scabrus* (Gadiformes: Macrouridae), Goban Spur (49° 30′ N, 12° 57′ W, depth 1500 m, 10.viii.1992). RRS *Challenger*.	AJ405276 AJ405262
Lepidapedon beveridgei Campbell & Bray, 1993	ex *Coryphaenoides* (*Nematonurus*) *armatus* (Gadiformes: Macrouridae), Porcupine Abyssal Plain (50° 10′ N, 14° 35′ W, depth 3965 m, 9.viii.1997). RRS *Challenger*.	AJ405277 AJ405263
Lepidapedon desclersae Bray & Gibson, 1995	ex *Mora moro* (Gadiformes: Moridae) Goban Spur (49° 38′ N, 11° 49′ W, depth 985 m, 14.viii.1997). RRS *Challenger*.	AJ405278 AJ405264
Lepidapedon discoveryi Bray & Gibson, 1995	ex *Coryphaenoides* (*Nematonurus*) *armatus* (Gadiformes: Macrouridae), Porcupine Abyssal Plain (50° 10′ N, 14° 35′ W, depth 3965 m, 9.viii.1997). RRS *Challenger*.	AJ405279 AJ405265
Lepidapedon elongatum (Lebour, 1908)	ex *Gadus morhua* (Gadiformes: Gadidae), North Sea (56° 25′ N, 02° 09′ W, depth 52 m, 22.v.1990). RV *Scotia*.	AJ405280 AJ405266
Lepidapedon gaevskayae Campbell & Bray, 1993	ex *Coryphaenoides* (*Nematonurus*) *armatus* (Gadiformes: Macrouridae), Rockall Trough (54° 43′ N, 12° 07′ W, depth 2890 m, 17.xi.1991). RRS *Challenger*.	AJ405281 AJ405267
Lepidapedon rachion (Cobbold, 1858)	ex *Gadus morhua* (Gadiformes: Gadidae), St Abbs (56° 05′ N, 02° 06′ W, depth 64 m, 6.vi.1993). RV *Scotia*.	AJ405274 AJ405260
Lepidapedon rachion (Cobbold, 1858)	ex *Melanogrammus aeglefinus* (Gadiformes: Gadidae), Bressar (60° 22′ N, 01° 13′ E, depth 142 m, 9.vi.1993). RV *Scotia*.	AJ405275 AJ405261
Lepidapedon sommervillae Bray & Gibson, 1995	ex *Coryphaenoides guentheri* (Gadiformes: Macrouridae), Goban Spur (49° 27′ N, 13° 19′ W, depth 2000 m, 7.viii.1992). RRS *Challenger*.	AJ405282 AJ405268
Lepidapedon zubchenkoi Campbell & Bray, 1993	ex *Coryphaenoides* (*Chalinura*) *leptolepis* (Gadiformes: Macrouridae), Goban Spur (50° 10′ N, 14° 53′ W, depth 4100 m, 5–6.viii.1992). RRS *Challenger*.	AJ405283 AJ405269

Odhner, 1905 and nine of the lepocreadiid genus *Lepidapedon* Stafford, 1904 were collected from various hosts and localities, along with various outgroup taxa (Table 4). Both genera are dominated by deep-sea forms. *Steringophorus* contains about 15 species (Bray, 1995b; Bray & Campbell, 1995; Zdzitowiecki, 1997b), 12 of which are deep-sea species. *Lepidapedon* is a bigger genus, with about 47 species (Bray & Gibson, 1995; Zdzitowiecki & Cielecka, 1997; Blend et al. 2000) again with the majority (about 33) found in deep-water, mostly in fishes of the gadiform family Macrouridae.

CASE-STUDIES ON THE PHYLOGENY OF TWO DIGENEAN GENERA

Material and methods

The original results in this paper are from collections made aboard the NERC research vessels *Challenger* and *Discovery* in 1989, 1992, 1994 and 1997. Shallow water forms are mainly from collections made aboard the SOAEFD research vessel *Scotia* in 1990, 1992, 1993, 1995 and 1996. When fish arrive on deck, collecting of parasites continues as for shallow water fishes, but due to the hazard of handling hot saline or fixatives in unstable conditions, worms for morphological study were fixed in Berland's fluid and preserved in 80% ethanol. Worms for molecular study were fixed and preserved in 100% ethanol and are listed in Table 4 along with GenBank/EMBL accession numbers for all sequences reported herein.

Gene amplification and sequencing. Prior to DNA extraction individual worms were rehydrated in two washes and one 1 h soak in TE (pH 8·0). Individuals were ground in 150 μl TE (pH 8·0), 0·5% SDS and digested for 3–4 h with the addition of 6 μl proteinase K (10 mg ml^{-1}) at 37 °C. Genomic DNA was phenol-chloroform extracted and precipitated over 15 mins at -20 °C in the presence of 0·1 volume sodium acetate, pH 4·5–6·0, and 2·5 volumes 100% ethanol. After washing in 70% ethanol DNA pellets were dried and redissolved in TE (pH 8·0).

Partial 28S rDNA was amplified from each extract with PCR (Saiki et al. 1988) using primers LSU5 and ECD2 listed in Littlewood, Rohde & Clough (1997) using the same PCR conditions. ND1 segments were amplified using primers NDJ1 (5'-AGATTCGTAAGGGGCCTAATA) from Morgan & Blair (1998) and ND1J2A primers (5'-CTTCAGCCTCAGCATAATC). Standard 25 μl PCR reactions were performed using PCR Beads (Pharmacia) and the cycling conditions were generally hot start (95 °C/5 min) followed by 30 cycles of 94 °C/1 min, 50 °C/1 min and 72 °C/1 min. At least two reactions were performed for each template.

Amplified products were cleaned directly with Qiaquick PCR purification kit (Qiagen) or were run out on a 1% TAE agarose gel, cut out, pooled and purified using Qiaquick gel extraction kit (Qiagen).

Gene fragments were directly sequenced using standard reaction mixes and procedures on a 373 or 377 ABI automated sequencer with the Big Dye Readymix reaction kit (ABI, Perkin Elmer). Products were sequenced with the primary PCR primers only. Both strands of the amplified DNA products were sequenced fully and contigs were assembled with Sequencher v. 3.0 (Gene Codes Corporation, MI).

Sequence alignment. Both 28S and ND1 sequences were each aligned for each taxon group (*Steringophorus* and *Lepidapedon*, including outgroups) using ClustalW (Thompson, Higgins & Gibson, 1994) using default settings. Alignments were checked by eye and with reference to the protein for ND1. Both genes were easy to align for both taxon groups, homology assessment was straightforward and very few indels appeared in the alignments. Full sequence alignments used in these analyses have been deposited with EMBL under accessions ds43164 and ds43242 and are available via anonymous FTP from FTP.EBI.AC.UK under directory pub/databases/embl/align.

Phylogenetic reconstruction. Phylogenies for each of the *Steringophorus* and *Lepidapedon* taxon sets were constructed independently, and all analyses were conducted with PAUP* (Swofford, 1998). For each set individual 28S and ND1 gene data were treated separately in the first instance. Gaps were treated as missing data in the analyses. Using maximum parsimony (MP), nucleotide data were analysed for both genes and also proteins for ND1 (using the platyhelminth code Table 14 of GenBank). MP was run with a branch and bound search followed by bootstrap resampling ($n = 1000$) and decay analysis (Bremer support estimated with Autodecay 4.0.1, Eriksson, 1998) to estimate branch support. The results from both genes were compared and a partition homogeneity test (ILD, Farris et al. 1994) as implemented in PAUP* comparing 28S nucleotides with ND1 nucleotides and 28S nucleotides with ND1 amino acids was computed to determine whether the data sets were combinable under the principles of conditional combination (Huelsenbeck, Bull & Cunningham, 1996; Cunningham, 1997). Minimum evolution (ME) trees were constructed for each data set using a maximum likelihood model with transition:transversion ratios, gamma statistics and proportion of invariable sites all estimated from a logdet neighbour-joining tree. ME solutions were identical to those found with MP and are not shown. ND1 data treated as amino acids had no effect on tree topology nor resolution.

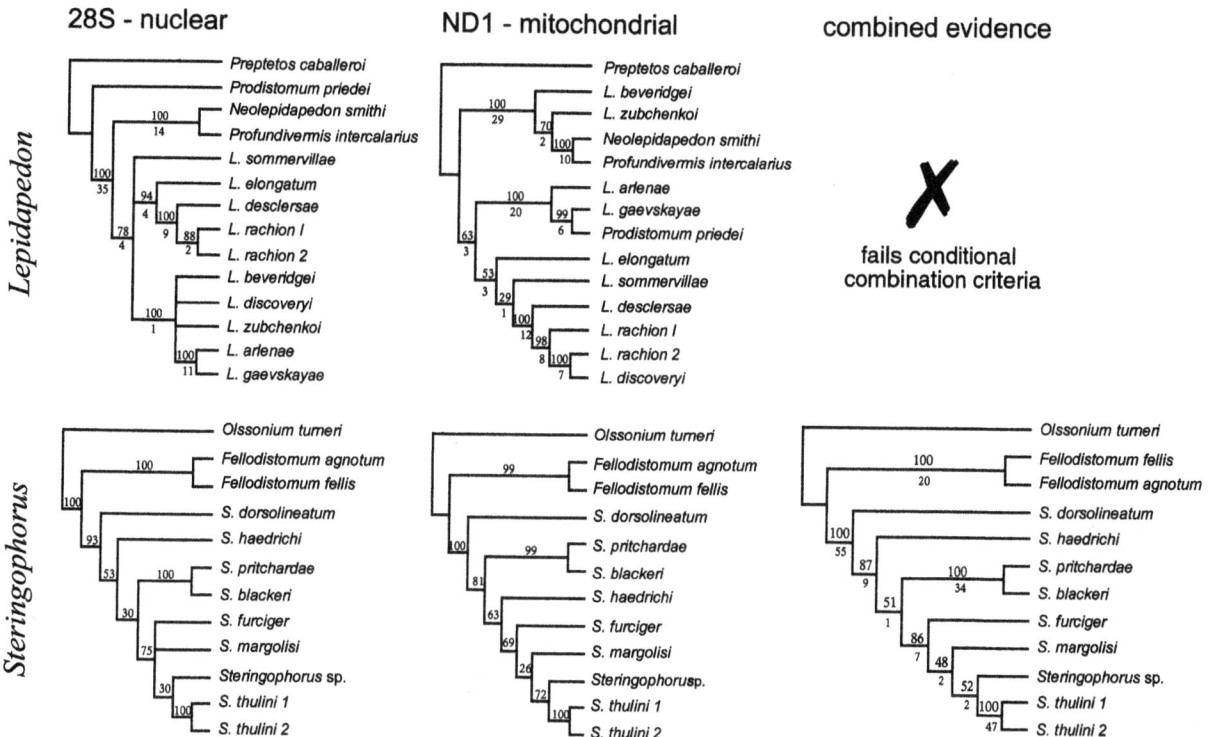

Fig. 2. Phylogenies of deep-sea digenean parasites using partial nuclear 28S rDNA and partial mitochondrial ND1 sequences. *Lepidapedon*: analyses of individual genes indicate incongruent tree topologies with only the solution from the nuclear data suggesting a monophyletic ingroup. Maximum parsimony (MP) results are shown for each; for 28S rDNA the strict consensus of 4 equally most parsimonious trees is shown (length = 236; CI = 0·704; RI = 0·776), and for ND1 the single most parsimonious solution (length = 493; CI = 0·649; RI = 0·754); 28S and ND1 contributed 94 and 169 parsimony informative sites respectively. *Steringophorus*: MP analyses of both genes yielded congruent topologies; data sets were combined after passing Templeton's test and the Kishino-Hasegawa test. 28S solution is strict consensus of 3 equally most parsimonious trees (length = 275; CI = 0·756; RI = 0·825), ND1 yielded a single most parsimonious solution (length = 904; CI = 0·558; RI = 0.484), as did the combined evidence (length = 904; CI = 0·595; RI = 0·577). 28S and ND1 contributed 122 and 244 parsimony informative sites respectively. Bootstrap (% of 1000 replicates) and Bremer support is shown above and below the nodes respectively; see text for further details.

Results and conclusions from molecular phylogenies

Lepidapedon – Phylogeny. Inferences on the phylogeny within the genus *Lepidapedon* are difficult to draw from our results using two genes (Figs 2, 3). The two genes are not congruent and the ILD test indicated conflict between them ($P < 0.01$). Consequently, we will discuss the phylogenies from each gene tree separately. The 28S tree implies monophyly for the genus, but that of ND1 not only includes the lepidapedines *Neolepidapedon smithi* and *Profundivermis intercalarias* within the genus, but also the lepocreadiine *Prodistomum priedei*. From the morphological point of view this latter attribution is difficult to understand or believe. The inclusion of the other lepidapedines within *Lepidapedon* is not so difficult to believe as, when Bray & Gibson (1991) attempted to produce a morphological phylogeny of the subfamily, they were unable to detect an autapomorphy for *Lepidapedon*. A few relationships are congruent on the trees. *N. smithi* and *P. intercalarius* are sister species in both trees. The close relationships of *L. arlenae* and *L. gaevskayae* are confirmed. The species *L. elongatum*, *L. desclersae* and *L. rachion* form a clade although, in the mitochondrial gene tree, the species *L. sommervillae* and *L. discoveryi* are included within the clade. We have no *a priori* reasons to accept or reject one gene tree over the other, but evidence from morphology would tend to favour the 28S topology. Such conflict highlights the need for careful interpretation of phylogenetic data and the need for congruence and dense sampling of genes and taxa prior to the over interpretation of results. Nucleotide trees do not necessarily track species trees with absolute fidelity (e.g. Page & Charleston, 1997). Furthermore, we should be open to the possibility that the selective pressure on the genes may be influenced not only by their function, but by the habitat of the organism, e.g. ND1 genes involved in the metabolic pathway may be under different selective pressures depending on the depth range in which they must function (e.g. see Somero, 1998, and Table 2).

Hosts. All members of the subfamily Lepidapedinae we have sampled here inhabit members of the order

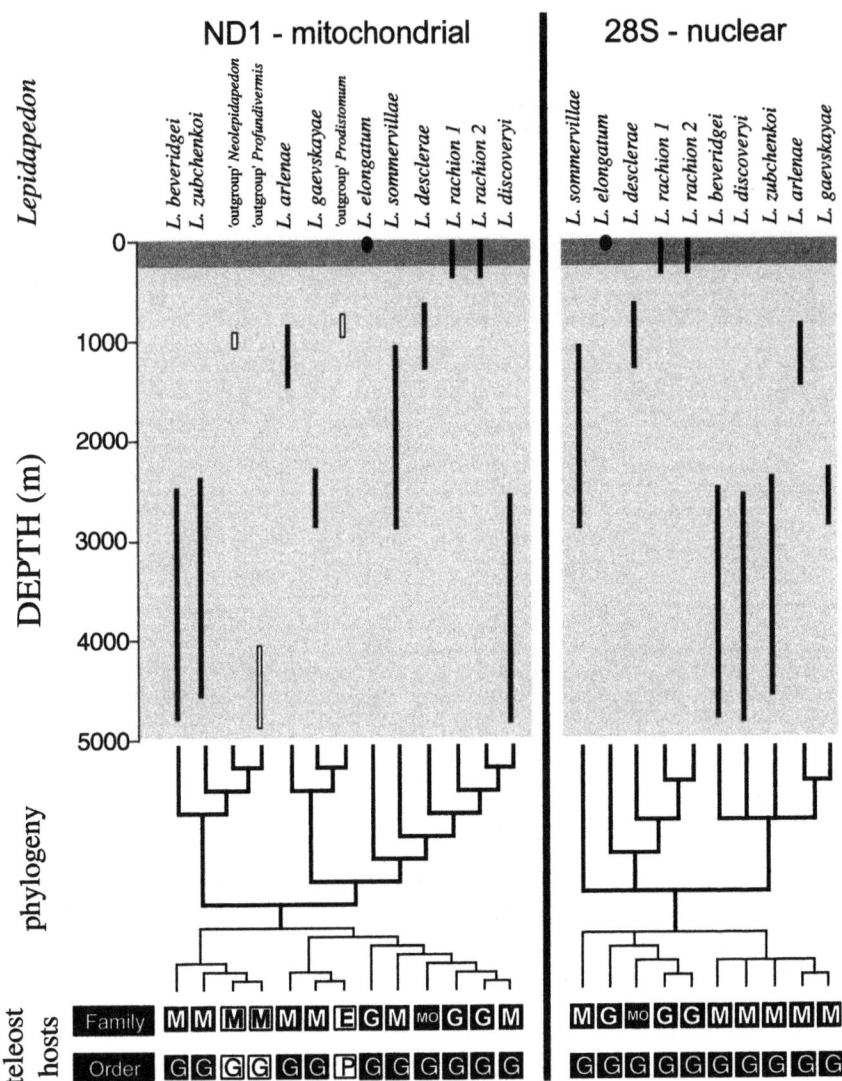

Fig. 3. Phylogeny, distribution and host associations of *Lepidapedon* spp. Phylogenies suggested by both 28S rDNA and ND1 are illustrated as each argue for markedly different solutions; outgroup details for taxa appearing in the ingroup are included for ND1. Fish host families and orders are abbreviated using first letters except MO which represents the Moridae; see Table 4 for full details.

Gadiformes. Both trees indicate that the family Macrouridae is probably the plesiomorphic host-group for the subfamily. Radiation into the Gadidae and Moridae has occurred in the clade (*L. elongatum* (*L. desclersae, L. rachion*)) (28S tree). According to the ND1 tree, this group is polyphyletic with two re-invasions of macrourids. Much of this discussion is predicated on fairly strict host-specificity in the group, which does appear to be the case, usually. The phylogenetic relationships of the three gadiform families discussed have been resolved as (Macrouridae (Moridae, Gadidae)) by Howes (1991) and as ((Moridae, Macrouridae) Gadidae) by Markle (1989), both utilizing morphological characters. Our results may indicate a slight preference for Howes' phylogeny, in that the gadid and morid worms appear related. Howes (1991) suggested that the 'phylogenetic trend amongst gadoids appears to have been from oceanic to shelf habitats', which parallels somewhat the perceived situation in *Lepidapedon*. The host-relations of the Lepidapedinae are, however, too complex to indicate straightforward co-evolution. Howes (1991) also suggested that attempts to designate various gadiforms as primitive or derived is misplaced, and that the depth ecotypes should be seen as resulting from vicariant events coinciding with such occurrences as the tectonic formation of a new ocean basin. Thus, 'the abyssal habits of macrouroids are not 'ancestral' to the shelf habitats of gadoids but coordinate with them'. The phylogeny of the Lepidapedinae does not reflect this vicariance event, but indicates subsequent host-switching within a recognizable host-clade (i.e. order Gadiformes). The nuclear gene tree appears to be more easily interpretable in terms of the host-parasite relationship.

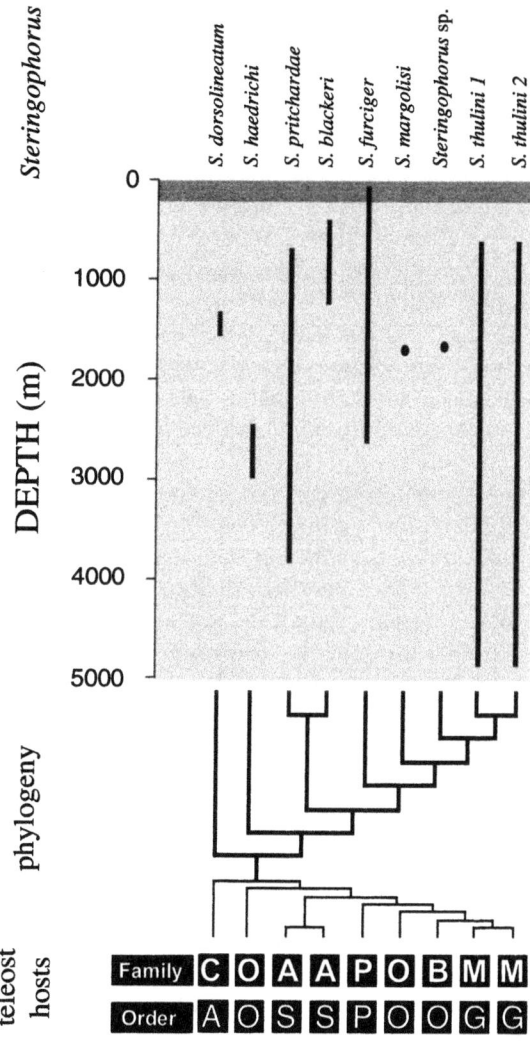

Fig. 4. Phylogeny, distribution and host associations of *Steringophorus* spp. Phylogeny is based on the combined evidence from 28S rDNA and mtND1. Fish host families and orders are abbreviated using first letters; see Table 4 for full details.

Steringophorus – Phylogeny: In contrast to *Lepidapedon*, the nuclear and mitochondrial gene trees are close to congruent for *Steringophorus* (Fig. 2), slight differences in the placing of *S. haedrichi* and the resolution of the relationships of *S. furciger* and *S. margolisi* being the only incongruence (ILD partition homogeneity test gave $P = 0.83$ and $P = 0.85$ for 28S+ND1 nucleotides and nucleotides +amino acids respectively strongly suggesting the two genes were arguing towards the same phylogenetic solution). It is, therefore, possible to discuss the single most parsimonious tree from combined evidence (Figs 2, 4). Evidence for the monophyly of *Steringophorus* is strong, as is the evidence for *S. dorsolineatum* as the most basal species.

Hosts. A problem associated with interpreting the host-associations in this genus is the variable specificity encountered. *S. furciger*, for example, is reported (according to Bray & Campbell, 1995) in 67 fish species of 41 genera, 17 families and five orders, including Gadiformes. Other species are restricted to orders (e.g. *S. thulini* to Gadiformes), families (e.g. *S. haedrichi* to Ophidiidae), genera (e.g. *S. dorsolineatum* to *Bathypterois*, *S. pritchardae* to *Alepocephalus*) or species (e.g. *S. blackeri* to *Xenodermichthys copei*). Nevertheless, there does appear to be a clade, (*S. furciger* (*S. margolisi* (*Steringophorus* sp., *S. thulini*))), associated with the related orders Ophidiiformes and Gadiformes (some authors place ophidiiforms within the Gadiformes – see Nelson, 1994). Considerable host-switching must have occurred. The small clade (*S. pritchardae*, *S. blackeri*) shows some evidence of co-evolution with alepocephalid salmoniforms.

Depth. As with host, discussion of depth in this genus is problematical, as species may occur over a wide depth range. *S. thulini* has been found between the deep abyssal region at 4865 m depth (Bray, 1995*b*) and the Mid-Skagerrak (i.e. about 600 m) (Bray & Gibson, 1980). Even *S. furciger*, which is a common parasite of shallow water flatfishes, has been reported at depths in excess of 2000 m (Bray & Campbell, 1995). Nevertheless, it is clear that *Steringophorus* is predominantly a deep-sea genus and only one species in our study, *S. furciger*, reaches into neritic waters.

General conclusions on molecular trees

Even taking into consideration the factors complicating interpretation of these trees, it is apparent that adaptation to the deep-sea is related to phylogeny, in that there are distinct lineages adapted to bathyal and abyssal conditions. Both of the genera studied have radiated predominantly in deeper waters off the continental shelf.

Depth. The relationship with depth is not well resolved, but it is clear that the whole subfamily Lepidapedinae is associated with deep-water of various depths. The mitochondrial gene tree shows no relationship between parasite phylogeny and depth, apart from the paraphyletic grouping *L. beveridgei* and *L. zubchenkoi* which inhabit deep bathyal and abyssal waters. It can be noted also that the clade including these two worms also includes *Profundivermis intercalarius*, a worm found so far only at abyssal depth (Bray & Gibson, 1991). Again, as with host-relationships, the nuclear gene tree is more readily interpretable in terms of bathymetric distribution. The clade (*L. elongatum* (*L. desclersae*, *L. rachion*)), shows a tendency to be found in shallower water, i.e. neritic or upper bathyal zones. Both trees indicate that the genus has radiated in deep water, i.e. waters off the continental shelf, with isolated clades moving into neritic waters.

In one case, *Lepidapedon*, this has accompanied a close association with one group of hosts, the Gadiformes, whereas in the other genus, *Steringophorus*, it has been associated with fishes from several deep-sea orders, including the gadiforms and the related ophidiiforms. The species of *Lepidapedon* studied are mainly parasites of macrourids, but they show no indication of co-speciation with their hosts. Where congeneric parasites infect the same host, the relationship is never that of sister-species, consequently there is no evidence of sympatric speciation within hosts.

DISCUSSION

Deep-sea digeneans

We have developed a database of published records of deep-sea fish digeneans, which currently contains 835 host-parasite records and 202 nominal species. This has no pretensions of being complete, and it should be noted that it is sometimes difficult to decide whether records are from taxa found deep enough to qualify for the database as the recording of depth data is the exception rather than the rule in the literature. Eighteen families were covered in our database (Table 5). There are about 150 digenean families in all and Gibson & Bray (1994) found that 53% of digenean families occurred in fishes. As far as host-parasite records are concerned, the major families are the Hemiuridae (19%), Lepocreadiidae (19%), Derogenidae (18%), Opecoelidae (12.5%) and Fellodistomidae (11%). The Zoogonidae and Lecithasteridae, with 6.6% and 5.75%, respectively, are the only other families with more than 2% of records. The diversity of deep-sea digeneans is, therefore, much lower than in shallow waters. Unfortunately, the disparity in collecting effort between these two environments does not lend itself to direct comparison by the use of diversity indices. Several major fish digenean families are not represented at all in deep-water (e.g. Apocreadiidae, Haploporidae, Haplosplanchnidae, Didymozoidae, Transversotrematidae) and other major families (e.g. Cryptogonimidae (0.4%), Bucephalidae (0.8%), Acanthocolpidae (1.2%), Faustulidae [recently split from the Fellodistomidae – see Hall, Cribb & Barker, 1999] (0.1%)) are poorly represented. The Monorchiidae (0.7%) is found in deep pelagic fishes (Bray & Gaevskaya, 1993). The high proportion of gadiforms sampled (55% of records), resulting from the preponderance of collecting in areas where macrourids are the predominant deep-sea fish family, may have baised these results. The most commonly reported genera are *Lepidapedon* (131 records, 16%), *Gonocerca* (105, 12.6%) and *Steringophorus* (66, 8%). Only derogenids (*Gonocerca phycidis*), fellodistomids (*Steringophorus thulini*) and lepocreadiids (*Lepidapedon discoveryi*, *L. zubchenkoi* and *Profundivermis intercalarius*) have been reported from the abyssal plain (c. 4850 m) (Bray & Gibson, 1991, 1995; Bray, 1991, 1995b). Only *P. intercalarius* appears to be restricted to the abyssal waters. Bray & Gibson's (1995) conclusion that *Lepidapedon* is 'probably the dominant digenean genus in deep water' was supported by Blend et al. (2000).

Campbell et al. (1980) reported that more than 70% of the digeneans they recovered from deep-water fishes in the NW Atlantic occurred in single host species. If this level of oioxenicity is typical, then deep-water digeneans appear to have a distinctly greater specificity than shallow-water forms. This tends to contradict the intuitive view, as expressed in fiction by Silverberg (1991): 'In an ocean the size of this one, the parasites can't afford to be host-specific. They'll take root in whatever they can'. Campbell et al. (1980) found that the benthic macrourids shared greater similarity of parasites among themselves than with other teleosts, but were not able to conclude whether this was due to co-evolution or to shared diet.

Digenean life-cycles

The life-cycles of digeneans in the deep-sea are not known, but it is possible to extrapolate from shallow-water relatives (Table 6). Members of all of the families listed, except for the Sanguinicolidae, use only molluscs as first intermediate hosts. Blood-flukes of the family Sanguinicolidae are also known to use molluscs in freshwater, but the only marine species whose life-cycle has been fully elucidated, *Aporocotyle simplex*, uses the terebellid polychaete *Artacama probiscidea* (Køie, 1982; Smith, 1997). The cercariae of *A. simplex* penetrate their final host, flatfishes, without the intervention of a second intermediate host. *Aporocotyle simplex* is the species reported from a macrourid, but as these worms are generally highly specific this is probably a misidentification (Bray, 1995a). The other families reported have basically a two or three-host life-cycle, utilizing a mollusc as the first intermediate host. The parthenitae of the Lepocreadiidae, Opecoelidae, Zoogonidae and the hemiuroid families Derogenidae, Hemiuridae, Bunocotylidae and Lecithasteridae are found in gastropods and those of the Bucephalidae, Fellodistomidae, Faustulidae and Gorgoderidae infect bivalves. The life-cycles of the Accacoeliidae and Hirudinellidae are not fully known. A wide variety of invertebrates serves as second (or occasionally third) intermediate hosts, which harbour metacercariae infective for the definitive host.

The life-cycle of *Lepidapedon elongatum*, a shallow-water species, has been described in some detail by Amosova (1955) and Køie (1985). The cercaria is atypical for a lepocreadiid, in that it lacks eye-spots and has a stumpy tail lacking setae. The

Table 5. Host-parasite records of digeneans of deep-sea fishes taken from the literature

Host Order	A1	A2	B1	B2	B3	C	D	F1	F2	G	H1	H2	L1	L2	M	O	S	Z	Total	Percentage
Anguilliformes				2					8		7		2			8		1	28	3·4
Aulopiformes							1		4		6					6			17	2·0
Beloniformes											1								1	0·1
Beryciformes									1		3		1			4		2	11	1·3
Chimaeriformes																6			6	0·7
Gadiformes	3	10		3	13		112	1	20		81	2	16	150	1	35	4	12	463	55·4
Gasterosteiformes																1			1	0·1
Gonorhynchiformes																1			1	0·1
Lophiiformes							2				4		4			2			12	1·4
Myctophiformes															1				1	0·1
Notacanthiformes			1				1		2	3	1		1			4		8	19	2·3
Ophidiiformes						3	10		9		8		4		1	6		1	40	4·8
Osmeriformes		1			3				6		2								16	1·9
Perciformes	4	1		2	2		3		10		9		1	6		7		12	53	6·3
Pleuronectiformes	1						10		9		14		7			7	1	5	58	6·9
Polymixiiformes											1					1			2	0·2
Salmoniformes									20		11		9					11	51	6·1
Scorpaeniformes	2						8		1		11		1	1		13		1	38	4·6
Stomiiformes											1		2		3	2			8	1·0
Syngnathiformes									1										1	0·1
Unknown		1							2									2	2	0·2
Zeiformes											2					1			6	0·7
Total	10	13	1	7	18	3	147	1	93	3	162	2	48	157	6	104	5	55	835	100
Percentage	1·2	1·6	0·1	0·8	2·2	0·4	17·6	0·1	11·1	0·4	19·4	0·2	5·7	18·8	0·7	12·5	0·6	6·6	100	

A1: Acanthocolpidae; A2: Accacoeliidae; B1: Bivesiculidae; B2: Bucephalidae; B3: Bunocotylidae; C: Cryptogonimidae; D: Derogenidae; F1: Faustulidae; F2: Fellodistomidae; G: Gorgoderidae; H1: Hemiuridae; H2: Hirudinellidae; L1: Lecithasteridae; L2: Lepocreadiidae; M: Monorchiidae; O: Opecoelidae; S: Sanguinicolidae; Z: Zoogonidae

Table 6. Summary of intermediate hosts of shallow-water relatives of deep-sea digeneans (for more detail see Table 2.3 of Williams & Jones, 1994)

Family	First intermediate host	Second or subsequent intermediate hosts
Sanguinicolidae†	Terebellid polychaete	None
Bucephalidae	Bivalves	Fish
Faustulidae	Bivalves	Amphipods
Fellodistomidae	Bivalves	None, or ophiuroid, or whelk or fish paratenic host
Gorgoderidae	Bivalves	Probably none in marine gorgoderids
Hemiuroidea*	Gastropods	Copepods, may use chaetognaths or fishes as paratenic hosts
Lepocreadiidae	Gastropods	Medusa, ctenophore, polychaete, bivalve, echinoid
Opecoelidae	Gastropods	Amphipods, decapods, mysids, fish
Zoogonidae	Gastropods	Brittle stars, polychaetes, bivalves, gastropods

† Marine only.
* Includes Derogenidae, Hemiuridae, Bunocotylidae and Lecithasteridae.

metacercariae encyst in a variety of annelids, molluscs and echinoderms, but infections in the latter two hosts are rare and probably short-lived. There are two major subfamilies in the Lepocreadiidae, the Lepidapedinae and the Lepocreadiinae. The latter occur widely in tropical and sub-tropical shallow-water fishes, whereas, in contrast, the lepidapedines are predominantly found in deep-water or cold-water fishes, mostly in Gadiformes (Bray & Gibson, 1997). The lack of eye-spots in lepidapedine cercariae and adults is a striking difference, perhaps indicating that the subfamily has evolved in the lightless deep-sea and those members which have moved onto the shelf have not reacquired eye-spots. Similar evidence has been used to argue that isopods of cold shallow areas have been derived from deeper waters (Hessler, Wilson & Thistle, 1979).

The other genus whose phylogeny is discussed above, *Steringophorus*, is a member of the subfamily Fellodistominae, the only fellodistomid subfamily found in deep-water. Aspects of the life-cycle of *Steringophorus furciger* have been elucidated by Chubrik (1966) and Køie (1979). The known first intermediate hosts are protobranch bivalves of the genus *Nuculana* (of the superfamily Nuculanoidea [= Nuculanacea], the commonest molluscs in the deep sea; Gage & Tyler, 1991, p. 135ff). Neither Chubrik (1966) nor Køie (1979) found a metacercaria, and Køie (1979) reported that small flatfishes consume the cercariae and become infected with adult *S. furciger*. She also pointed out that piscivorous hosts presumably acquire this parasite from fish prey. This type of life-cycle has also been demonstrated for the fellodistomine genus *Monascus* (by Køie, 1979) and inferred for the fellodistomine genus *Steringtrema* (by Bray, 1985). The hosts we have found for *Steringophorus* spp. include not only piscivorous species, e.g. *Coryphaenoides*, but also feeders on cnidarians and ctenophores, e.g. *Alepocephalus*, on crustaceans, e.g. *Xenodermichthys* and on benthopelagic plankton, e.g. *Bathypterois*. From this we might infer that *Steringophorus* uses a wide range of second intermediate hosts. A parallel situation has been discovered in *Monascus*, where the second intermediate host is now known to be not only fishes (as demonstrated by Køie, 1979), but also medusae (Girola, Martorelli & Sardella, 1992) and chaetognaths (Martorelli & Cremonte, 1998).

Knowledge of the life cycle of the digeneans of fishes may also be a valuable source of general information on feeding strategies of deep-sea fishes, especially those prone to stomach eversion during capture (Merrett & Haedrich, 1997).

Putative intermediate host frequency at depth

There would seem to be no shortage of candidates for intermediate hosts for digeneans in the deep-sea, at least to 2500 m depth. Grassle & Maciolek (1992), working in the NW Atlantic, showed that annelids, molluscs and arthropods (mainly crustaceans) were highly diverse in continental slope samples (1500–2500 m). They extrapolated from their findings and estimated that there are over 1 million and possibly more than 10 million deep-sea macrofaunal species. Subsequent authors have variously re-examined their findings suggesting that the fauna may be about 0·5 million (May, 1992) or 5 million (Poore & Wilson, 1993). The density of this macrofauna is high, with Grassle & Maciolek (1992) reporting 798 species in 21 m^2 surface area. Of those taxa likely to act as digenean intermediate hosts, they mention 106 mollusc species from 43 families (including Gastropoda 28 species/18 families, Bivalvia 45/18, Scaphopoda 9/4), 385 annelid species from 49 families (Polychaeta 367/49, Oligochaeta 18/2) and 185 arthropod species in 40 families (including Cumacea 25/4, Tanaidacea 45/8, Isopoda 59/11, Amphipoda 55/16). Copepods are also diverse in deep-water and are mostly found in the upper layer of the sediment (Vincx *et al.* 1994). Rex (1973) found

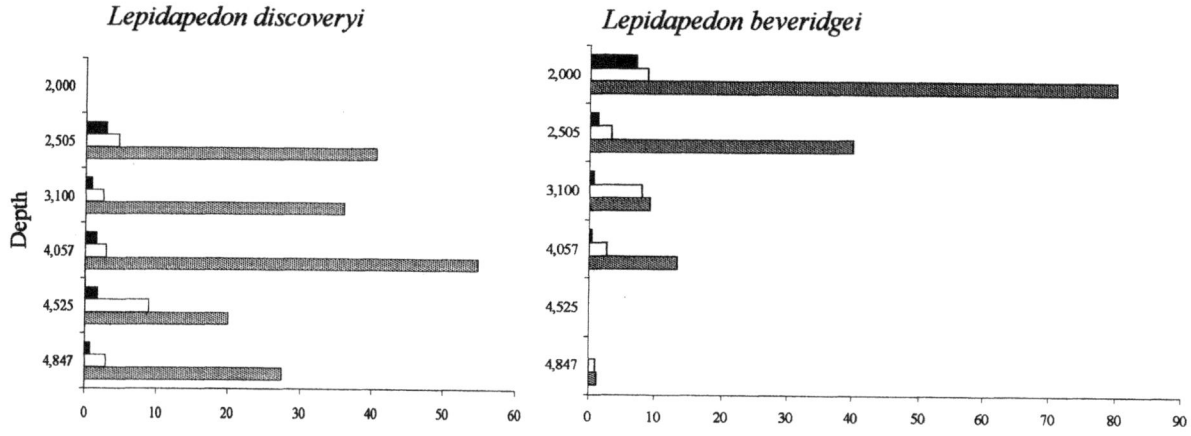

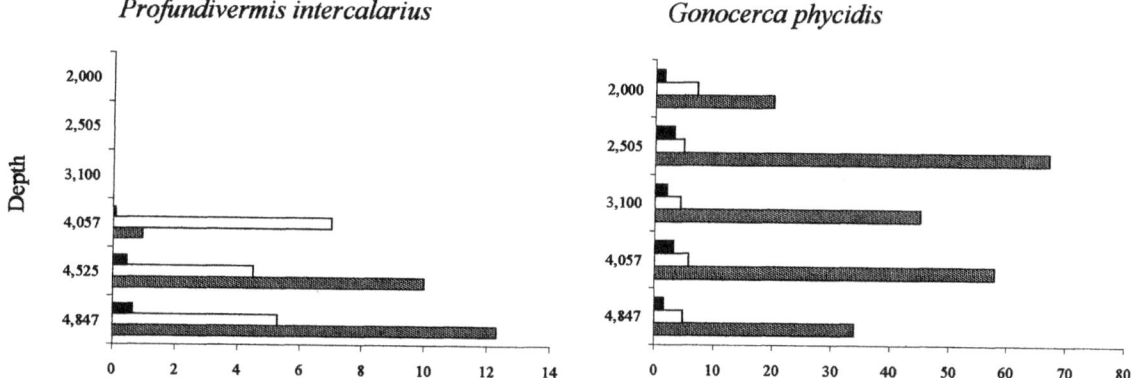

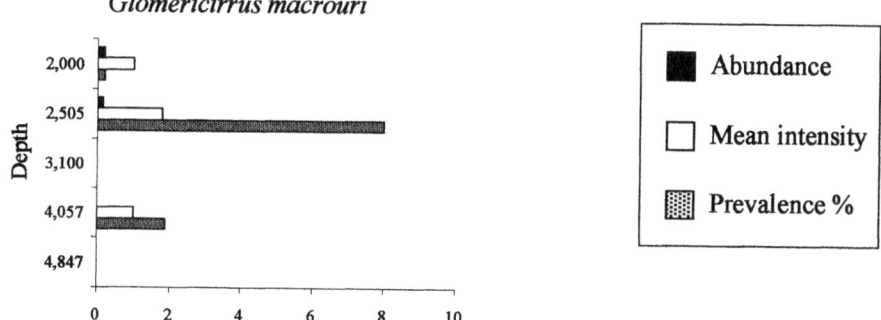

Fig. 5. Graphs showing prevalence, intensity and abundance of five digeneans at various depths in *Coryphaenoides* (*Nematonurus*) *armatus* in the NE Atlantic: depths are given as median of depth range (Note differences in the depths in samples of stomach worms *Gonocerca phycidis* and *Glomericirrus macrouri* – stomachs were not sampled at all sites). Sample sizes: 2000 m = 5; 2505 m = 75; 3100 m = 11; 4057 m = 105; 4525 m = 20; 4847 m = 81.

that in the NW Atlantic gastropod diversity was low on the continental shelf, high on the continental slope and rise and decreased out onto the abyssal plain.

Zoning at depth

As noted earlier, there appears to be no straightforward zonation of fauna delimited by depth (Haedrich, 1997). Each species has its own bathymetric distribution which overlaps with that of other species in a complex way. Little is known of the bathymetric limits of individual deep-sea species, but we present preliminary results of sampling the digenean fauna of *Coryphaenoides* (*Nematonurus*) *armatus* at various depths (Fig. 5). Caution should be exercised in that samples are small and the shallowest sample represents very few individuals. These results agree with Haedrich's (1997) observations, in that there is no real evidence of zoned communities of parasites. Some species occur more-or-less throughout the depth range (i.e. *Lepidapedon discoveryi, Gonocerca phycidis*), whilst others are restricted to relatively shallower depths (*L. beveridgei* and *Glomericirrus macrouri*) and to deeper regions (*Profundivermis intercalarius*). Another note of caution must be placed on the data relating to the stomach-inhabiting forms, *Gonocerca phycidis* and *Glomericirrus macrouri*, in that some loss of parasites is likely to have occurred with eversion of the fish's stomach during surfacing. *C. (N.) armatus* itself is known to have a particularly large bathymetric range (Merrett & Haedrich, 1997). Campbell et al. (1980) suggested that decreased 'incidences' of lepidapedines in *C. (N.) armatus* were directly related to decline in the occurrence of polychaete prey. The effect relating to *Glomericirrus macrouri* [as *G. ulmeri*] and isopods (their presumed second intermediate hosts) was most dramatic in that *G. macrouri* was much commoner in smaller fish. Some of the bathymetric gradients in our results may result from this sort of trophic relationship, as the deeper the fish is found, the larger it is likely to be ('Heincke's Law'). This argument could be advanced to explain the distributions of *L. beveridgei* and *Glomericirrus macrouri*, but does not apply to the other three species, in particular *P. intercalarius*, which do not disappear in deeper water. Campbell et al. (1980) found a steep decline in the numbers of parasites (Monogenea, Digenea, Nematoda) with depth, until they were absent at 5000 m, this being related, they suggested, to the change to a more pelagic diet in older *C. (N.) armatus*. Our results do not confirm these findings, in that three digenean species were found at abyssal depths with prevalences between 12 and 34%.

Constraints of physical conditions

Tyler (1995) listed important aspects of the conditions on the deep-sea floor, particularly in relation to their influence on the biota. Some of these are summarized in Table 2. Those which may influence digenean biology are briefly discussed below.

Temperature. The similarity of the deep-sea and the Antarctic cold shallow water fauna bear witness to the importance of adaptation to low temperatures in the inhabitants of the deep-sea. In the case of the digeneans many common deep-water genera, e.g. *Lepidapedon, Steringophorus, Gonocerca, Glomericirrus, Neolepidapedon* and *Lecithophyllum* occur in Antarctic shelf fishes (Zdzitowieki, 1997 a, b).

Pressure. If pressure was a major factor in the distribution of digeneans, we would not expect one species to spread over a large bathymetric range, particularly into shallow water. In fact, species may appear to be adapted a wide range, e.g. *Gonocerca phycidis*, which is reported in macrourids from just below 200 m to abyssal depths (4850 m) (Bray, 1995 a). *G. phycidis* is also common in Antarctic shelf fishes, particularly nototheniids (Zdzitowiecki, 1997 a). We do not know whether digeneans survive in the hadal zone. Apart from a record of an unnamed digenean from *Macrourus acrolepis*, captured between 2500 and 7000 m in the Pacific Ocean, we have no information on digeneans from deeper than the abyssal plain in the NE Atlantic ($\sim$ 5000 m) (Gusev, 1957). The eurybathy of many deeper-water digeneans and their occurrence in Antarctic shelf waters suggests that pressure effects are not as important as temperature in their distribution.

Light. Many digenean cercariae have eye-spots, testifying to the use of light for orientation and host finding (Smyth & Halton, 1983). In deep-water, below the illuminated zone, other cues are presumably used, and the lack of eye-spots in members of the Lepidapedinae may reflect their long association with deep-water. This is in comparison with the other major lepocreadiid subfamily, the Lepocreadiinae, whose species are predominantly shallow water forms, which often retain eye-spot pigment into the adult stage. It may also be noteworthy that the cercaria of *Steringophorus furciger* apparently lacks eye-spots (Køie, 1979).

Seabed composition. The soft sediment found in much of the deep-sea bed may be related to the relative abundance of digeneans using bivalves in the deep-sea.

Currents. As Tyler (1995) pointed out, it is difficult to get data on a small enough scale to assess the effects of currents, but it is thought that scavenging fish and amphipods detect prey items up-current to them. This may have some bearing on digenean strategies for transmission.

CONCLUDING QUESTIONS

Is the deep-water fauna smaller than the shallow?

This is difficult to say even in general terms, as the sampling is so skewed towards shallow water and a few deep-water areas. The area covered by deep-water is so vast that it seems possible that the fauna is not significantly smaller in biomass terms, even though biomass per square metre of deep-sea 'macrobenthos' (not including fishes) is reckoned by Rowe (1983) to decline exponentially with depth. Fish biomass is also considered to decline with depth, although Merrett & Haedrich (1997) confided that they were 'still far from wholeheartedly believing in our deep-sea estimates of even biomass'.

Is the deep-water fauna less diverse than shallow?

Earlier results, especially those of Campbell et al. (1980) and those in our own database (see Table 4), in which only 18 of over 60 fish digenean families are recorded in deep-water, indicate that the fauna is, indeed, less diverse in the deep-sea.

Are evolutionary transitions from shallow to deep (or vice versa) relatively rare?

This is probably the case, in that such a small percentage of the families of digeneans found in fishes have extended into the deep-sea. Those that have made the transition often inhabit related hosts, which inhabit similar depths, such that depth and host effects are not separable.

Have most deep-sea digeneans radiated in the deep-sea?

This appears to be the case in *Lepidapedon* and *Steringophorus*. In other cases it appears that isolated individual species have invaded the bathyal zones, but not the abyssal parts of the deep-sea. Examples are the lone deep-sea faustulid, *Echinobreviceca coelorhynchae* (see Dronen, Blend & McEachran, 1994) in a bathyal macrourid and the lone lepocreadiine lepocreadiid *Prodistomum priedei*, which has apparently invaded the shallow bathyal regions along with its perciform host (Bray & Merrett, 1998).

How much do we really know about deep-sea digeneans?

The answer to this must be 'very little'. Firstly, the amount of sampling has been small, and it has been patchy. Nearly all has occurred in the north Atlantic and north-eastern Pacific. There has been none, or practically none, in the Indian Ocean, any part of the tropics and the southern Ocean. There has been none around vents nor in trenches. We know no deep-sea life-cycles and all our ideas are predicated on the presumed similarity of the life-cycles of related deep-sea and shallow forms. The assessment of deep-sea parasite diversity and an understanding of deep-sea parasitology remain distant goals.

FINAL CONCLUSIONS

The conflict we have found amongst gene trees highlights the problems associated with reconstructing phylogenies (Page & Charleston, 1997), but we believe that new insights have been forthcoming in this study. All the results show that the groups of species surveyed have radiated in deeper 'off-shelf' waters with occasional invasions of the neritic zone, rather than having two vicariant clades, one neritic and one deeper, or the shallow waters forms being basal. It is clear that if progress is to be made in deep sea parasite studies, the emphasis will need to be on systematics and phylogenetics of both parasite and host; systematics to understand biodiversity in the deep and phylogenetics as a framework for understanding patterns and processes in the evolution and radiation of parasites and their hosts under this extreme environment.

ACKNOWLEDGEMENTS

Thanks are due to the crew and officers of the NERC Research Vessels *Challenger* and *Discovery* and the SOAEFD Research Vessel *Scotia*, and to Dr Monty Priede, University of Aberdeen, Dr Nigel Merrett and Ms Mary Spencer Jones, The Natural History Museum, London, and Dr Alasdair McVicar, SOAEFD, Aberdeen, for help with collection of material. DTJL and EAH were funded by a Wellcome Trust Senior Research Fellowship to DTJL (043965/Z/95/Z). BW completed the 28S phylogeny of *Lepidapedon* as part of her MSc dissertation.

REFERENCES

AMOSOVA, I. S. (1955). [On the occurrence of metacercariae of digenetic trematodes in certain polychaetes in the Barents Sea.] *Zoologicheskii Zhurnal* **34**, 286–290 (In Russian).

ANGEL, M. V. (1997). What is the deep sea? In *Deep-sea Fishes* (ed. Randall, D. J. & Farrell, A. P.), pp. 1–41. San Diego, Academic Press.

BELL, F. J. (1887). Description of new species of *Distomum*. *Annals and Magazine of Natural History* (5th series) **19**, 116–117.

BLEND, C. K., DRONEN, N. O. & ARMSTRONG, H. W. (2000). Six new species of *Lepidapedon* Stafford, 1904 (Digenea: Lepocreadiidae) from deep-sea macrourid fishes from the Gulf of Mexico and Caribbean Sea, with revised keys to the species of the genus. *Systematic Parasitology* **45**, 29–51.

BRAY, R. A. (1985). Some helminth parasites of marine fishes of South Africa: Families Gorgoderidae, Zoogonidae, Cephaloporidae, Acanthocolpidae and Lepocreadiidae (Digenea). *Journal of Natural History* **19**, 377–405.

BRAY, R. A. (1991). The helminth parasites of the deep-sea macrourid *Coryphaenoides* (*Nematonurus*) *armatus*. In *Abstracts of reports. IIIrd International Symposium "Problems of Fish Parasitology"* (Petrozavodsk). Petrozavodsk and Leningrad, Biological Institute KSC and Zoological Institute Academy of Sciences USSR, 11–12.

BRAY, R. A. (1995a). Annotated checklist of digenean parasites of Macrouridae (Teleostei, Gadiformes). *Acta Parasitologica* **40**, 169–192.

BRAY, R. A. (1995b). Observations on the genus *Steringophorus* Odhner, 1905 (Digenea: Fellodistomidae) in deep-sea fishes from the northeastern Atlantic, with the description of *Steringophorus margolisi* n. sp. In *Parasites of Aquatic Organisms: a Festschrift dedicated to Dr. Leo Margolis, O.C., Ph.D., F.R.S.C.* (ed. Arthur, J. R.). *Canadian Journal of Fisheries and Aquatic Sciences* **52** (Suppl. 1), 71–77.

BRAY, R. A. & CAMPBELL, R. A. (1995). Fellodistomidae and Zoogonidae (Digenea) from deep-sea fishes of the NW Atlantic Ocean. *Systematic Parasitology* **31**, 201–213.

BRAY, R. A. & GAEVSKAYA, A. V. (1993). *Bathymonorchis polyipni* (Reimer, 1985) n. g., n. comb. (Digenea: Monorchiidae) from bathypelagic fishes of the eastern mid-Atlantic Ocean. *Systematic Parasitology* **26**, 91–95.

BRAY, R. A. & GIBSON, D. I. (1980). The Fellodistomidae (Digenea) of fishes from the north-east Atlantic. *Bulletin of the British Museum (Natural History) (Zoology Series)* **37**, 199–293.

BRAY, R. A. & GIBSON, D. I. (1991). The Lepocreadiidae (Digenea) of fishes from the north-east Atlantic: *Profundivermis intercalarius* n. gen., n.sp. from the marine fish *Coryphaenoides* (*Nematonurus*) *armatus* (Hector) (Macrouridae) from the Porcupine Abyssal Plane. *Systematic Parasitology* **18**, 121–125.

BRAY, R. A. & GIBSON, D. I. (1995). The Lepocreadiidae (Digenea) of fishes of the north-east Atlantic: a review of the genus *Lepidapedon* Stafford, 1904. *Systematic Parasitology* **31**, 81–132.

BRAY, R. A. & GIBSON, D. I. (1997). The Lepocreadiidae Odhner, 1905 (Digenea) of fishes from the north-east Atlantic: summary paper, with keys and checklists. *Systematic Parasitology* **36**, 223–228.

BRAY, R. A. & MERRETT, N. R. (1998). *Prodistomum priedei* n. sp. (Digenea: Lepocreadiidae) from the deepwater cardinalfish *Epigonus telescopus* (Perciformes: Epigonidae) in the northern Atlantic Ocean. *Systematic Parasitology* **41**, 71–77.

BROAD, W. J. (1997). *The Universe Below. Discovering the Secrets of the Deep-sea.* New York, Touchstone

CAMPBELL, R. A. (1977). *Degeneria halosauri* (Bell, 1887) gen. et comb. nov. (Digenea: Gorgoderidae) from the deep-sea teleost *Halosauropsis macrochir*. *Journal of Parasitology* **63**, 76–79.

CAMPBELL, R. A. (1983). Parasitism in the deep sea. In *The Sea. Volume 8* (ed. Rowe, G. T.), pp. 473–552. New York, John Wiley & Sons, Inc.

CAMPBELL, R. A. (1990). Deep water parasites. *Annales de Parasitologie Humaine et Comparée* **65** (suppl. I), 65–68.

CAMPBELL, R. A., HAEDRICH, R. L. & MUNROE, T. A. (1980). Parasitism and ecological relationships among deep-sea benthic fishes. *Marine Biology* **57**, 301–313.

CHUBRIK, G. K. (1966). [Fauna and ecology of trematode larvae from molluscs in the Barents and White Seas.] *Trudy Murmanskii Morskoi Biologicheskii Institute* **10**(14), 78–166 (In Russian).

COLLARD, S. B. (1970). Some aspects of host-parasite relationships in mesopelagic fishes. In *A Symposium on Diseases of Fishes and Shellfishes* (ed. Snieszko, S. F.), pp. 41–56. Washington, D.C., American Fisheries Society, Special publication 5.

CRIBB, T. H., BRAY, R. A., LITTLEWOOD, D. T. J., PICHELIN, S. P. & HERNIOU, E. A. (In press). The Digenea. In *Interrelationships of the Platyhelminthes* (ed. Littlewood, D. T. J. & Bray, R. A.). London, Taylor and Francis.

CUNNINGHAM, C. W. (1997). Can three incongruence tests predict when data should be combined? *Molecular Biology and Evolution* **14**, 733–740.

DRONEN, N. O., BLEND, C. K. & MCEACHRAN, J. D. (1994). *Echinobreviceca coelorhynchae* n. gen., n. sp. (Echinobrevicecinae n. subf.), fellodistomid from *Coelorhynchus coelorhynchus* (Macrouridae) from the Gulf of Mexico. *Journal of Parasitology* **80**, 309–311.

EKMAN, S. (1953). *Zoogeography of the Sea.* London, Sidgwick & Jackson.

ERIKSSON, T. (1998). *Autodecay version 4.0* (program distributed by the author). Department of Botany, Stockholm University, Stockholm.

FARRIS, J. S., KÄLLERSJÖ, M., KLUGE, A. G. & BULT, C. (1994). Testing significance of incongruence. *Cladistics* **10**, 315–319.

GAGE, J. D. & TYLER, P. A. (1991). *Deep-sea Biology: A Natural History of Organisms at the Deep-sea Floor.* Cambridge, Cambridge University Press.

GARTNER, J. V. JR & ZWERNER, D. E. (1989). The parasite faunas of meso- and bathypelagic fishes of Norfolk Submarine Canyon, western North America. *Journal of Fish Biology* **34**, 79–95.

GIBSON, D. I. & BRAY, R. A. (1994). The evolutionary expansion and host-parasite relationships of the Digenea. *International Journal for Parasitology* **24**, 1213–1226.

GIROLA, C. V., MARTORELLI, S. R. & SARDELLA, N. H. (1992). Presencia de metacercarias de *Monascus filiformis* (Digenea, Fellodistomidae) en hidromedusas del Océano Atlántico Sur. *Revista Chilena de Historia Natural* **65**, 409–415.

GRASSLE, J. F. & MACIOLEK, N. J. (1992). Deep-sea species richness: regional and local diversity estimates from quantitative bottom samples. *American Naturalist* **139**, 313–341.

GUSEV, A. V. (1957). [Parasitological investigations on some deepwater fishes of the Pacific Ocean.] *Trudy Instituta Okeanologii* **27**, 362–366 (In Russian).

HAEDRICH, R. L. (1997). Distribution and population ecology. In *Deep-sea Fishes* (ed. Randall, D. J. & Farrell, A. P.), pp. 79–114. San Diego, Academic Press.

HALL, K. A., CRIBB, T. H. & BARKER, S. C. (1999). V4 region of small subunit rDNA indicates polyphyly of the Fellodistomidae (Digenea) which is supported by morphology and life-cycle data. *Systematic Parasitology* **43**, 81–92.

HARRISON, C. M. H. (1996). On the first halosaur leptocephalus: from Madeira. *Bulletin of the British Museum (Natural History) (Zoology)* **14**, 441–486.

HESSLER, R. R., WILSON, G. D. & THISTLE, D. (1979). The deep-sea isopods: a biogeographic and phylogenetic overview. *Sarsia* **64**, 67–76.

HOWES, G. J. (1991). Biogeography of gadoid fishes. *Journal of Biogeography* **18**, 595–622.

HUELSENBECK, J. P., BULL, J. P. & CUNNINGHAM, C. W. (1996). Combining data in phylogenetic analysis. *Trends in Ecology and Evolution* **11**, 152–158.

KØIE, M. (1979). On the morphology and life-history of *Monascus* [= *Haplocladus*] *filiformis* (Rudolphi, 1819) Looss, 1907 and *Steringophorus furciger* (Olsson, 1868) Odhner, 1905 (Trematoda, Fellodistomidae). *Ophelia* **18**, 113–132.

KØIE, M. (1982). The redia, cercaria and early stages of *Aporocotyle simplex* Odhner, 1900 (Sanguinicolidae)

– a digenetic trematode which has a polychaete annelid as the only intermediate host. *Ophelia* 21, 115–145.

KØIE, M. (1985). On the morphology and life-history of *Lepidapedon elongatum* (Lebour, 1908) Nicoll, 1910 (Trematoda: Lepocreadiidae). *Ophelia* 24, 135–153.

KUNZIG, R. (1999). *The Restless Sea*. New York: W. W. Norton.

LINKLATER, E. (1972). *The Voyage of the Challenger*. London, John Murray.

LINTON, E. (1898). Notes on trematode parasites of fishes. *Proceedings of the United States National Museum* 20, 507–548.

LITTLEWOOD, D. T. J., ROHDE, K. & CLOUGH, K. A. (1997). Parasite speciation within and between host species? – phylogenetic evidence from site-specific polystome monogeneans. *International Journal for Parasitology* 27, 1289–1297.

LUMB, S. M., BRAY, R. A. & ROLLINSON, D. (1993). Partial small subunit (18S) rRNA gene sequences from fish parasites of the families Lepocreadiidae and Fellodistomidae (Digenea) and their use in phylogenetic analyses. *Systematic Parasitology* 26, 141–149.

MANTER, H. W. (1934). Some digenetic trematodes from deep-water fish of Tortugas, Florida. *Papers from Tortugas Laboratory* 28, 257–346.

MANTER, H. W. (1954). Some digenetic trematodes from fishes of New Zealand. *Transactions of the Royal Society of New Zealand* 82, 475–568.

MARKLE, D. F. (1989). Aspects of character homology and phylogeny of the Gadiformes. In *Papers on the Systematics of Gadiform Fishes* (ed. Cohen, D. M.), pp. 59–88. Los Angeles: Natural History Museum of Los Angeles County, Science series, 32.

MARTORELLI, S. R. & CREMONTE, F. (1998). A proposed three-host life history of *Monascus filiformis* (Rudolphi, 1819) (Digenea: Fellodistomidae) in the southwest Atlantic Ocean. *Canadian Journal of Zoology* 76, 1198–1203.

MAY, R. M. (1992). Bottoms up for the oceans. *Nature* 357, 278–279.

MCCAULEY, J. E. (1968). Six species of *Lepidapedon* Stafford, 1904 (Trematoda: Lepocreadiidae) from deep-sea fishes. *Journal of Parasitology* 54, 496–505.

MERRETT, N. R. & HAEDRICH, R. L. (1997). *Deep-sea Demersal Fish and Fisheries*. London, Chapman & Hall.

MORGAN, J. A. T. & BLAIR, D. (1998). Mitochondrial ND1 gene sequences used to identify echinostome isolates from Australia and New Zealand. *International Journal for Parasitology* 28, 493–502.

MOSELEY, H. N. (1880). Deep-sea dredging and life in the deep-sea. III. *Nature* 21, 591–593.

NELSON, J. S. (1994). *Fishes of the World* (Third Edition). New York, John Wiley & Sons, Inc.

NOBLE, E. R. (1973). Parasites and fishes in a deep-sea environment. *Advances in Marine Biology* 11, 121–195.

ODHNER, T. (1911). Zum natürlichen System der digenen Trematoden. II. *Zoologischen Anzeiger* 37, 237–253.

PAGE, R. D. M. & CHARLESTON, M. A. (1997). From gene to organismal phylogeny: reconciled trees and the gene tree/species tree problem. *Molecular Phylogenetics and Evolution* 7, 231–240.

POORE, G. C. B. & WILSON, G. D. F. (1993). Marine species richness. *Nature* 361, 597–598.

REX, M. A. (1973). Deep-sea species diversity: decreased gastropod diversity at abyssal depths. *Science* 181, 1051–1053.

ROWE, G. T. (1983). Biomass and production of the deep-sea macrobenthos. In *The Sea. Volume 8* (ed. Rowe, G. T.), pp. 97–121. New York, John Wiley & Sons, Inc.

SAIKI, R. F., GELFAND, D. H., STOFFEL, S., SCHARF, S. J., HIGUCHI, R., HORN, G. T., MULLIS, K. B. & ERLICH, A. H. (1988). Primer-directed enzymatic amplification of DNA with a thermostable DNA polymerase. *Science* 239, 489–491.

SILVERBERG, R. (1991). *The Face of the Waters*. London: GraftonBooks.

SMITH, J. W. (1997). The blood flukes (Digenea: Sanguinicolidae and Spirorchiidae) of cold-blood vertebrates: Part 1. A review of the literature published since 1971, and bibliography. *Helminthological Abstracts* 66, 255–294.

SMYTH, J. D. & HALTON, D. W. (1983). *The Physiology of Trematodes*. Second edition. Cambridge: Cambridge University Press.

SOMERO, G. N. (1992). Adaptation to high hydrostatic pressure. *Annual Review of Physiology* 54, 557–577.

SOMERO, G. N. (1998). Adaptation to cold and depth: contrasts between polar and deep-sea animals. In *Cold Ocean Physiology* (ed. Pörtner, H. O. & Playle, R. C.), pp. 33–57. Cambridge, Cambridge University Press.

SWOFFORD, D. L. (1998). *PAUP*: Phylogenetic Analysis Using Parsimony (and other methods), version 4.0*. Massachussetts: Sinauer Associates.

THOMPSON, J. D., HIGGINS, D. G. & GIBSON, T. J. (1994). CLUSTAL-W – improving the sensitivity of progressive multiple sequence alignment through sequence weighting, position-specific gap penalties and weight matrix choice. *Nucleic Acids Research* 22, 4673–4680.

TYLER, P. A. (1995). Conditions for the existence of life at the deep-sea floor: an update. *Oceanography and Marine Biology* 33, 221–244.

VINCX, M., BETT, B. J., DINET, A., FERRERO, T., GOODAY, A. J., LAMBSHEAD, P. J. D., PFANNKUCHE, O., SOLTWEDEL, T. & VANREUSEL, A. (1994). Meiobenthos of the deep northeast Atlantic. *Advances in Marine Biology* 30, 1–88.

VON LINSTOW, O. (1888). Report on the Entozoa collected by H. M. S. *Challenger* during the years 1873–1876. *Report of the Scientific Results of the Voyage of H. M. S. Challenger 1873–1876* 23, 1–18.

WAGENER, G. R. (1852). Enthelminthica No. III. *Archiv für Anatomie, Physiologie und wissenschaftliche Medicin*, 555–569.

WEITZMAN, S. H. (1997). Systematics of deep-sea fishes. In *Deep-sea Fishes* (ed. Randall, D. J. & Farrell, A. P.), pp. 43–77. San Diego, Academic Press.

WILLIAMS, H. H. & JONES, A. (1994). *Parasitic Worms of Fish*. London, Taylor & Francis.

WOODWARD, A. S. (1898). The antiquity of the deep-sea fish-fauna. *Natural Science* 12, 257–260.

YAMAGUTI, S. (1938). *Studies on the Helminth Fauna of*

Japan. Part 21. Trematodes of Fishes, IV. Kyoto, Satyû Yamaguti.

ZDZITOWIECKI, K. (1997*a*). Antarctic Digenea parasites of fishes. *Synopses of the Antarctic Benthos* **8**, 1–156.

ZDZITOWIECKI, K. (1997*b*). Digenea of fishes of the Weddell Sea. V. Two new species of the genus *Steringophorus* (Fellodistomidae). *Acta Parasitologica* **42**, 144–148.

ZDZITOWIECKI, K. & CIELECKA, D. (1997). Digenea of fishes of the Weddell Sea. I. Parasites of *Macrourus whitsoni* (Gadiformes, Macrouridae). *Acta Parasitologica* **42**, 23–30.

ZUBCHENKO, A. V. (1981*a*). Use of parasitological data in studies of the local groupings of rock grenadier, *Coryphaenoides rupestris* Gunner. In *Abstracts. Symposium on Parasitology and Pathology of Marine Organisms* (ed. Bauer, O. N.), pp. 25–32. Leningrad: Nauka, Leningrad Branch (In Russian: English translation (1985) *NOAA Technical Reports NMFS* **25**, 19–23).

ZUBCHENKO, A. V. (1981*b*). Parasitic fauna of some Macrouridae in the northwest Atlantic. *Journal of Northwest Atlantic Fishery Science* **2**, 67–72.

ZUBCHENKO, A. V. (1984). Ecological peculiarities of parasite fauna of some Alepocephalidae. In *Ecological-parasitological Investigations of Northern Seas* (ed. Polyansky, Y. I., Galaktionov, K. V. & Dobrovolsky, A. A.), pp. 77–81. Apatity, Kola Branch of Academy of Sciences of the USSR, Murmansk Biological Institute (In Russian).

For EU product safety concerns, contact us at Calle de José Abascal, 56–1°,
28003 Madrid, Spain or eugpsr@cambridge.org.

www.ingramcontent.com/pod-product-compliance
Lightning Source LLC
LaVergne TN
LVHW082244060526
838200LV00046B/2049